Zoophysiology *Volume 28*

AF786018

Zoophysiology

Volumes already published in the series:

Volume 1: *P.J. Bentley*
Endocrines and Osmoregulation (1971)

Volume 2: *L. Irving*
Arctic Life of Birds and Mammals Including Man (1972)

Volume 3: *A.E. Needham*
The Significance of Zoochromes (1974)

Volume 4/5: *A.C. Neville*
Biology of the Arthropod Cuticle (1975)

Volume 6: *K. Schmidt-Koenig*
Migration and Homing in Animals (1975)

Volume 7: *E. Curio*
The Ethology of Predation (1976)

Volume 8: *W. Leuthold*
African Ungulates (1977)

Volume 9: *E.B. Edney*
Water Balance in Land Arthropods (1977)

Volume 10: *H.-U. Thiele*
Carabid Beetles in Their Environments (1977)

Volume 11: *M.H.A. Keenleyside*
Diversity and Adaptation in Fish Behaviour (1979)

Volume 12: *E. Skadhauge*
Osmoregulation in Birds (1981)

Volume 13: *S. Nilsson*
Autonomic Nerve Function in the Vertebrates (1983)

Volume 14: *A.D. Hasler*
Olfactory Imprinting and Homing in Salmon (1983)

Volume 15: *T. Mann*
Spermatophores (1984)

Volume 16: *P. Bouverot*
Adaption to Altitude-Hypoxia in Vertebrates (1985)

Volume 17: *R.J.F. Smith*
The Control of Fish Migration (1985)

Volume 18: *E. Gwinner*
Circannual Rhythms (1986)

Volume 19: *J.C. Rüegg*
Calcium in Muscle Activation (1986)

Volume 20: *J.-R. Truchot*
Comparative Aspects of Extracellular Acid-Base Balance (1987)

Volume 21: *A. Epple* and *J.E. Brinn*
The Comparative Physiology of the Pancreatic Islets (1987)

Volume 22: *W.H. Dantzler*
Comparative Physiology of the Vertebrate Kindney (1988)

Volume 23: *G.L. Kooyman*
Diverse Divers (1989)

Volume 24: *S.S. Guraya*
Ovarian Follicles in Reptiles and Birds (1989)

Volume 25: *G.D. Pollak* and *J.H. Casseday*
The Neural Basis of Echolocation in Bats (1989)

Volume 26: *G.A. Manley*
Peripheral Hearing Mechanisms in Reptiles and Birds (1989)

Volume 27: *U.M. Norberg*
Vertebrate Flight (1990)

Volume 28: *M. Nikinmaa*
Vertebrate Red Blood Cells (1990)

M. Nikinmaa

Vertebrate Red Blood Cells

Adaptations of Function to Respiratory Requirements

With 47 Figures

Springer-Verlag Berlin Heidelberg New York
London Paris Tokyo Hong Kong

Dr. Mikko Nikinmaa
Division of Physiology
Department of Zoology
University of Helsinki
Arkadiankatu 7
00100 Helsinki, Finland

ISBN-13:978-3-642-83911-5 e-ISBN-13:978-3-642-83909-2
DOI: 10.1007/978-3-642-83909-2

Library of Congress Cataloging-in-Publication Data. Nikinmaa, M. (Mikko), 1954– Vertebrate red blood cells : adaptations of function to respiratory requirements / M. Nikinmaa. p. cm. – (Zoophysiology ; v. 28) Includes bibliographical references. 1. Erythrocytes. 2. Vertebrates–Physiology. I. Title. II. Series. [DNLM: 1. Erythrocytes– metabolism. 2. Oxygen–blood. 3. Oxygen–pharmacokinetics. W1 Z0615M v. 28 / WH 150 N692v] QP96.N55 1989 596'.0113–dc20 DNLM/DLC 89-21820

Typesetting: International Typesetters Inc., Makati, Philippines

2131/3145-543210 – Printed on acid-free paper

Dedicated to Barbro, Kristian and Markus

Preface

This book reviews the respiratory function of vertebrate red cells. I have defined the phrase "respiratory function" broadly to include, in addition to the actual oxygen and carbon dioxide transport, erythropoiesis, haemoglobin synthesis, red cell structure, the deformability of red cells in circulation, ion and substrate transport across the cell membrane, cellular metabolism, and control of cellular volume and pH. All of these aspects of the red cell function may affect gas transport between the respiratory epithelia and the tissues.

Throughout the book, I have tried to relate our current knowledge about the nucleated red cell function to the wealth of information about the function of mammalian red cells. However, whenever possible, I have placed the emphasis on the nucleated red cell function for two reasons. First, the erythrocytes of 90% of vertebrate species are nucleated, and, second, nucleated red cell function has not been reviewed earlier in a single volume. This being the case, I have tried to make the reference list as complete as I could with regard to nucleated red cells. I hope that the approach adopted is useful for both comparative and human physiologists.

Many people have contributed to the making of this book directly or indirectly. Antti Soivio started me in this field. Prof. Henrik Wallgren has always encouraged fresh scientific ideas in his department. My present ideas of red cell function have been influenced by work carried out with Prof. Roy E. Weber, Prof. Wray H. Huestis, Prof. David J. Randall, Frank Jensen, Aimo Oikari, Heikki Tuurala, Eira Railo, Bruce L. Tufts, Joseph J. Cech Jr, John F. Steffensen, Peter G. Bushnell, Annika Salama, Niina Marttila, Leona Mattsoff, Kirsti Tiihonen, Marita Paajaste and many other co-workers and friends.

Prof. David J. Randall, the editor responsible for this volume, not only suggested that I write the book, but also had many valuable comments on the manuscript. Parts of the manuscript were also read by Holmer Sordyl, Robert G. Boutilier and Bruce L. Tufts. They also gave me access to their unpublished material. Marita Paajaste, Leona Mattsoff, Kirsti Tiihonen and Annika Salama have helped me with different aspects of the work, not least in compiling the reference list.

This work has been financially supported by the University of Helsinki and the Finnish National Research Council for Science.

Helsinki, Winter 1989/90 MIKKO NIKINMAA

Contents

Chapter 1

Introduction

Up to the present, most of the available reviews on red cell function have been heavily biased towards mammalian, especially human, red cells. As a result, most physiologists' picture of the red cell is overly simplistic – the nucleated red cells of all nonmammalian vertebrates are metabolically more versatile than mammalian red cells. They produce energy aerobically, their membrane transport pathways are more varied than those of the human erythrocytes and their function is often under hormonal control. The major aim of the present book is to relate our current knowledge about nucleated red cell function to the wealth of information about the function of mammalian red cells. In many instances this amounts to pointing out gaps in our knowledge, in others to presenting personal views on the basis of a limited amount of experimental data. Current knowledge about red cell function in many vertebrate groups is minimal. For example, very little information is available on the red cells of reptiles. Also, the information on fish is limited to a few species, especially rainbow trout.

The second purpose of the book is to relate how different aspects of red cell function are involved in the regulation of gas transport. Although red cells have been intensively used in studies of amino acid, nucleoside and ion transport across the red cell membrane, these results have seldom been used by respiration physiologists. Similarly, the structure of the red cell and the control of red cell structure have been intensively studied, but the relation of these studies to oxygen transport – or to ion transport – is, as yet, relatively unexplored. Furthermore, erythropoiesis and the developmental changes in haemoglobin components have been studied from two angles, oxygen transport and developmental biology, with relatively little overlap.

The contents of the book can be summarized by a brief description of the factors influencing circulatory gas transport. The relationship between the oxygen uptake by the animal (M_{O2}), the circulatory flow rate (cardiac output; Vb) and the oxygen concentration in the arterial (Ca_{O2}) and mixed venous (Cv_{O2}) blood can be described by the Fick equation

$$M_{O2} = Vb \cdot (Ca_{O2} - Cv_{O2}).$$

The oxygen concentration of blood depends on the partial pressure of oxygen (P_{O2}), the number of red cells per unit volume of blood and their haemoglobin concentration, and the oxygen affinity of haemoglobin.

The number of red cells per unit volume of blood depends ultimately on the balance between erythropoiesis and red cell breakdown. The erythropoietic patterns of different vertebrates and the senescence of red cells are described in Chapter 2. Chapter 2 also describes how, in the short term, the number of circulating red cells can be adjusted via a liberation of erythrocytes from storage organs or via changes in plasma volume.

The amount of functional haemoglobin within the cell depends on the age of the red cells, on the oxidative stresses experienced by the cells and on the reduction of oxidized haemoglobin back to functional state. Factors influencing haemoglobin synthesis are the subject of Chapter 3, and reduction of methaemoglobin is described in Chapter 10.

The oxygen affinity of haemoglobin depends on the intrinsic properties of the haemoglobin molecules, on the temperature and on the properties of the microenvironment of haemoglobin, i.e. the properties of the cytoplasm of the red cell. The major factors influencing the composition of the cytoplasm of the red cell are the energy metabolism of the cells, including the transport of substrates into the cell, and the ion transport across the red cell membrane. The oxygenation properties of haemoglobin, and the control of oxygen affinity by environmental factors, are described in Chapter 10. The transport of substrates into the red cells and the energy metabolism of red cells are discussed in Chapter 6. It appears that, in contrast to human erythrocytes, most nucleated erythrocytes do not use glucose as the primary fuel of energy metabolism, and that their energy metabolism is predominantly aerobic. Chapter 7 describes the major ion transport pathways, and Chapter 8 outlines how the different membrane transport pathways are involved in the control of red cell volume and pH. Again, in contrast to human red cells, which are practically unable to regulate their volume after osmotic disturbances, most nucleated red cells respond to both hypertonic and hypotonic shocks by an activation of secondarily active ion transport pathways, which utilize either the sodium or the potassium gradient to regulate the cellular volume. Sodium dependent acid extrusion is also used by many nucleated red cells in the control of cellular pH.

The cardiac work required to pump a given volume of blood through the vasculature depends on the vascular architecture and the viscosity of blood. The major determinants of the blood viscosity are the aggregation and deformability of the red cells, discussed in Chapter 5. Both of these factors depend on the structure of the red cell, outlined in Chapter 4. The structure of nucleated red cells, which is more complex than that of mammalian red cells, affects their flow properties. For example, they are more resistant to shear deformation than mammalian red cells. The dynamic molecular structure of the red cell membrane depends on the energy metabolism of the cell. It is also affected by the volume and the shape of the red cell. The deformability of the red cells also influences their microcirculatory behaviour, thereby affecting tissue oxygenation.

Thus, the properties of red cells affect all the components of the Fick equation. Furthermore, all the aspects of red cell function, including the structure and deformability of the cells, are involved in the oxygen transport function of the red cell.

The properties of red cells also influence carbon dioxide transport, as discussed in Chapter 9. Carbonic anhydrase, packaged within the red cells, acts in the tissues to speed up the hydration of carbon dioxide to bicarbonate, whereby the carbon dioxide gradient in the tissues can be maintained, and in the lungs to speed up the dehydration of bicarbonate, whereby carbon dioxide excretion can be facilitated. The concentrations and properties of intra- and extracellular buffers influence the amount of bicarbonate formed from carbon dioxide in the tissues and the amount of bicarbonate dehydrated to carbon dioxide in the gas exchange organs. Furthermore, the different time courses of intra- and extracellular interconversion between bicarbonate and carbon dioxide, and bicarbonate transport across the red cell membrane play an important role, both in the control of carbon dioxide transport and the red cell pH.

Chapter 2

Erythropoiesis and the Control of Circulating Red Cell Number

2.1 Erythropoietic Sites

The major erythropoietic sites in different vertebrate groups are given in Table 2.1. Basically, two sets of erythrocytes, primitive and definitive, are produced in all vertebrates. Primitive erythrocytes are produced in the yolk sac: blood islands are formed in the splanchnic mesoderm (Jordan 1933; Stohlman 1970).

Table 2.1. The major erythropoietic tissues in vertebrates

Group	Embryo	Adult	Comments
Cyclostomes	Yolk sac Spiral valve of intestine Nephric fold Circulation	Fat column Circulation	Erythropoiesis stops in lampreys during spawning migration
Elasmobranchs	Yolk sac Circulation	Spleen Circulation	
Teleosts	Yolk sac Circulation	Spleen Kidney	Maturation of definitive (adult) erythrocytes occurs in circulation
Amphibians	Yolk sac Intertubular regions of mesonephric kidney Liver Circulation	Spleen Bone marrow Circulation Heart	Bone marrow erythropoietically active in metamorphosis and after hibernation
Reptiles	Yolk sac Circulation	Spleen Bone marrow	Bone marrow major site in lizards
Birds	Yolk sac	Bone marrow Thymus	
Mammals	Yolk sac Liver Spleen	Bone marrow Spleen	Embryonic erythropoiesis in yolk sac, foetal in liver or spleen; spleen erythropoiesis active in several adult hibernators and rodents

Based on the observations of Jordan and Speidel (1930); Jordan (1933, 1938); Andrew (1965); Kendall and Ward (1974); Mattison and Fänge (1977); Broyles et al. (1981); Potter et al. (1982); Fänge (1987); Frangioni and Borgioli (1988)

In this stage, early erythroid precursors are liberated into the circulation, where they continue cell divisions, take up iron and produce haemoglobin (see e.g. Jordan and Speidel 1930; Jordan 1938; Yamamoto and Iuchi 1975; Mattison and Fänge 1977; Potter et al. 1982). During the relatively long larval life of cyclostomes and amphibians, other tissues also start producing erythrocytes. In cyclostomes, a diffuse, erythropoietic spleen is formed in the spiral valve of the intestine (Jordan 1933). Red cells are also produced in the nephric fold (Youson 1981; Potter et al. 1982). In amphibians, larval red cells are produced in the intertubular regions of mesonephric kidneys and in the liver (Broyles et al. 1981). The primitive erythrocytes of all vertebrate groups are easily distinguishable from definitive erythrocytes: in mammals they are nucleated and megaloblastic, in contrast to the nonnucleated definitive erythrocytes, and in other vertebrate groups they are round, in contrast to the oval shape of definitive erythrocytes (see e.g. Iuchi 1985). In nonmammalian vertebrates, embryonic erythropoiesis is followed by adult-like erythropoiesis, whereas in mammals the yolk sac erythropoiesis is followed early in the gestation by hepatic erythropoiesis, during which definitive erythrocytes are produced by liver and spleen (Jordan 1933; Lucarelli et al. 1968; Krantz and Jacobson 1970). Late in gestation and after birth, red cells are produced in the bone marrow.

2.2 Mammalian Erythropoiesis

In adult animals, erythrocytes are continuously produced in the erythropoietic tissues to replace aged cells. Red cell production increases markedly in situations of tissue hypoxia: in anaemia, carbon monoxide poisoning and hypobaric hypoxia (for an early review see Grant and Root 1952). On the other hand, erythropoiesis subsides after transfusion (see e.g. Krantz and Jacobson 1970). The most important factor affecting erythropoiesis is the ratio of oxygen supply to oxygen demand (see e.g. Fried et al. 1957; Krantz and Jacobson 1970; Jelkmann 1986).

Tissue hypoxia stimulates the production of erythropoietin, a glycoprotein hormone, mainly in the kidney (for reviews see Krantz and Jacobson 1970; Sherwood 1984; Jelkmann 1986). In the foetal stage, erythropoietin is produced mainly by extrarenal sites, e.g. the liver (e.g. Hågå and Kristiansen 1981). The oxygen tension of renal cells appears to be critical for erythropoietin production. It has been suggested that erythropoietin would be produced either in the glomerular mesangial or proximal tubule cells (for a review see Jelkmann 1986). Recent results of Lacombe et al. (1988), however, indicate that the sites of erythropoietin synthesis are most likely the peritubular cells: erythropoietin mRNA synthesis occurred exclusively in these cells in the murine hypoxic kidney. Intrarenally, prostaglandins affect erythropoietin synthesis (e.g. Gross and Fisher 1980), and several hormones systematically exert an influence on erythropoietin production (see Fig. 2.1. and Jelkmann 1986).

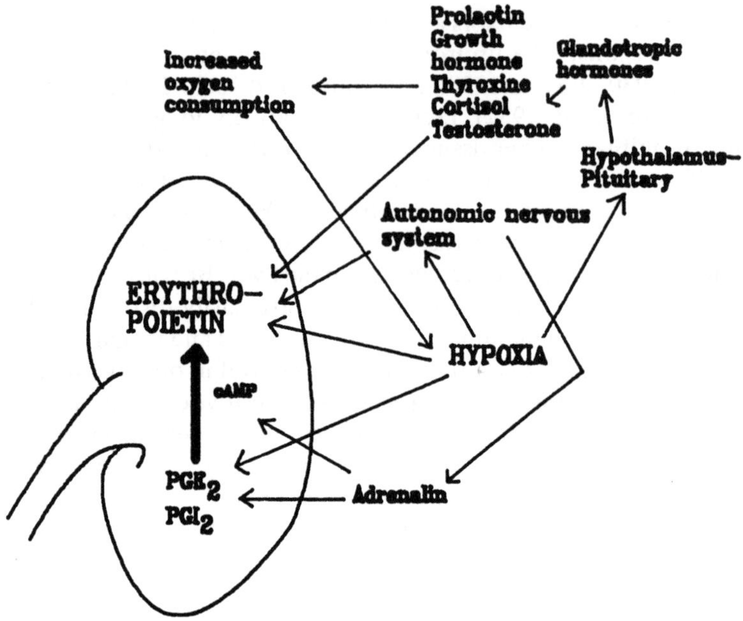

Fig. 2.1. A schematic representation of factors that appear to be involved in the regulation of renal erythropoietin production. (After Jelkmann 1986)

The sequence of events in erythrocyte formation is summarized in Fig. 2.2. Initially, it involves the proliferation of the most primitive haemopoietic stem cells. These cells appear to produce progenitors both to the myelopoietic and to the lymphopoietic cell lines (Abramson et al. 1977). The stem cells of myeloid line (colony forming unit-spleen; CFU-S) can differentiate to either erythrocytes, granulocytes or megakaryocytes. The proliferation (and self-renewal) of CFU-Ss appears to be under the influence of multilineage haemopoietic growth factors (multiHGF which include interleukins 1–6; see Iscove et al. 1985; Whetton and Dexter 1986; Dexter et al. 1988). In addition, CFU-S proliferation is affected by β_1 adrenergic and cholinergic agonists, testosterone and prostaglandin E_2 (for a review see Jelkmann 1986.). It is not clear how CFU-Ss are committed to the different cell lines. Till et al. (1964) proposed that the stem cell commitment would be a random event. Trentin (1970) formulated the haemopoietic microenvironment concept, which suggests that anatomical niches in marrow and spleen instruct stem cells as to their direction of differentiation. Interestingly, Jordan (1933) suggested that the formation of either granulocytes or erythrocytes would depend on the position of the progenitors in the vasculature of the haematopoietic tissue. Goldwasser (1975) has suggested that the stem cells would be committed to the different cell lines by inducers (humoral factors) during distinct periods of their cell cycle. Another possibility is that the stem cell differentiation is a result of either orderly (Nicola and Johnson 1982) or random (Ogawa et al. 1983, 1985) loss of lineage potential.

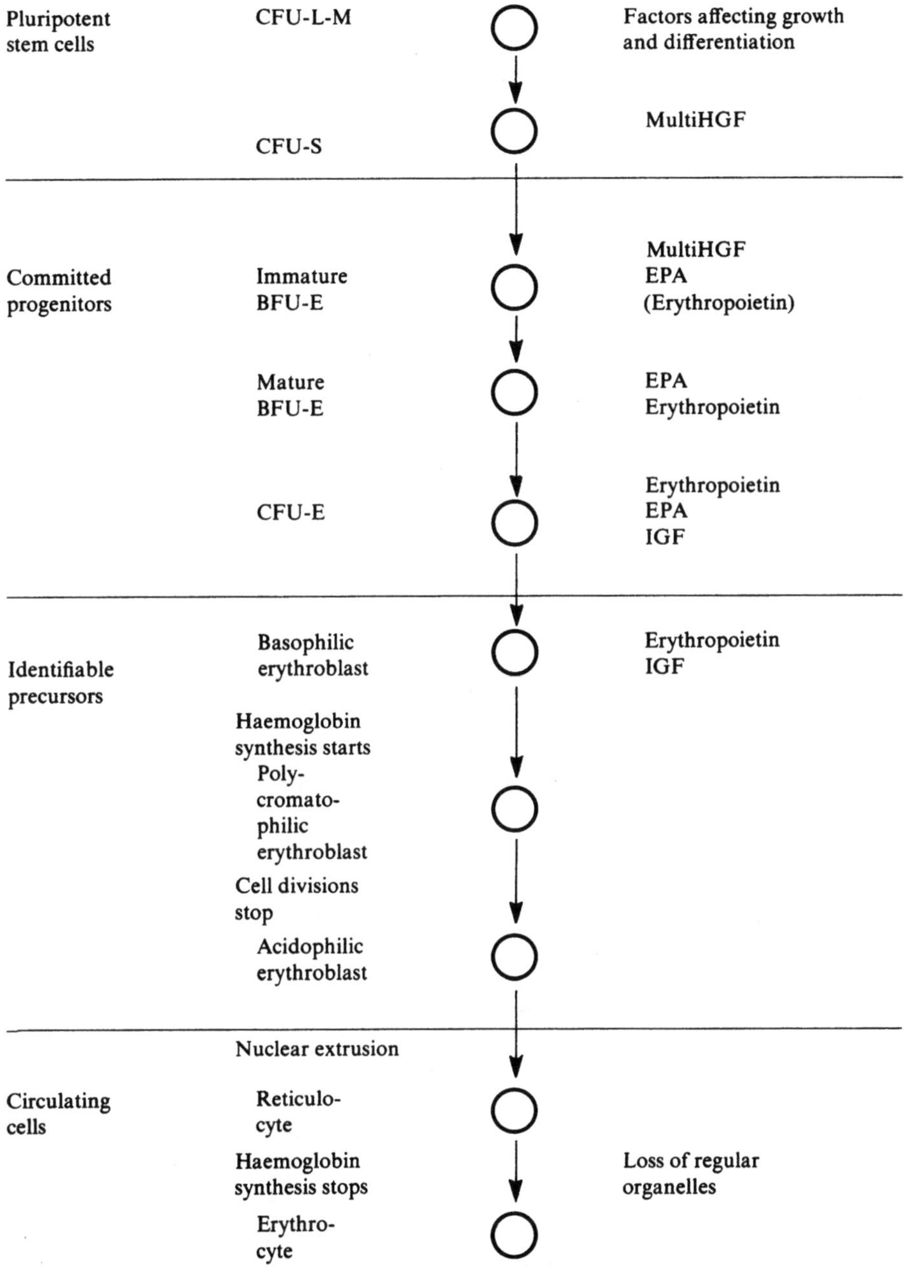

Fig. 2.2. A schematic representation of mammalian erythropoiesis. Based mainly on Jelkmann (1986), with additional information from Axelrad et al. (1974); Iscove et al. (1985); Gregory (1976); Kurtz et al. (1982); Gasson et al. (1985); Harrison et al. (1974); Denton and Arnstein (1973); Repasky and Eckert (1981) and Rapoport (1985)

The first cells committed to erythroid development are called erythroid burst-forming units (BFU-E). These cells produce large bursts of haemoglobin-synthesizing cells when cultured with high doses of erythropoietin (Axelrad et al. 1974). The BFU-Es are initially relatively insensitive to erythropoietin (cf. Iscove 1977; Iscove and Guilbert 1978), but are sensitive to "burst-promoting activity" (BPA). The presence of T-lymphocytes, and possibly monocytes/macrophages (or a medium conditioned by these), are required for burst-promoting activity (see e.g. Cline and Golde 1979; Zanjani and Kaplan 1979; Dainiak and Cohen 1985). With maturation of BFU-Es, their sensitivity also to erythropoietin increases (Gregory 1976).

The erythroid colony-forming units (CFU-E) appear to be the major target of erythropoietin: the erythropoietin concentration needed for their maximal in vitro growth is up to two orders of magnitude lower than for mature BFU-Es (Gregory 1976) and their growth is not affected by BPA (Gregory and Eaves 1977). Despite the importance of erythropoietin in the proliferation of CFU-Es, other factors influencing their proliferation have recently been found. Fagg's (1981) experiments suggested that a factor distinct from erythropoietin may stimulate late erythroid differentiation. Kurtz et al. (1982) showed that insulin-like growth factor I (IGF I) induced CFU-E proliferation. Gasson et al. (1985) purified erythroid potentiating activity (EPA) which also stimulated, in contrast to BPA, the growth of CFU-Es.

The CFU-Es produce clusters of erythroblasts, in which haemoglobin synthesis is initiated (see Harrison et al. 1974) and continues up to the reticulocytic stage. Cell divisions stop in polychromatophilic erythroblasts (Denton and Arnstein 1973). The denucleation of erythroblasts occurs as a result of nuclear extrusion (Simpson and Kling 1967; Repasky and Eckert 1981).

Reticulocytes lose their cellular organelles during maturation to erythrocytes. According to Rapoport (1985), the key step in the maturation of reticulocytes is the destruction of mitochondria, since the reduction of ATP production leads to a loss of protein synthesis (e.g. globin synthesis) and other ATP-requiring processes. Rapoport et al. (1985) have shown that the initial step in the degradation of mitochondria is the formation of lipoxygenase mRNA, followed by massive synthesis of the enzyme which lyses mitochondrial membranes. Thereafter other proteases and phospholipases complete mitochondrial degradation.

The erythroid differentiation from the stem cell to the erythrocyte involves ca. 13 cell divisions during normal red cell production. Of these, only three to six divisions occur in recognizable erythroid precursors, erythroblasts (Monette 1983). The total amplification capacity of the erythroid cell line, with an average cell cycle time of 12–13 h, appears to be much larger than this, up to 20 cell divisions. Under normal steady-state conditions mature BFU-Es account for most of the erythroid progenitor cell amplification. Under conditions in which erythropoiesis is stimulated, the increase in erythroid cell production is a consequence of added divisions in the earlier, more primitive cellular compartments, e.g. CFU-Ss (see Monette 1983).

2.3 Erythropoiesis in Nonmammalian Vertebrates

In the following sections, the factors influencing erythropoiesis in nonmammalian vertebrates are discussed. In the absence of contrasting information, it can be assumed that the process of erythropoiesis in nonmammalian vertebrates is largely similar to that of mammals (see Fig. 2.2). The major difference between mammals and other vertebrate classes is the fact that the nucleus is retained in all vertebrate red cells apart from mammalian red cells. Upon erythroblast-reticulocyte transformation, however, the nucleus of nonmammalian red cells becomes inert and highly condensed. The condensation of DNA in the terminal differentiation of nonmammalian erythroid cells may be due to the dephosphorylation of the major erythroid-specific histone (H5). After dephosphorylation, H5 interacts strongly with DNA, and may bring about the condensation of the chromatin (see e.g. Rapoport 1986).

2.3.1 Fish

Very little is known about the control of erythropoiesis in agnathans and elasmobranchs. In the hagfish (*Myxine* and *Eptatretus*) bleeding increases the number of erythroblasts, and occasional bursts of erythropoietic activity occur at other times (Mattison and Fänge 1977). In the lamprey (*Lampetra*) erythropoiesis stops during the upstream spawning migration (Percy and Potter 1977; Youson et al. 1979; Potter et al. 1982).

In teleost fish, erythropoiesis is stimulated by a decrease in the tissue oxygen supply to demand ratio, as in mammals. Erythropoiesis increases in fish made anaemic either by bleeding (e.g. Weinberg et al. 1976; McLeod et al. 1978) or by phenylhydrazine treatment (e.g. Chudzik and Houston 1983). Erythropoiesis is also stimulated in fish subjected to hypoxia (e.g. McLeod et al. 1978; Härdig et al. 1978). In addition, an increase in environmental temperature activates erythropoiesis both in normal (Hevesy et al. 1964) and anaemic (Chudzik and Houston 1983) fish, because of the increased tissue oxygen demand.

As in mammals, androgens and the activity of the hypothalamus seem to affect erythropoiesis in teleost fish. Hypophysectomy decreases red cell counts markedly. This anaemia can be prevented by chronic administration of adrenocorticotropic hormone, methyltestosterone or thyroid-stimulating hormone (Slicher 1961, cited by Slicher and Pickford 1968). These results are somewhat equivocal, because hypophysectomy and the above hormones may affect the apparent number of red cells in unit volume of blood by influencing the plasma volume. However, later observations by Pickering (1986) and Pottinger and Pickering (1987) have confirmed the effect of androgens on the erythropoiesis. Red cell numbers are elevated in sexually maturing male brown trout (*Salmo trutta*). The elevation coincides with the peak of 11-ketotestosterone concentration.

Circumstantial evidence suggests that thyroid hormones may also affect erythropoiesis in teleosts. There is a strong, partly temperature-independent seasonality in the haematological values of fish (see e.g. Denton and Yousef 1975).

Lane (1979) observed that in trout the proportion of immature erythrocytes decreased markedly in winter and increased in spring. Härdig and Höglund (1984) observed that the proportion of immature erythrocytes in young salmon started to increase in early spring, before an increase in the ambient temperature. The increase in the proportion of immature erythrocytes correlated best with increasing light. Notably, a peak in plasma thyroxine concentration occurs in young salmonids in the early spring (Virtanen, personal communication).

In view of the above, erythropoiesis in teleosts appears to be controlled largely by the same factors as in mammals. The existence of fish erythropoietin was first postulated by Zanjani et al. (1969) who found that a plasma factor from anaemic fish (blue gourami, *Trichogaster trichopterus*) increased erythropoiesis in normal animals. Later, Weinberg et al. (1976) showed that human erythropoietin at a high concentration stimulated erythropoiesis in starved kissing gourami (*Helostoma temmincki*). However, despite these observations, fish erythropoietin has not, as yet, been found and purified.

2.3.2 Amphibians

It appears that the stimuli affecting red cell production in amphibians are similar to those in teleosts and mammals. In the urodeles *Triton cristatus*, made anaemic by starvation, and *Necturus*, subjected to hypoxia, erythropoietic activity increases (cf. Jordan 1938). In another urodele, *Desmognathus ptroca*, human urinary erythropoietin causes a marked increase in the ^{59}Fe uptake in the spleen and circulating red cells (Gordon 1960). Plasma from bled frogs increased erythropoiesis in normal frogs (Rosse et al. 1963). Furthermore, testosterone and tri-iodotyronine treatments (capable of increasing erythropoiesis in mammals) also increased erythropoiesis in the frog, *Rana pipiens* (Bossak et al. 1948; Meints and Carver 1973). However, in contrast to the salamander, *Desmognathus ptroca*, frogs appear to be insensitive to human erythropoietin. In addition, the plasma factor from anaemic frogs does not influence erythropoiesis in polycythemic mice (Rosse et al. 1963).

2.3.3 Reptiles

Because the erythropoietic responses of mammals, teleosts and amphibians are largely similar, the observation of a complete lack of effect of environmental hypoxia on the erythropoiesis of the turtle (Altland and Parker 1955) was somewhat surprising. However, Meints et al. (1975) have since shown that bleeding and hypoxia elicit an erythropoietic response in the turtle, *Chrysemys picta*. The authors concluded that a reduction in the ratio of oxygen supply to oxygen demand is an important stimulus for erythropoiesis in turtles as well. Hormonal effects on erythropoiesis in reptiles have also been observed. Human erythropoietin and thyroxine stimulated erythropoiesis in the snake, *Xenochropsis piscator*, made anaemic by starvation (Pati and Thapliyal 1984). On the other hand, testosterone had no effect.

2.3.4 Birds

Rosse and Waldmann (1966) observed that both in the Japanese quail (*Coturnix coturnix*) and the chicken (*Gallus*) bleeding and acute hypoxia caused an increase in the carbon-14 labelled thymidine uptake into red cells. Furthermore, hypoxic quail serum and anaemic chicken serum were capable of stimulating erythropoiesis in the quail. So far, the erythropoiesis-stimulating factors purified from anaemic chicken serum have turned out to be similar or identical to transferrin; erythropoietin from chicken has not been purified (see Coll and Ingram 1978; Ingram 1985). Rosse and Waldmann (1966) also observed that human anaemic serum did not stimulate erythropoiesis in quail. In contrast, Thapliyal et al. (1982) reported that human urinary erythropoietin increased the erythrocyte number and haemoglobin concentration of spotted munia (*Lonchura punctulata*). These differences may be due to different concentrations of erythropoietin used. The effects of other hormones on the erythropoiesis of birds are also somewhat equivocal. Taber et al. (1943) and Domm et al. (1943) could increase the red cell counts of chicken with testosterone, whereas Thapliyal et al. (1982) reported that testosterone had no effect in spotted munia. However, both studies only report the red cell counts and/or haemoglobin concentrations. These values can be markedly affected by other effects of the hormone treatments, e.g. changes in plasma volume (see Sect. 2.5).

2.4 Senescence of Red Blood Cells

Circulating red cell number is determined by the balance between erythropoiesis and red cell breakdown. In some mammals, as in the mouse (e.g. Burwell et al. 1953), red cells are removed from the circulation in a random fashion. In others, as in man, the red cells are cleared from circulation in an age-dependent manner. Thus, in these animals the senescence of red blood cells, and their consecutive removal from circulation, are an important determinant of circulating red cell number. This is clearly seen in hibernating mammals. Although erythropoiesis is markedly slowed down in hibernation (e.g. Lyman et al. 1957; Brock 1960) there is either no change (e.g. Kramm et al. 1975) or an increase (e.g. Rosokivi and Suomalainen 1973) in the haemoglobin concentration and haematocrit value of a hibernating as compared to a nonhibernating hedgehog (*Erinaceus europeus*). This is the result of the markedly longer life span of red cells from hibernating animals than of those from non-hibernating animals (Brock 1960, 1964; Larsen 1968).

The senescence of red blood cells has recently been reviewed by Clark (1988). Various functional changes appear to be associated with senescence. First, the red cell potassium content decreases, and the red cells shrink (Astrup 1974; Cohen et al. 1976). The reduced water content causes an increase in intracellular haemoglobin concentration, and reduces red cell deformability (see Sect. 5.2.4.4). The reduced deformability may lead to the removal of the senescent red cells from circulation in the spleen. Second, calcium appears to accumulate within the cell (e.g. Shiga et al. 1985). An increase in intracellular concentration of calcium may lead to the potassium loss via the Gardos channel (see Sect. 7.2) and consecutive cell shrinkage. This, in turn, would cause the reduced deformability of the cells, and their removal from circulation in the spleen. Third, the sialic acid content on the red cell surface may decrease (see discussion in Clark 1988). Sordyl (unpublished data) observed that the old population of rainbow trout red cells had a lower concentration of sialic acid in the membrane than young cells. However, at present the role of sialic acid in the removal of red cells from circulation is unclear. For example, Sordyl (unpublished data) observed that rainbow trout red cells, from which sialic acid had been removed, deformed in the spleen. Fourth, specific structures on the red cell membrane form binding sites for serum antibodies. Macrophages then recognize the bound antibodies and phagocytose the cells (see e.g. Singer et al. 1986). One possible candidate for the antibody-binding sites on red cell membrane are aggregated band 3 proteins (see Sect. 4.2.2.4 and Low et al. 1985). Alternatively, the antibody-binding site may be a breakdown product of band 3 protein (Kay et al. 1988b).

In many instances the number of circulating red cells can decrease because of increased breakdown of red cells. For example, the macrocytic erythrocytes, produced in response to acute blood loss in the rabbit, are destroyed much faster than normocytic erythrocytes (Card and Valberg 1967). Red cell breakdown increases in the early stages of physical training in man (see Röcker et al. 1983), and paper and pulp mill effluents appear to increase the breakdown of erythrocytes in salmonid fish (McLeay 1973; Bushnell et al. 1985; Mattsoff and Nikinmaa 1987).

2.5 Erythropoiesis-Independent Factors Increasing Circulating Red Cell Number

The hypoxic stimulus increases erythropoietin titers in 1–3 days in man, and in 6–24 h in dogs and rodents (see e.g. Jelkmann 1986). The consequent increase in circulating red cell numbers in man at altitude is seen within 1–5 weeks (see Grant and Root 1952). Meints et al. (1975) observed an erythropoietic response in turtles in a 17-day exposure to hypoxia. In the Florida gar, 12–18 days were required for an increase in immature circulating erythrocytes (McLeod et al. 1978). In salmon, the proportion of immature erythrocytes in hypoxia increased from day 3 to day 20 of the exposure (Härdig et al. 1978). Because the production of new erythrocytes via erythropoiesis takes up to several weeks, the stimulation of erythropoiesis does not play a role in the short-term (minutes-hours) adjustments of blood oxygen carrying capacity.

The spleen functions as a storage organ for red cells in many vertebrates. The volume of the spleen is controlled mainly via the autonomic nervous system, but also by circulating catecholamines (see Davies and Withrington 1973; Nilsson

1983 for reviews). Red cells are liberated from the spleen when the capsular smooth muscle contracts. In mammals, the development of capsular smooth muscle is pronounced in cats and dogs. During muscle contraction, their spleen volume can decrease as much as 50–70 ml (see Davies and Withrington 1973). On the other hand, the capsular musculature in man is poorly developed. Thus, the spleen volume decreases only slightly, up to 5 ml, during muscle contraction. The splenic contraction in cats and dogs occurs as a response to haemorrhage, asphyxia, hypoxia, anaesthesia and emotion (see Davies and Withrington 1973). In hibernating hedgehogs, marked changes occur in the haematocrit value of the spleen and blood during periodic arousals and following bouts of hibernation. Soivio (1967) and Rosokivi and Suomalainen (1973) observed a marked increase in both the haemoglobin concentration and haematocrit value of blood on arousal, and a marked drop in both values when the animals re-entered hibernation. Opposite changes were observed in the haematocrit value of the spleen, indicating that liberation of red cells from and uptake into the spleen significantly contribute to the rapid changes in circulating red cell numbers in hibernating mammals.

Nerve stimulation also causes splenic contraction in nonmammalian vertebrates. The volume of toad (*Bufo marinus*) spleen decreases markedly upon nerve stimulation (Nilsson 1978). In the same species, exercise causes a marked increase in both haematocrit value and haemoglobin concentration of blood (Tufts et al. 1987a). Stevens (1968) observed that erythrocytes are liberated from the spleen of trout during exercise. In exercised and hypoxic yellowtail (*Seriola quinqueradiata*), the liberation of erythrocytes from the spleen could account for up to 40–50% of the total change in the blood oxygen carrying capacity (Yamamoto et al. 1980, 1983). Yamamoto (1987) has since extended his studies to several cyprinid species. Exercise-induced contraction of the spleen significantly increased the number of circulating erythrocytes in all the species studied. Nilsson and Grove (1974) observed that the stimulation of the splanchnic nerve, and addition of adrenaline or noradrenaline in the perfusion medium, caused a release of red cells from the perfused spleen of the cod (*Gadus morhua*). Similarly, splenic contraction as a result of nerve stimulation also occurs in the elasmobranchs, *Squalus acanthias* and *Scyliorhinus canicula* (Nilsson et al. 1975).

Changes in the plasma volume also influence the apparent number of circulating red cells (measured as rbc/unit volume). A short-term (< 10 min) physical exercise in man can decrease the plasma volume by more than 10% (see Röcker et al. 1983), leading to an increase in blood haematocrit value and haemoglobin concentration. The plasma volume of the rat (*Rattus norvegicus*) decreased during a 4-day food deprivation, and simultaneously the measured haematocrit value increased (Wright et al. 1977). Similarly, the plasma volume decreased and the haematocrit value increased in water-deprived chickens (Koike et al. 1983). Temperature affects the blood oxygen capacity of some sea snakes: blood oxygen capacity increased from 10° to 25–30°C and thereafter decreased (Pough and Lillywhite 1984). Hibernating sandvipers (*Cerastes cerastes* and *Cerastes vipera*) show marked decreases in haematocrit values as compared to nonhibernating specimens (Al-Badry and Nuzhy 1983). Although in this case a decrease in erythropoiesis may contribute, it is likely that an increase in plasma volume occurs, since the concentrations of plasma proteins, free amino acids, total lipids and glucose decrease. In fish, marked changes in plasma volume occur as responses to environmental changes. The plasma volume of hypoxic rainbow trout and yellowtail decreases significantly (Soivio, Nikinmaa and Egginton, unpublished data, see Nikinmaa 1981; Yamamoto et al. 1983; Sordyl, unpublished data). Yamamoto et al. (1980) also observed a reduction of plasma volume in exercised yellowtail. On the other hand, Sleet and Weber (1983) found that the blood volume of buffalo sculpins (*Enophrys bison*) increased 48 h after arterial cannulation.

The site of blood sampling affects the apparent number of circulating red cells (per unit volume). The haematocrit value of capillary blood is generally lower than that of the major vessels. This is due to at least the following factors: (1) the flow of red cells is faster than that of plasma in the capillaries (Farhaeus effect; see Skalak and Chien 1981), (2) there are stagnant plasma "pockets" in the capillaries (e.g. Duling and Desjardins 1987), and (3) the blood may be separated to high-haematocrit and low-haematocrit blood at bifurcations of blood vessels (e.g. Johnson 1971). Such separation of plasma and red cells at the capillary openings, "plasma skimming", occurs at least in the gill vasculature (Olsen 1984), and at the entrance of secondary vascular system (Steffensen et al. 1986) in fish. As a result of the plasma skimming in the gill vasculature (Fig. 2.3), prebranchial blood often has a lower haemoglobin concentration than postbranchial blood (e.g. Soivio et al. 1981; Nikinmaa 1981).

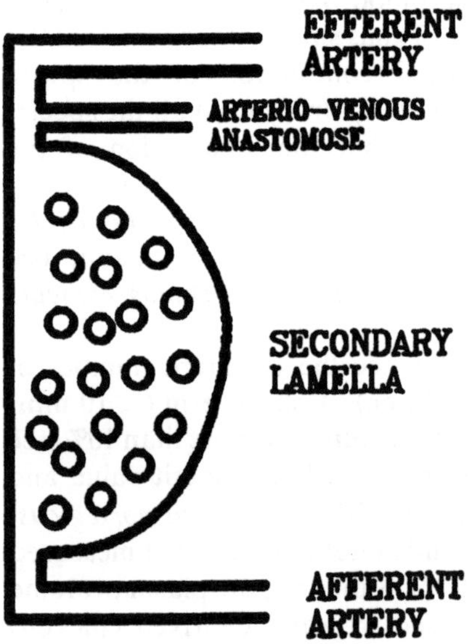

Fig. 2.3. A schematic picture of plasma skimming in the gills of teleost fish. Blood entering the arterio-venous anastomoses on the efferent side of the gills has a low haematocrit value, whereby the red cell number in unit volume of the efferent artery exceeds that of the afferent artery

Chapter 3

The Biosynthesis and Structure of Haemoglobin

Circulating erythrocytes of mammals are incapable of haemoglobin synthesis. Haemoglobin is produced mainly by erythroblasts and reticulocytes in the bone marrow. In addition, circulating reticulocytes synthesize some haemoglobin. The circulating immature erythrocytes of nonmammalian vertebrates actively take up iron and produce haemoglobin. Upon maturation haemoglobin synthesis slows down. The synthesis of haem and globin chains are independent events. After the synthesis of the haem group and globin chains, they are assembled to form the characteristic quarternary structure of haemoglobin.

A detailed account of the molecular structure of haemoglobin has been given, e.g. by Dickerson and Geis (1983) and Bunn and Forget (1986). The primary functional unit of the haemoglobin molecule is haem group, in which ferrous iron is surrounded by porphyrin ring (Fig. 3.1). The haem group is surrounded by globin chain, which forms a hydrophobic pocket around the haem.

In most vertebrates, there is marked heterogeneity of haemoglobin components. Furthermore, marked changes in the haemoglobin components (haemoglobin switching) occur during the ontogeny of most vertebrates. Haemoglobin heterogeneity can arise from allelic or nonallelic variation in the genes coding for haemoglobin chains, from posttranslational modifications of the globin chains (e.g. acetylation or glycolysation), errors in gene expression,

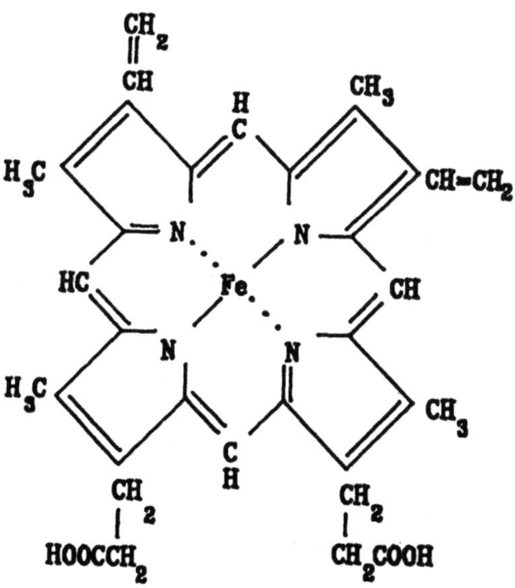

Fig. 3.1. Structure of the haem group

variation in the subunit assembly or variations in the number of subunits in the quarternary structure of haemoglobin (see e.g. Garrick and Garrick 1983; Bunn and Forget 1986).

3.1 Uptake of Iron into Erythroid Cells

In most physiological conditions, iron is in the ferric form. At neutral pH most ferric salts hydrolyse to form insoluble ferric hydroxide, and thereby iron as such can be very poorly transported in plasma. To facilitate iron transport, an iron-binding protein, transferrin, is present in all vertebrates (see Williams 1982; Huebers and Finch 1987). Transferrin is a single chain polypeptide (MW 80000) with two iron-binding sites per molecule (Aisen and Listowsky 1980; Aisen 1982; Huebers and Finch 1987). The strong reversible binding of iron to transferrin requires the presence of an anion, in body fluids normally bicarbonate (see Aisen 1982).

Approximately three quarters of the plasma iron turnover is directed towards the developing erythroid cells (see e.g. Paterson et al. 1984). The mechanism of iron transport into the cell is schematically depicted in Fig. 3.2. In the first step of cellular iron uptake, the iron-transferrin complex (diferric transferrin) binds to the transferrin receptors on the cell membrane in all vertebrates. Several recent reviews (e.g. Newman et al. 1982; Seligman 1983; Trowbridge et al. 1984; May and Cuatrecasas 1985; Testa 1985; Huebers and Finch 1987) give detailed accounts of the different aspects of the structure and function of the transferrin receptor in mammals. In short, the molecule is a membrane spanning, dimeric glycoprotein with probably identical subunits (MW 90 000), both capable of binding one transferrin molecule.

In mammals, the number of transferrin receptors in the cell membrane depends on the stage of erythroid development (see e.g. Testa 1985). The receptors are present in all erythroid progenitors from CFU-Ss, but increase in number from immature BFU-Es to CFU-Es (Sieff et al. 1982; Lesley et al. 1983). In the erythroblasts, the number of transferrin receptors increases from early to intermediate normoblasts, and thereafter starts to decrease (Iacopetta et al. 1982). The receptors are present in reticulocytes, although their number decreases markedly during reticulocyte maturation (Frazier et al. 1982). Mature erythrocytes do not contain transferrin receptors. Transferrin receptors also disappear from the cell surface of definitive series of chicken erythrocytes during maturation, but are retained on the surface of primitive erythrocytes throughout their life span (Schmidt et al. 1986). In other nonmammalian vertebrates, iron is rapidly taken up and haemoglobin synthesized by circulating immature erythrocytes (see Hevesy et al. 1964; Härdig 1978). Transferrin receptors are present in these circulating erythroid cells, as shown by Lim and Morgan (1984) for phenylhydrazine-treated toad, *Bufo marinus*. However, even in this case transferrin and iron uptake by mature erythrocytes is minimal.

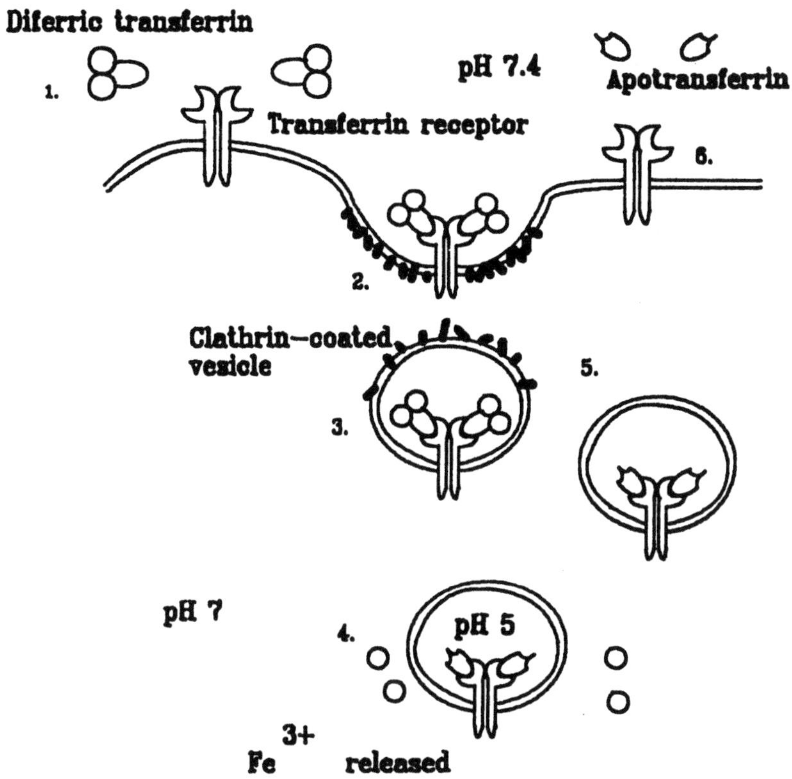

Fig. 3.2. Iron transport into erythroid cells. *1* Fe^{3+}-transferrin complex binds to the transferrin receptor. *2* The iron-transferrin-transferrin receptor complex is invaginated and *3* endocytosed in a clathrin-coated vesicle. *4* The endocytotic vesicle is acidified, whereby the iron is liberated into the cytoplasm. *5* The apotransferrin-transferrin receptor complex is recycled back to the cell membrane. *6* Apotransferrin is released from transferrin receptor when encountering the neutral pH of the plasma

Studies with human leukemic cell lines, Hela cells and fibroblasts have indicated that the number of transferrin receptors on the cell surface may be modulated by the intracellular concentration of chelatable iron (see Testa 1985; Rouault et al. 1985). Delivery of iron in the form of soluble salts, as diferric transferrin or as hemin, results in a decrease in the receptor biosynthesis, whereas chelation of intracellular iron with desferrioxamine enhances receptor biosynthesis (see Rouault et al. 1985).

After the binding of diferric transferrin to the transferrin receptors, the complex is endocytosed in vesicles, which are acidified down to a pH of ca. 5 (Tycho and Maxfield 1982; Paterson et al. 1984). It is possible that the endocytotic vesicles, observed in the haemoglobin-producing circulating erythroid cells e.g. of cyclostomes and elasmobranchs (see Mattison and Fänge 1977; Zapata 1980), are vesicles containing diferric transferrin-transferrin receptor complex. However, to date, the presence of transferrin or transferrin receptor in these vesicles has not been investigated.

The vesicular pH is probably lowered by proton pump, as shown by the effects of proton ionophores and metabolic inhibitors on the intravesicular pH (Paterson et al. 1984; see Mellman et al. 1986 for a review on the acidification of the endocytic pathways). The low pH of the endocytic vesicles causes iron to dissociate from the transferrin-transferrin receptor complex. Treatments which increase intravesicular pH markedly slow down iron release and decrease iron uptake into the cell (Paterson et al. 1984).

The iron-free transferrin (apotransferrin) remains bound to the receptor in the acid environment, and the complex is recycled to the cell membrane, where, upon encountering neutral pH, apotransferrin is released from the receptor (Dautry-Varsat et al. 1983).

The mechanism of iron transport to mitochondria after its dissociation from transferrin in the acid vesicles is somewhat unclear. Ferritin or unidentified iron-binding proteins have been presented as intermediates in the transport (see Testa 1985).

3.2 Haem Synthesis

For detailed reviews on the subject, see e.g. Kaplan (1970), Maines and Kappas (1977) and Ibrahim et al. (1983). Figure 3.3 outlines the pathway of haem synthesis and the intracellular location of the different enzymes.

The formation of δ-aminolevulic acid (ALA) from glycine and succinyl coenzyme A, in a reaction catalyzed by ALA synthase, appears to be the rate-limiting enzymatic reaction in haem synthesis, at least in hepatocytes (cf. Granick and Urata 1963; Ibrahim et al. 1983). The activity of this mitochondrial enzyme is largely controlled by haem. In undifferentiated stem cells of the erythroid line, the presence of haem induces the enzyme, whereas in more differentiated erythroid cells, erythroblasts and reticulocytes, haem inhibits the enzyme (Ibrahim et al. 1978, 1983). In addition, ALA synthase can be induced by 5-β-H-steroids, both in erythroid cells and hepatocytes (Levere et al. 1967; Sassa and Granick 1971), and by some drugs, like 3,5-diethoxycarbonyl-1,4-dihydrocollidine and allylisopropylacetamide in hepatocytes (see Ibrahim et al. 1983).

In addition to ALA synthase, some steps in the pathway of iron from extracellular transferrin to protoporphyrin may limit the rate of haem synthesis in erythroid cells (see Ponka and Schulman 1985). When the transferrin route for iron uptake is bypassed by the use of pyridoxal isonicotinoylhydrazone (Ponka and Schulman 1985), iron uptake and haem synthesis can be speeded up. The exact controlling step is not known, but may involve either the activity or processing of transferrin receptors (see Pelicci et al. 1982), the rate of release of iron from transferrin (see Paterson et al. 1984), the reduction of ferric iron, transport of iron from transferrin to ferrochelatase or the activity of ferrochelatase (Bottomley 1968; Rutherford et al. 1979). The mitochondrial enzyme ferrochelatase catalyzes the reaction between ferrous iron and protoporphyrin. It is inhibited by oxidized haem (hemin; Bottomley 1968).

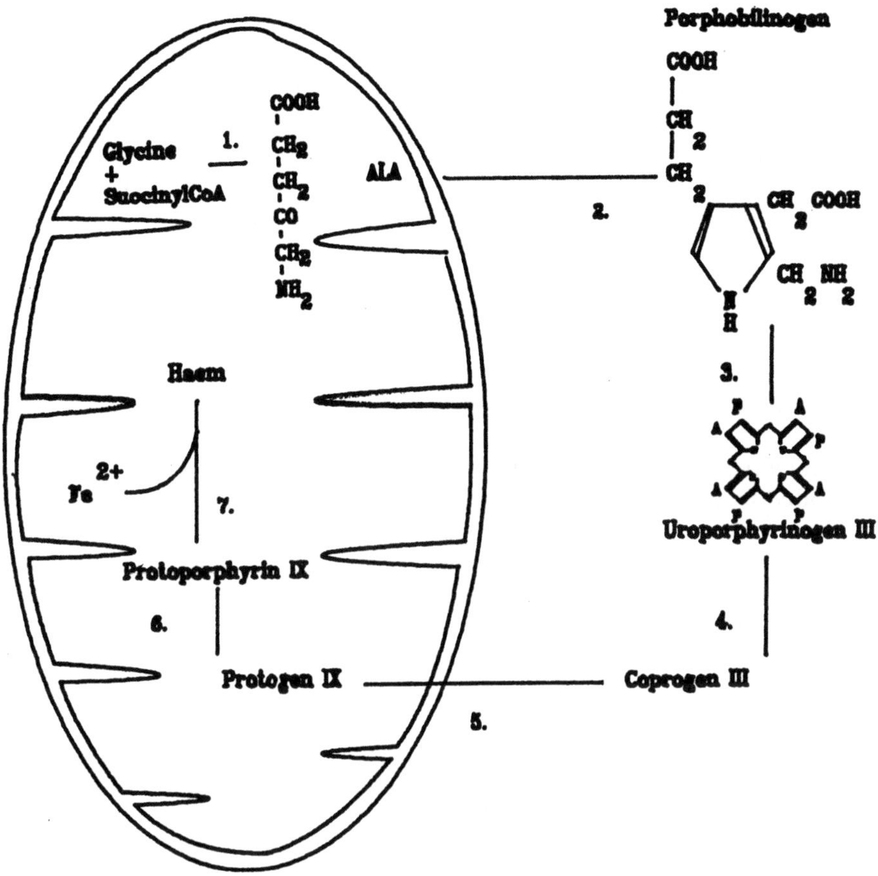

Fig. 3.3. Biosynthesis of haem. *1* The mitochondrial enzyme δ-aminolevulinic acid synthase catalyzes the formation of δ-aminolevulinic acid (ALA) from glycine and succinyl coenzyme A. *2* ALA is transported across the mitochondrial membrane, and dehydrated to porphobilinogen (PBG) in a reaction catalyzed by ALA dehydratase. *3* Four PBG molecules are condensed to form uroporphyrinogen III (urogen III) in reactions catalyzed by PBG deaminase and urogen III cosynthetase. *4* Urogen III decarboxylase converts the acetyl side chains of urogen III to methyl groups to yield coproporphyrinogen III (coprogen III). *5* Coprogen III is transported to mitochondria, where coprogen oxidase decarboxylates the propionate side chains of rings A and B to vinyl groups, yielding protoporphyrinogen IX (protogen IX). *6* Protoporphyrinogen oxidase removes six hydrogen atoms to yield protoporphyrin IX. *7* Ferrochelatase inserts ferrous ion into protoporphyrin IX to yield haem

3.3 Globin Synthesis

3.3.1 Structure of Globin Genes

In man, the three genes coding for α-like globins (ξ, α_1 and α_2) are found in chromosome 16, and the five genes coding for β-like globins (ε, Gγ, Aγ, δ and β) in chromosome 11 (for details of globin gene structure in mammals see Maniatis et al. 1980; Efstratiadis et al. 1980; Forget 1983; Garrick and Garrick 1983; Collins and Weissman 1984; Bunn and Forget 1986). In most mammals studied, there are several genes coding for both α and β globin chains: adult goats have eight β globin genes and four α globin genes, adult dogs two β and two α genes, adult rats eight β and six α globin genes etc (see Garrick and Garrick 1983). In addition, embryonic and foetal globin genes are present in most mammals. The exceptions appear to be camels and the donkey, for which no evidence is available for globin chain heterogeneity.

Going from the 5′ end of the DNA chain, the human globin genes are arranged in the order coding for embryonic, foetal and adult haemoglobins. The genes for adult globin chains are separated from the foetal ones by pseudogenes. However, this sequence of the genes is not necessary for the sequential, developmental expression of the genes during ontogeny (see Sect. 3.5.2), as indicated by a different sequence in both the goat (e.g. Lingrel et al. 1983) and the chicken (e.g. Dolan et al. 1981). In both these species, the adult β genes are surrounded by the embryonic/foetal genes. In the frog, *Xenopus laevis,* both α and β chains occur in a single chromosome in two clusters (Hosbach et al. 1983), which may have arisen by tetraploidization (Jeffreys et al. 1980). In the direction of transcription, there are first two larval α genes, one adult α gene, one adult β gene and two larval β genes. The structure and organization of reptilian and fish globin genes have not been studied. However, the marked structural and functional heterogeneity of teleost haemoglobins indicates that gene duplications have occurred in the genes coding for both α-like and β-like haemoglobins. For example, Tsuyuki and Ronald (1971) have observed that *Oncorhynchus nerka* haemoglobins have three different α chains and five different β chains. Powers (1980) has reviewed the haemoglobin systems of *Fundulus heteroclitus* and *Catostomus clarkii.* In *Fundulus,* two different α and two different β chains have been observed. In *Catostomus clarkii,* there are at least two different α chains, and four different β chains. Furthermore, the embryonic haemoglobins appear to be distinct from adult haemoglobins in many cases (see Sect. 3.5.2 and Iuchi 1985).

The globin genes are separated from each other by flanking sequences which are never read out as messenger RNA. Within every globin gene there are two introns, base sequences transcribed to messenger RNA, but "spliced" away from the mRNA during post-translational processing, and not used to direct protein synthesis in the ribosomes. The three regions of DNA that are translated into globin chains are called exons. This exon-intron-exon-intron-exon sequence of globin genes appears to be the same in all vertebrates (see e.g. Collins and Weissman 1984). Interestingly, the introns occur at homologous positions in both α and β globin genes (see Bunn and Forget 1986). They subdivide the globin gene into exons which encode distinct functional (or structural) regions of the protein. The central exon encodes the entire haem pocket, and also the $\alpha_1\beta_2$ intersubunit contacts (Eaton 1980; Craik et al. 1980, 1981).

3.3.2 Globin Chain Synthesis and Its Regulation

The transcription of globin genes is initiated at the "cap" site, which constitutes the 5′ end of the mature cytoplasmic globin mRNA. Three preserved DNA sequences occur in the promoter region relatively close to the cap site: ATAA sequence, constituting the "ATA box" ca. 30 nucleotides from the cap site, the CCAAT sequence 70–80 nucleotides from the cap site and GGGG(C/T)G or C(A/T)CCCCC sequence 80–100 nucleotides from the cap site (see Bunn and Forget 1986). The "ATA box" is necessary, and often sufficient, to direct globin transcription in cell-free systems, whereas CCAAT is required for efficient transcription of globin genes transferred into tissue culture cells (see e.g. Dierks et al. 1981; Grosweld et al. 1982; Collins and Weissman 1984; Bunn and Forget 1986). In addition to the three regions mentioned above, three other regions of the promoter region are important for the induction of transcription of erythroid cells. Two of these regions bind the erythroid specific nuclear factor (NF-E1) and the third a ubiquitous CCAAT-box factor (CP1; see deBoer et al. 1988). The site of transcription termination appears to be unknown at the moment (see e.g. Proudfoot 1984; Wickerson and Stephenson 1984; Gil and Proudfoot 1984).

Very rapidly after transcription, the 5′ end of the mRNA is "capped", and the 3′ end polyadenylated at the AATAAA base sequence in the 3′ noncoding portion of the mRNA, whereafter the introns are sequentially spliced (see e.g. Bunn and Forget 1986).

Transcription of the globin genes to messenger RNA appears to be an important regulatory event in globin gene expression (e.g. Sahr and Goldwasser 1983). The specific factors that activate globin gene transcription appear to be poorly known. Nijhof et al. (1987) have shown, however, that erythropoietin may induce globin mRNA transcription. Undermethylation and increased DNAse I sensitivity are often seen in actively transcribed globin genes. The globin genes are methylated at specific sites in tissues that do not produce haemoglobin, e.g. sperm and brain tissue, whereas in erythroid cells certain sites in or near the expressed globin genes are un- or undermethylated in chicken (Haigh et al. 1982; Peschle et al. 1983). In man, active globin genes are undermethylated (Van der Ploeg and Flavell 1980). Mavilio et al. (1983) observed a strong correlation between the undermethylation in the close flanking sequences of globin genes and their expression. The globin genes of the chicken erythroid cells, present at 20–23 h of development, are chromosomically and transcriptionally inert (Groudine and Weintraub 1981), but during the next 10 hours the genes become undermethylated and sensitive to DNAse I. Similarly, the genes expressing globin in human erythroid cells become sensitive to DNAse I (Groudine et al. 1983). The active globin genes thus appear to be sensitive to DNAse I and undermethylated. However, Adams and Burdon (1985) emphasize that "it is easy to overinterpret these results and claim a good correlation between undermethylation and gene expression, but there are too many partial or complete exceptions to the rule for more than limited confidence to be placed in this simple idea".

The region on the 3′ side of the chicken β globin gene contains an enhancer region, which functions in transcriptionally active chicken erythrocytes but not in chicken fibroblasts (see e.g. Evans et al. 1988). Erythroid cells contain a protein that binds to two sites on the enhancer region. Mutation of these sites so that the protein (erythroid specific factor 1) cannot bind to the enhancer region leads to a loss of enhancing ability (Evans et al. 1988). These data suggest that the erythroid specific factor 1 may play an important role in globin gene transcription.

The globin messenger RNA of erythroid cells appears to be very stable, since globin synthesis by yolk sac erythroid cells continues at a high rate even if RNA synthesis is inhibited by actinomycin D, whereas the synthesis of nonhaem proteins decreases markedly (Fantoni et al. 1968; Marks et al. 1968). Hemin affects the transcription of globin genes, at least in the human and murine erythroleukemic cell lines K562 and T3-C12 respectively: globin mRNAs are markedly accumulated in the presence of hemin (Ross and Sautner 1976; Charnay and Maniatis 1983).

Haem and hemin affect the translation of mRNA to globin polypeptide chains. The translation is initiated by the eukaryotic peptide chain initiation factor (eIF-2) (see e.g. Gupta et al. 1983; Gross and Kaplansky 1983; London et al. 1983; Siekierka and Ochoa 1983). In the absence of haem/hemin, the haem/hemin-controlled translational repressor phosphorylates eIF-2 at specific sites, whereby chain initation is inhibited (e.g. Gross and Rabinovitz 1972; Gross and Kaplansky 1983; London et al. 1983; Siekierka and Ochoa 1983). Thus, in the absence of hemin, polyribosomes disaggregate and aggregate when hemin is added (Grayzel et al. 1966; Gross and Rabinovitz 1972). The effects of haem/hemin on both the transcription and translation of globin genes indicate that haem synthesis may control the production of globin in the cells (see e.g. Ibrahim et al. 1983).

3.4 Subunit Assembly of Haemoglobin

The subunit assembly of haemoglobin has been reviewed recently by Bunn (1987). A slight mismatch in globin chain synthesis can be tolerated within the erythroid cells, because the proteolytic enzymes attack free globin chains but not assembled dimers or tetramers. The rate-limiting step of haemoglobin assembly is the formation of $\alpha\beta$-dimer. This reaction is nearly irreversible, whereas the combination of dimers to form tetramers is reversible (see Bunn 1987). The association of α and β chains to dimers depends on the electrostatic attraction between the positively charged α subunit (at pH 7.2 2.4 positive charges) and the negatively charged β subunit (at pH 7.2 2.5 negative charges; see Bunn and McDonald 1983; Bunn 1987). The importance of the electrostatic attraction in the association has been studied by Mrabet et al. (1986) using globin chains with different net surface charges. The proportion of dimers with the different β chains was a function of their surface charge, showing the importance of electrostatic attraction.

3.5 Haemoglobin Structure in Vertebrates

The globin chains have 140–160 amino acids, and have a molecular weight between 15 000 and 17 000 (human α chain has 141 amino acids, and a molecular weight of 15 126, and human β chain has 146 amino acids and a molecular weight of 15 867). The primary structure of globin chains is known for a wide variety of vertebrates. Table 3.1 gives the amino acid sequences of (haemo)globin chains for selected vertebrates. Although there are pronounced differences in the sequences, discussed below, some general features of the haemoglobin structure are apparent. In all cases studied, the globin chains contain eight helical segments (A-H) which are folded upon each other to form a globular tertiary structure. The folds can be abrupt, i.e. involving only one amino acid residue (between the helical segments A and B, B and C, and D and E) or more gradual, involving several amino acids in nonhelical segments (between helical segments C and D, E and F, F and G, and G and H).

Table 3.1. Amino acid sequences of the haemoglobin chains of various species[a]

			α chain				
	Panda	Pheasant	*Caiman*	*Carassius*	*Squalus*	*Petromyzon*	
						Pro	NA
						Ile	
						Val	
						Asp	
						Thr	
						Gly	
						Ser	
						Val	
1	Val	Val	Val	Ac-Ser	Val	Ala	
						Pro	
2	Leu	Leu	Leu	Leu	Leu	Leu	
3	Ser	Ser	Ser	Ser	Ser	Ser	A
4	Pro	Ala	Glu	Asp	Ala	Ala	
5	Ala	Ala	Glu	Lys	Ala	Ala	
6	Asp	Asp	Asp	Asp	Asp	Glu	
7	Lys	Lys	Lys	Lys	Lys	Lys	
8	Thr	Asn	Ser	Ala	Thr	Thr	
9	Asn	Asn	His	Val	Ala	Lys	
10	Val	Val	Val	Val	Ile	Ile	
11	Lys	Lys	Lys	Lys	Lys	Arg	
12	Ser	Gly	Ala	Ala	His	Ser	
13	Thr	Ile	Ile	Leu	Leu	Ala	
14	Trp	Phe	Trp	Trp	Thr	Trp	
15	Asp	Thr	Gly	Ala	Gly	Ala	
16	Lys	Lys	Lys	Lys	Ser	Pro	
17	Leu	Ile	Val	Ile	Leu	Val	
18	Gly	Ala	Ala	Gly	Arg	Tyr	
19	Gly	Gly	Gly	Ser	Thr	Ser	AB
20	His	His	His	Arg	Asn	Asn	B

Table 3.1. (Continued)

	α chain						
	Panda	Pheasant	*Caiman*	*Carassius*	*Squalus*	*Petromyzon*	
21	Ala	Ala	Leu	Ala	Ala	Tyr	
22	Gly	Glu	Glu	Asp	Glu	Glu	
23	Glu	Glu	Glu	Glu	Ala	Thr	
24	Tyr	Tyr	Tyr	Ile	Trp	Ser	
25	Gly	Gly	Gly	Gly	Gly	Gly	
26	Gly	Ala	Ala	Ala	Ala	Val	
27	Glu	Glu	Glu	Glu	Glu	Asp	
28	Ala	Ala	Ala	Ala	Ser	Ile	
29	Leu	Leu	Leu	Leu	Leu	Leu	
30	Glu	Glu	Glu	Gly	Ala	Val	
31	Arg	Arg	Arg	Arg	Arg	Lys	
32	Thr	Met	Met	Met	Met	Phe	
33	Phe	Phe	Phe	Leu	Phe	Phe	
34	Ala	Ile	Cys	Thr	Ala	Thr	
35	Ser	Thr	Ala	Val	Thr	Ser	
36	Phe	Tyr	Tyr	Tyr	Thr	Thr	C
37	Pro	Pro	Pro	Pro	Pro	Pro	
38	Thr	Ser	Gln	Gln	Ser	Ala	
39	Thr	Thr	Thr	Thr	Thr	Ala	
40	Lys	Lys	Lys	Lys	Lys	Gln	
41	Thr	Thr	Ile	Thr	Thr	Glu	
42	Tyr	Tyr	Tyr	Tyr	Tyr	Phe	
43	Phe	Phe	Phe	Phe	Phe	Phe	CD
44	Pro	Pro	Pro	Ser	Ser	Pro	
45	His	His	His	His	Lys	Lys	
46	Phe	Phe	Phe	Trp	Phe	Phe	
				Ser	Thr	Lys	
47	Asp	Asp	Asp	Asp	Asp	Gly	
48	Leu	Leu	Met	Leu	Phe	Leu	
49	Ser	Ser	Ser	Ser	Ser	Thr	
						Thr	
						Ala	
						Asp	
						Gln	
						Leu	
50	Pro	His	His	Pro	Ala	Lys	D
51	Gly	Gly	Asn	Gly	Asn	Lys	
52	Ser	Ser	Ser	Ser	Gly	Ser	E
53	Ala	Ala	Ala	Ala	Lys	Ala	
54	Gln	Gln	Gln	Pro	Arg	Asp	
55	Val	Ile	Ile	Val	Val	Val	
56	Lys	Lys	Arg	Lys	Lys	Arg	
57	Ala	Gly	Gly	Lys	Ala	Trp	
58	His	His	His	His	His	His	
59	Gly	Gly	Gly	Gly	Gly	Ala	
60	Lys	Lys	Lys	Lys	Gly	Glu	
61	Lys	Lys	Lys	Thr	Lys	Arg	
62	Val	Val	Val	Ile	Val	Ile	
63	Ala	Val	Phe	Met	Leu	Ile	
64	Asp	Ala	Ala	Gly	Asn	Asn	
65	Ala	Ala	Ala	Ala	Ala	Ala	
66	Leu	Leu	Leu	Val	Val	Val	

Table 3.1. (Continued)

			α chain				
	Panda	Pheasant	*Caiman*	*Carassius*	*Squalus*	*Petromyzon*	
67	Thr	Ile	His	Gly	Ala	Asn	
68	Leu	Glu	Asp	Asp	Asp	Asp	
69	Ala	Ala	Ala	Ala	Ala	Ala	
70	Val	Val	Val	Val	Thr	Val	
71	Gly	Asn	Asn	Ser	Asp	Ala	
72	His	His	His	Lys	His	Ser	EF
73	Leu	Ile	Ile	Ile	Leu	Met	
74	Asp	Asp	Asp	Asp	Asp	Asp	
75	Asp	Asp	Asp	Asp	Asn	Asp	
						Thr	
						Glu	
						Lys	
76	Leu	Ile	Leu	Leu	Val	Met	
77	Pro	Thr	Ala	Val	Ala	Ser	
78	Gly	Gly	Gly	Gly	Gly	Met	
79	Ala	Thr	Ala	Ala	His	Lys	
80	Leu	Leu	Leu	Leu	Leu	Leu	F
81	Ser	Ser	Cys	Ser	Asp	Arg	
82	Ala	Lys	Arg	Ala	Pro	Asp	
83	Leu	Leu	Leu	Leu	Leu	Leu	
84	Ser	Ser	Ser	Ser	Ala	Ser	
85	Asp	Asp	Asp	Glu	Val	Gly	
86	Leu	Leu	Leu	Leu	Leu	Lys	
87	His	His	His	His	His	His	
88	Ala	Ala	Ala	Ala	Gly	Ala	
89	His	His	His	Phe	Thr	Lys	FG
90	Lys	Lys	Asn	Lys	Thr	Ser	
91	Leu	Leu	Leu	Leu	Leu	Phe	
92	Arg	Arg	Arg	Arg	Cys	Gln	
93	Val	Val	Val	Ile	Val	Val	
94	Asp	Asp	Asp	Asp	Asp	Asp	G
95	Pro	Pro	Pro	Pro	Pro	Pro	
96	Val	Val	Val	Ala	His	Gln	
97	Asn	Asn	Asn	Asn	Asn	Tyr	
98	Phe	Phe	Phe	Phe	Phe	Phe	
99	Lys	Lys	Lys	Lys	Pro	Lys	
100	Leu	Leu	Phe	Ile	Leu	Val	
101	Leu	Leu	Leu	Leu	Leu	Leu	
102	Ser	Gly	Ser	Ala	Thr	Ala	
103	His	Gln	Gln	His	Gln	Ala	
104	Cys	Cys	Cys	Asn	Cys	Val	
105	Leu	Phe	Ile	Val	Ile	Ile	
106	Leu	Leu	Leu	Ile	Leu	Ala	
107	Val	Val	Val	Val	Val	Asp	
108	Thr	Val	Val	Val	Thr	Thr	
109	Leu	Val	Phe	Ile	Leu	Val	
110	Ala	Ala	Gly	Gly	Ala	Ala	
111	Cys	Ile	Val	Met	Ala	Ala	
112	His	His	His	Leu	His		
113	His	His	His	Phe	Leu		GH
114	Pro	Pro	Pro	Pro			
115	Ala	Ser	Cys	Gly	Thr		

Table 3.1. (Continued)

	α chain						
	Panda	Pheasant	*Caiman*	*Carassius*	*Squalus*	*Petromyzon*	
116	Glu	Ala	Ser	Asp	Glu		
117	Phe	Leu	Leu	Phe	Leu		
118	Thr	Thr	Thr	Thr	Lys		H
119	Pro	Pro	Pro	Pro	Pro		
120	Ala	Glu	Glu	Glu	Glu		
121	Val	Val	Val	Val	Thr	Gly	
122	His	His	His	His	His	Asp	
123	Ala	Ala	Ala	Met	Cys	Ala	
124	Ser	Ser	Ser	Ser	Ala	Gly	
125	Leu	Leu	Leu	Val	Leu	Phe	
126	Asp	Asp	Asp	Asp	Asp	Glu	
127	Lys	Lys	Lys	Lys	Lys	Lys	
128	Phe	Phe	Phe	Phe	Phe	Leu	
129	Phe	Leu	Leu	Phe	Leu	Met	
130	Ser	Cys	Cys	Gln	Cys	Ser	
131	Ala	Ala	Ala	Asn	Glu	Met	
132	Val	Val	Val	Leu	Val	Ile	
133	Ser	Gly	Ser	Ala	Ala	Cys	
134	Thr	Thr	Ala	Leu	Thr	Ile	
135	Val	Val	Met	Ala	Ala	Leu	
136	Leu	Leu	Leu	Leu	Leu	Leu	
137	Thr	Thr	Thr	Ser	Gly	Arg	
138	Ser	Ala	Ser	Glu	Ser	Ser	
139	Lys	Lys	Lys	Lys	His	Ala	HC
140	Tyr	Tyr	Tyr	Tyr	Tyr	Tyr	
141	Arg	Arg	Arg	Arg	Arg		

	β chain						
	Panda	Pheasant	*Caiman*	*Rana*	*Carassius*	*Squalus*	
1	Val	Val	Ser		Val	Val	NA
2	His	His	Pro		Glu	His	
3	Leu	Trp	Phe		Trp	Trp	
4	Thr	Ser	Ser		Thr	Thr	A
5	Gly	Ala	Ala		Asp	Gly	
6	Glu	Glu	His		Ala	Glu	
7	Glu	Glu	Glu	Gly	Glu	Glu	
8	Lys	Lys	Glu	Ser	Arg	Lys	
9	Ala	Gln	Lys	Asp	Ser	Ala	
10	Ala	Leu	Leu	Leu	Ala	Leu	
11	Val	Ile	Ile	Val	Ile	Val	
12	Thr	Thr	Val	Ser	Ile	Asn	
13	Gly	Gly	Asp	Gly	Gly	Ala	
14	Leu	Leu	Leu	Phe	Leu	Val	
15	Trp	Trp	Trp	Trp	Trp	Trp	
16	Ser	Gly	Ala	Gly	Gly	Thr	
17	Lys	Lys	Lys	Lys	Lys	Lys	
18	Val	Val	Val	Val	Leu	Trp	
19	Asn	Asn	Asp	Asp	Asn	Asp	B
20	Val	Val	Val	Ala	Pro	His	

Table 3.1. (Continued)

	β chain						
	Panda	Pheasant	*Caiman*	*Rana*	*Carassius*	*Squalus*	
21	Asp	Ala	Ala	His	Asp	Gln	
22	Glu	Asp	Ser	Lys	Glu	Ala	
23	Val	Cys	Cys	Ile	Leu	Val	
24	Gly	Gly	Gly	Gly	Gly	Val	
25	Gly	Ala	Gly	Gly	Pro	Ala	
26	Glu	Glu	Asp	Glu	Gln	Lys	
27	Ala	Ala	Ala	Ala	Ala	Ala	
28	Leu	Leu	Leu	Leu	Leu	Leu	
29	Gly	Ala	Ser	Ala	Ala	Glu	
30	Arg	Arg	Arg	Arg	Arg	Arg	
31	Leu	Leu	Met	Leu	Cys	Leu	
32	Leu	Leu	Leu	Leu	Leu	Phe	
33	Val	Ile	Ile	Val	Ile	Val	
34	Val	Val	Ile	Val	Val	Val	
35	Tyr	Tyr	Tyr	Tyr	Tyr	Tyr	C
36	Pro	Pro	Pro	Pro	Pro	Pro	
37	Trp	Trp	Trp	Trp	Trp	Trp	
38	Thr	Thr	Lys	Thr	Thr	Thr	
39	Gln	Gln	Arg	Gln	Gln	Lys	
40	Arg	Arg	Arg	Arg	Arg	Thr	
41	Phe	Phe	Tyr	Tyr	Tyr	Tyr	
42	Phe	Phe	Phe	Phe	Phe	Phe	CD
43	Asp	Ala	Glu	Thr	Ala	Val	
44	Ser	Ser	His	Thr	Thr	Lys	
45	Phe	Phe	Phe	Phe	Phe	Phe	
46	Gly	Gly	Gly	Gly	Gly	Asn	
47	Asp	Asn	Lys	Asn	Asn	Gly	
48	Leu	Leu	Leu	Leu	Leu	Lys	
49	Ser	Ser	Ser	Gly	Ser	Phe	
50	Ser	Ser	Tyr	Ser	Ser	His	D
51	Pro	Pro	Asp	Ala	Pro	Ala	
52	Asp	Thr	Gln	Asp	Ala		
53	Ala	Ala	Asp	Ala	Ala		
54	Val	Ile	Val	Ile	Ile		
55	Met	Leu	Leu	Cys	Met		
56	Gly	Gly	His	His	Gly	Ser	
57	Asn	Asn	Asn	Asn	Asn	Asp	E
58	Pro	Pro	Glu	Ala	Pro	Ser	
59	Lys	Met	Lys	Lys	Lys	Thr	
60	Val	Val	Ile	Val	Val	Val	
61	Lys	Arg	Arg	Leu	Ala	Gln	
62	Ala	Ala	Glu	Ala	Ala	Thr	
63	His	His	His	His	His	His	
64	Gly	Gly	Gly	Gly	Gly	Ala	
65	Lys	Lys	Lys	Glu	Arg	Gly	
66	Lys	Lys	Lys	Lys	Thr	Lys	
67	Val	Val	Val	Val	Val	Val	
68	Leu	Leu	Leu	Leu	Met	Val	
69	Asn	Thr	Ala	Ala	Gly	Ser	
70	Ser	Ser	Ser	Ala	Gly	Ala	
71	Phe	Phe	Phe	Ile	Leu	Leu	
72	Ser	Gly	Gly	Gly	Glu	Thr	

Table 3.1. (Continued)

	β chain						
	Panda	Pheasant	*Caiman*	*Rana*	*Carassius*	*Squalus*	
73	Glu	Asp	Glu	Glu	Arg	Val	
74	Gly	Ala	Ala	Gly	Ala	Ala	
75	Leu	Val	Val	Leu	Ile	Tyr	
76	Lys	Lys	Lys	Lys	Lys	Asn	
77	Asn	Asn	His	His	Asn	His	EF
78	Leu	Leu	Leu	Pro	Met	Ile	
79	Asp	Asp	Asp	Glu	Asp	Asp	
80	Asn	Asn	Asn	Asn	Asn	Asp	
81	Leu	Ile	Ile	Leu	Ile	Val	
82	Lys	Lys	Lys	Lys	Lys	Lys	
83	Gly	Asn	Gly	Ala	Ala	Pro	
84	Thr	Thr	His	His	Thr	His	
85	Phe	Phe	Phe	Tyr	Tyr	Phe	F
86	Ala	Ser	Ala	Ala	Ala	Val	
87	Lys	Gln	His	Lys	Pro	Glu	
88	Leu	Leu	Leu	Leu	Leu	Leu	
89	Ser	Ser	Ser	Ser	Ser	Ser	
90	Glu	Glu	Lys	Glu	Val	Lys	
91	Leu	Leu	Leu	Tyr	Met	Lys	
92	His	His	His	His	His	His	
93	Cys	Cys	Phe	Ser	Ser	Tyr	
94	Asp	Asp	Glu	Asn	Glu	Glu	FG
95	Lys	Lys	Lys	Lys	Lys	Glu	
96	Leu	Leu	Phe	Leu	Leu	Leu	
97	His	His	His	His	His	His	
98	Val	Val	Val	Val	Val	Val	
99	Asp	Asp	Asp	Asp	Asp	Asp	G
100	Pro	Pro	Cys	Pro	Pro	Pro	
101	Glu	Glu	Glu	Ala	Asp	Glu	
102	Asn	Asn	Asn	Asn	Asn	Asn	
103	Phe	Phe	Phe	Phe	Phe	Phe	
104	Lys	Arg	Lys	Arg	Arg	Lys	
105	Leu	Leu	Leu	Leu	Leu	Leu	
106	Leu	Leu	Leu	Leu	Leu	Leu	
107	Gly	Gly	Gly	Gly	Ala	Ala	
108	Asn	Asp	Asp	Asn	Asp	Asn	
109	Val	Ile	Ile	Val	Cys	Cys	
110	Leu	Leu	Ile	Phe	Ile	Leu	
111	Val	Ile	Ile	Ile	Thr	Glu	
112	Cys	Ile	Val	Thr	Val	Val	
113	Val	Val	Val	Val	Cys	Glu	
114	Leu	Leu	Leu	Leu	Ala	Ley	
115	Ala	Ala	Gly	Ala	Ala	Gly	
116	His	Ala	Met	Arg	Met	His	
117	His	His	His	His	Lys	Ala	
118	Phe	Phe	His	Phe	Phe	Leu	GH
119	Gly	Ser	Pro	Gln	Gly	His	
120	Lys	Lys	Lys	His	Pro	Lys	
121	Glu	Asp	Asp	Glu	Ser	Glu	
					Gly		
122	Phe	Phe	Phe	Phe	Phe	Phe	
123	Thr	Thr	Thr	Thr	Asn	Thr	H

Table 3.1. (Continued)

	α chain						
	Panda	Pheasant	*Caiman*	*Carassius*	*Squalus*	*Petromyzon*	
124	Pro	Pro	Leu	Pro	Ala	Pro	
125	Gln	Glu	Gln	Glu	Asp	Glu	
126	Val	Cys	Thr	Leu	Val	Val	
127	Gln	Gln	His	Gln	Gln	Gln	
128	Ala	Ala	Ala	His	Glu	Ala	
129	Ala	Ala	Ala	Ala	Ala	Ala	
130	Tyr	Trp	Phe	Leu	Trp	Trp	
131	Gln	Gln	Gln	Glu	Gln	Ser	
132	Lys	Lys	Lys	Ala	Lys	Lys	
133	Val	Leu	Leu	His	Phe	Phe	
134	Val	Val	Val	Phe	Leu	Ser	
135	Ala	Arg	Arg	Cys	Ser	Asn	
136	Gly	Val	His	Ala	Val	Val	
137	Val	Val	Val	Val	Val	Val	
138	Ala	Ala	Ala	Gly	Val	Val	
139	Asn	His	Ala	Asp	Ser	Asp	
140	Ala	Ala	Ala	Ala	Ala	Ala	
141	Leu	Leu	Leu	Leu	Leu	Leu	
142	Ala	Ala	Ser	Ala	Cys	Ser	
143	His	Arg	Ala	Lys	Arg	Lys	
144	Lys	Lys	Glu	Ala	Gln	Gly	HC
145	Tyr	Tyr	Tyr	Tyr	Tyr	Tyr	
146	His	His	His	His	His	His	

[a]Amino acid sequences of a mammal (lesser panda, *Ailurus fulgens*; Tagle et al. 1986), a bird (pheasant, *Phasianus colchinus*; Braunitzer and Godovac 1982), a reptile (caiman, *Caiman crocodylus*; Leclerq et al. 1981), a frog (*Rana esculenta*; Chauvet and Acher 1972), a teleost (*Carassius auratus*; Braunitzer and Rodewald 1980), an elasmobranch (*Squalus acanthias*; Aschauer et al. 1985) and a cyclostome (*Petromyzon marinus*; Hombrados et al. 1983) are given. For the species with multiple haemoglobins, only the major components are given. The haemoglobin of *Petromyzon* is included in the α chains. The numbering gives the amino acid positions of mammalian haemoglobin chain. The last column indicates the helical (A-H) and nonhelical (NA-HC) segments of the haemoglobin chains.

The contacts between haem and the globin chain in two widely separated vertebrates, man and the sea lamprey (*Petromyzon*), are given in Table 3.2. The contacts are similar in both cases. Thus, the binding site for oxygen is practically invariant throughout vertebrates. In all the globins studied, the haem group is bound covalently to a histidine residue in the F-helix (F8). The valine residue (FG5) interacts with the vinyl group of the ring C of haem. The leucine residue F4 ties the side chain of His F8 onto the F helix, limiting the conformational freedom of the histidine side chain. Phenylalanine CD1 and valine E11 probably play a role in keeping the haem in place and, together with histidine E7, cause the steric constraint which forces the iron-bound oxygen to assume a bent structure (see e.g. Winterhalter and Di Iorio 1984). In human oxyhaemoglobin, the iron-oxygen bond angle is 156°, and the oxygen molecule forms a hydrogen bond with histidine E7 (Shaanan 1982), which is stronger in α than in β chain.

Table 3.2. The amino acid residues interacting with haem group in human α and β chains (Fermi et al. 1984) and in the major haemoglobin of the cyclostome, *Petromyzon marinus* (Honzatko et al. 1985)[a]

α	β	*Petromyzon*
	Leu(B13)	
Tyr(C7)	Phe(C7)	
Phe(CD1)	Phe(CD1)	Phe(CD1)
His(CD3)		
Phe(CD4)		
His(E7)	His(E7)	His(E7)
Lys(E10)	Lys(E10)	Arg(E10)
Val(E11)	Val(E11)	Ile(E11)
	Ala(E14)	
Leu(F4)	Phe(F4)	Leu(F4)
		Ser(F5)
Leu(F7)	Leu(F7)	Lys(F7)
His(F8)	His(F8)	His(F8)
		Ala(F9)
Leu(FG3)	Leu(FG3)	Phe(FG3)
Val(FG5)	Val(FG5)	Val(FG5)
Asn(G4)	Asn(G4)	Tyr(G4)
Phe(G5)	Phe(G5)	Phe(G5)
Leu(G8)	Leu(G8)	Leu(G8)
Leu(H19)	Leu(H19)	
		Tyr(HC2)

[a] In man the residues less than 0.4 nm from the haem group are included, in the cyclostome, residues either within 0.35 nm from the haem group, or within 0.4 nm from His(F8) or the ligand of haem are included. Invariant residues are underlined.

The haemoglobin of most vertebrates is a tetrameric molecule, consisting of two α-like and two β-like chains. The three-dimensional quarternary structures of tetrameric horse and human deoxy- and oxyhaemoglobins have been elucidated and refined (for recent refinements see Shaanan 1983; Fermi et al. 1984, reviews on earlier work include those by Ten Eyck 1972; Perutz 1978; Dickerson and Geis 1983). On the basis of the quarternary structure and the primary amino acid sequence of the α and β chains of the haemoglobin, it has been possible to deduce the amino acid residues which are involved in the formation of the $\alpha_1\beta_1$ (packing) and $\alpha_1\beta_2$ (sliding) contacts in man. The contacts between the mammalian chains are depicted in Tables 3.3, 3.4 and 3.5. The $\alpha_1\beta_1$ contacts are similar in both deoxy- and oxyconformation of haemoglobin. In contrast, the number of contacts between α_1 chain and β_2 chain decreases markedly upon oxygenation, from ca. 40 to ca. 20. In addition, the contacts between α_1 and α_2 chains are only present in deoxyconformation.

Of the haemoglobins studied, at least the haemoglobins of the dogfish (*Squalus acanthias*) and the water snake (*Liophis miliaris*) break up to $\alpha\beta$-dimers upon oxygenation (Fyhn and Sullivan 1975; Matsuura et al. 1987). Human oxyhaemoglobin dissociates to two $\alpha\beta$-dimers at neutral pH at high (e.g. 2 M)

Table 3.3. The $\alpha_1\beta_1$ contacts of human deoxy- and oxyhaemoglobin[a]

α_1	Amino acids of the β_1 chain in contact
Arg(31)B12	Gln(127)H5
	Pro(125)H3
	Phe(123)H1
	Phe(122)GH5
Leu(34)B15	Ala(128)H6
	Pro(125)H3
	Pro(124)H2
Ser(35)B16	Gln(131)H9
	Ala(128)H6
	Gln(127)H5
Phe(36)C1	Gln(131)H9
His(103)G10	Gln(131)H9
	Asn(108)G10
Leu(106)G13	Tyr(35)C1
Val(107)G14	Gln(127)H5
	Ala(115)G17
	Cys(112)G14
Ala(110)G17	His(116)G18
	Ala(115)G17
	Cys(112)G14
Ala(111)G18	Gly(119)GH2
	Ala(115)G17
Pro(114)GH2	His(116)G18
Phe(117)GH5	His(116)G18
	Arg(30)B12
Thr(118)H1	Arg(30)B12
Pro(119)H2	Met(55)D6
	Val(33)B15
	Arg(30)B12
His(122)H5	Cys(112)G14
	Tyr(35)C1
	Val(34)B16
	Arg(30)B12
Ala(123)H6	Val(34)B16
Asp(126)H9	Tyr(35)C1
	Val(34)B16

[a]For each amino acid of the α_1 chain, the amino acid residues of the β_1 chain in distance less than 0.4 nm are given. Data are from Fermi and Perutz (1981).

ionic strength solutions of sodium chloride, whereas deoxyhaemoglobin fails to do so, but can be dissociated in 1 M NaI (Kellett 1971). The haemoglobins of teleost fish appear to be more resistant to dimerization than human deoxyhaemoglobin: trout (Brunori et al. 1973) and carp (Atha and Riggs 1982) oxyhaemoglobins do not dissociate to dimers in 1 M NaI. At present, the differences in the quarternary structure causing the different susceptibilities to dimerization do not appear to be known, because detailed X-ray crystallographic

Table 3.4. The amino acids participating in $\alpha_1\beta_2$ and $\alpha_1\alpha_2$ contacts in human deoxyhaemoglobin[a]

$\alpha_1\beta_2$ contact	
α_1	Amino acids of the β_2 chain in contact
Pro(37)C2	His(146)HC3
Thr(38)C3	Pro(100)G2
	Asp(99)G1
Lys(40)C5	His(146)HC3
Thr(41)C6	Tyr(145)HC2
	Asp(99)G1
	Val(98)FG5
	His(97)FG4
	Phe(41)C7
	Arg(40)C6
Tyr(42)C7	Asn(102)G4
	Asp(99)G1
	Val(98)FG5
	His(97)FG4
	Phe(41)C7
	Arg(40)C6
Pro(44)CD2	His(97)FG4
Ala(88)F9	Trp(37)C3
	Pro(36)C2
Leu(91)FG3	Arg(40)C6
Arg(92)FG4	Asn(102)G4
	Glu(43)CD2
	Phe(41)C7
	Arg(40)C6
	Trp(37)C3
Asp(94)G1	Leu(105)G7
	Glu(101)G3
	Asp(99)G1
	Trp(37)C3
Pro(95)G2	Trp(37)C3
Val(96)G3	Glu(101)G3
Asn(97)G4	Asp(99)G1
Tyr(140)HC2	Trp(37)C3
	Pro(36)C2
Arg(141)HC3	Pro(36)C2
	Tyr(35)C1
	Val(34)B16

[a] For each amino acid of the α_1 chain, the amino acids of the other chain in a distance less than 0.4 nm are given. Data are from Fermi and Perutz (1981).

Table 3.4. *(Continued)*

$\alpha_1\alpha_2$ contact α_1	Amino acids of the α_2 chain in contact
Val(1)NA1	Arg(141)HC3
Asp(126)H9	Arg(141)HC3
Lys(127)H10	Arg(141)HC3
Arg(141)HC3	Lys(127)H10
	Asp(126)H9
	Val(1)NA1

Table 3.5. The amino acids forming the $\alpha_1\beta_2$ contact in horse methaemoglobin

α_1	Amino acids of the β_2 chain in contact
Thr(38)C3	Asp(99)G1
	His(97)FG4
Thr(41)C6	His(97)FG4
	Arg(40)C6
Tyr(42)C7	Asp(99)G1
	Phe(41)C7
	Arg(40)C6
	Trp(37)C3
Arg(92)FG4	Asp(99)G1
	Asp(43)CD2
	Gln(39)C5
	Trp(37)C3
Asp(94)G1	Asn(102)G4
	Glu(101)G3
	Trp(37)C3
Pro(95)G2	Trp(37)C3
Val(96)G3	Asn(102)G4
	Glu(101)G3
	Asp(99)G1
	Trp(37)C3
Asn(97)G4	Asp(99)G1
Tyr(140)HC2	Trp(37)C3

[a] The contacts of horse and human oxyhaemglobin are nearly identical to the methaemoglobin contacts. For each amino acid of the α_1 chain, the amino acids of the β_2 chain in a distance less than 0.4 nm are given. Data are from Fermi and Perutz (1981).

data are not available for nonmammalian haemoglobins. However, the differences must occur in the $\alpha_1\beta_2$ contact region, or in $\alpha_1\alpha_2$ contact, since the dimers produced from human carboxyhaemoglobin appear to be $\alpha_1\beta_1$ dimers. Furthermore, the stabilization of the deoxytetramer involves the interaction between the C-terminal arginine of one α subunit, and the aspartate H9 of the other α subunit ($\alpha_1\alpha_2$ interaction; see Hewitt et al. 1972).

The tetrameric haemoglobins of some birds, reptiles and amphibians aggregate to higher polymers upon deoxygenation (see Riggs 1988). The structural features behind the aggregation have been studied using bullfrog (*Rana catesbeiana*) haemoglobins. Bullfrogs have two major haemoglobin components (B and C) with identical β chains, but with different α chains (Tam et al. 1986). The aggregates formed appear to be BC_2 trimers of tetramers, and result from an interaction between the different α chains (Tam and Riggs 1984).

The haemoglobins of cyclostomes, *Lampetra* and *Petromyzon*, appear to be in monomer/dimer/tetramer equilibria in physiological conditions (e.g. Briehl 1963; Riggs 1972; Dohi et al. 1973). Honzatko et al. (1985), and Honzatko and Hendrickson (1986) have investigated the molecular basis for the aggregation of lamprey haemoglobins. The amino acid residues corresponding to the $\alpha_1\beta_1$ contact of mammalian haemoglobins are much more polar in lamprey than in mammals (see e.g. Dickerson and Geis 1983), suggesting that these regions are exposed to aqueous environment and do not interact. In contrast, the amino acids of lamprey haemoglobin, which correspond to the $\alpha_1\beta_2$ contact region of mammals, have similar polarity as in mammals. Furthermore, hydrogen bonds can be formed between Glu 50 and Glu 110, Asp 112 and Tyr 115, Gln 114 and Tyr 115, and Glu 114 and Glu 114 of two monomers (Honzatko and Hendrickson 1986). Thus, dimers corresponding to $\alpha_1\beta_2$ are plausible for the lampreys. However, if this were the only possible contact region, formation of tetramers would be impossible. The other energetically favourable regions of interaction are between the E and F helices of one monomer and the E and F helices of another monomer, and between A and B helices of one monomer and the A and B helices of another monomer (Honzatko and Hendrickson 1986). Such subunit contacts do not occur in vertebrate haemoglobins, but are the characteristic feature of the tetrameric clam haemoglobin (Royer et al. 1985). However, in the absence of crystallographic data on the structure of aggregated lamprey haemoglobins, the models for interaction between lamprey haemoglobin monomers are necessarily tentative.

In most tetrameric haemoglobins, there is a pocket between the β subunits, to which organic phosphates (2,3-diphosphoglycerate, ATP, GTP or IPP) bind. Several positive charges line the anion binding site in human haemoglobin, i.e. the amino terminus, histidine NA2, lysine EF6 and histidine H21 of both β chains (see e.g. Arnone 1972). The binding site has a suitable geometry for organic phosphate binding in the deoxy conformation of haemoglobin only. The organic phosphate binding sites in different vertebrates are discussed in detail in Section 10.3.4.

Although the general features of the haemoglobin structure are similar in most vertebrates, there is marked variation in the structures of individual globin chains of different vertebrates (see Table 3.1). Most of the amino acid differences are functionally neutral, but a few amino acid changes at key positions may cause drastic alterations in the functional properties of the haemoglobins (cf. Perutz and Brunori 1982). In the following sections, the differences in the haemoglobin structure of adult animals, and the ontogenetic changes in the globin gene expression and the haemoglobin structure are outlined. The effects of structural changes on the function of haemoglobin are discussed in detail in Chapter 10.

3.5.1 Haemoglobin Heterogeneity in Adult Animals

Agnatha

The hagfish *Myxine glutinosa* has three haemoglobin components (Paleus et al. 1971). The major haemoglobin has been sequenced (see Hombrados et al. 1983), and has only 31 amino acids in common with both the human α chain and the human β chain. Also, there are 88 amino acid differences between *Myxine* and *Petromyzon* haemoglobin (see below). Among the substitutions is the replacement of His E7 with glutamine.

In *Petromyzon*, three different globin chains have been sequenced. Two of these (PMII and PMIII) according to the nomenclature of Hombrados et al. 1987) have very similar structures, differing from each other only by four amino acids, and each of the differences only requires a single-base interconversion in the genetic code (Hombrados et al. 1987). Of the amino acid differences, Thr/Ser in position 5, Ser/Thr in position 33, Ala/Val in position 86 and Arg/Gly in position 99, only the last one involves a marked change in polarity and the size of the side chain. In contrast, the third haemoglobin (PMI) differs from the major component by 27 amino acids. However, even in this case, only two amino acid substitutions (Met/Ala and Leu/Ala) require two base changes in the DNA. The difference between PMI and the other haemoglobins represents a clear divergence of the globin chains already in cyclostomes. The divergence between the globins of *Petromyzon* becomes even more apparent when they are compared to the *Lampetra* haemoglobins. The major haemoglobins of *Lampetra fluviatilis* and *Petromyzon* only differ from each other by three amino acids, and each of the differences corresponds to a single-base mutation in the DNA. In *Lampetra fluviatilis*, three to six electrophoretically different haemoglobins have been observed (Potter and Brown 1975). The two major haemoglobin components differ from each other only in that one of the components has a free amino terminus, and the other has N-formylproline as the terminal group (Fujiki et al. 1970). Both *Lampetra* and *Petromyzon* globins have only two histidine residues, each interacting with the haem group (see Hombrados et al. 1983). It is notable from the phylogenetic point of view that *Lampetra* and *Petromyzon* haemoglobins differ as much from *Myxine* haemoglobins as from human α-chain, suggesting that these animals are evolutionally widely divergent (see Hombrados et al. 1983, 1987).

Elasmobranchs

Fyhn and Sullivan (1975) studied the electrophoretic mobility of the haemoglobins of 14 species of elasmobranchs, and found heterogeneity in each species. Similarly, Mumm et al. (1978) observed five haemoglobin components in the blood of the common sting-ray (*Dasyatis sabina*), and Weber et al. (1983a) observed haemoglobin heterogeneity in the blood of the dogfish (*Squalus acanthias*).

The amino acid sequence of haemoglobins has been studied in two species, *Squalus acanthias* (Aschauer et al. 1985) and *Heterodontus portusjacksoni* (Nash et al. 1976; Fisher et al. 1977). Distinct α and β chains are found in both species.

Squalus acanthias has two α chains, which have 141 amino acids, and two β chains, which have 142 amino acids. The two different α chains differ from each other only by one amino acid, just as the two different β chains do (Aschauer et al. 1985). The α chain of *Heterodontus* has 148 amino acids, and the β chain 142 amino acids. The α chains of *Squalus* differ from the α chain of *Heterodontus* by 73 amino acid exchanges, and from the human α chain by 72 amino acid exchanges. With regards to the β chains, the number of amino acid exchanges between *Squalus* and *Heterodontus* is 70, and between *Squalus* and man 82. These differences suggest that *Squalus* and *Heterodontus* are phylogenetically quite far apart.

Dipnoi

Rodewald et al. (1984) have studied the haemoglobins of the South American lungfish (*Lepidosiren paradoxus*), which has only one haemoglobin component. In the African lungfish *Protopterus amphibius* two haemoglobin components appear to be present, whereas only one component is seen in *Protopterus annectens* (Weber et al. 1977).

The α chain of *Lepidosiren paradoxus* has 143 amino acids and the β chain 147 amino acids. The α chain differs from that of human haemoglobin by 83 amino acids, and the β chain by 75 amino acids. Both chains have a similar or greater histidine content than human haemoglobin (10 histidine residues in the α chain and 12 in the β chain; Rodewald et al. 1984). The β F9 position is occupied by a serine residue.

Teleosts

Teleost haemoglobins are characterized by marked polymorphism. In practically all species studied, electrophoretically different haemoglobins have been described in adult animals (e.g. Wilkins and Iles 1966; Westman 1970; Tsuyuki and Ronald 1971; Powers 1972; Hjorth 1974; Sharp 1975; Everaarts 1978; Wilhelm and Reischl 1981; Southard et al. 1986). The functional properties of the different haemoglobins are often similar, as in carp, *Cyprinus carpio* (Gillen and Riggs 1972; Weber and Lykkeboe 1978), *Catostomus insignis* (Powers 1972), plaice, *Pleuronectes platessa* and flounder, *Platicthys flesus* (Weber and DeWilde 1976). However, in several species like the eels, *Anguilla* (Yamaguchi et al. 1962; Gillen and Riggs 1973; Weber et al. 1976a), *Catostomus clarkii* (Powers 1972), rainbow trout (Brunori 1975; Weber et al. 1976b), the Amazonian fish *Mylossoma* (Martin et al. 1979) and *Pterygoplichthys pardalis* (Brunori et al. 1979), the different forms of haemoglobin differ from each other functionally. Furthermore, the haemoglobin components of fish change with age (e.g. Westman 1970) and as a response to environmental changes (e.g. Houston and Cyr 1974; Houston et al. 1976; Weber et al. 1976b; Tun and Houston 1986).

The amino acid sequences of globin chains are available only for a few teleosts. Braunitzer and Rodewald (1980) have sequenced the haemoglobin chains of *Carassius auratus*, which only has one haemoglobin component. The α chain has 66 amino acid substitutions as compared to human α chain, and the β chain 72 differences. Typically, both goldfish chains have fewer histidine residues than the corresponding human chains. In contrast, the β chain of goldfish has 4 cysteine residues, whereas human β chain has only two. Two of these cysteine residues appear to be exposed to the environment (Cys B12 (β31) and Cys H21 (β143)), which may make goldfish haemoglobin more susceptible to oxidation than human haemoglobin. Cys F9 of human β chain has been replaced by a serine residue.

The carp (*Cyprinus carpio*) has two major and one minor haemoglobin components. The α chain, sequenced by Hilse and Braunitzer (1968), appears to be the same in all the components. The α chain differs from the α chain of goldfish by 16% (23–24 amino acid substitutions), and, like the α chain of goldfish, has fewer histidine residues than human α chain. Carp haemoglobins have two β chains (β_A and β_B), sequenced by Grujic-Injac et al. (1980), which differ from each other at four positions (β13 Ala/Gly; β31 Phe/Tyr; β122 Gly/Ala; β143 Lys/Cys), and closely resemble the β chain of goldfish; β_A differs from it by 13 amino acids and β_B by 10. In comparison with human haemoglobin, the histidine content of carp β chains is smaller, and the Cys F9 has been replaced by serine. Similar to goldfish β chain, the β_B chain of carp has two external cysteine residues, whereas the β_A chain has only one.

Of the four major haemoglobin components (HbI-IV) of the trout, *Salmo irideus* (e.g. Brunori 1975), the complete amino acid sequence is available for the cathodically migrating HbI (Bossa et al. 1978; Barra et al. 1983). The β chain of trout HbI differs from the β chain of goldfish by 51 amino acids, i.e. by ca. 35%, and from that of man by 68 amino acids, i.e. by ca. 47%. Interestingly, Barra et al. (1983) report that the β chain trout HbI differs more (i.e. by ca. 42%) from the β chain of the anodically migrating trout HbIV than from the β chain of goldfish. In contrast, the β chains of trout HbIV and goldfish differ from each other only by ca. 25%.

Similarly, the α chain of trout HbI differs by 45 amino acids from goldfish α chain (Bossa et al. 1978; Braunitzer and Rodewald 1980). Again, on the basis of preliminary data on the α chain of trout HbIV, it appears that the difference between trout HbI and trout HbIV is greater than that between trout HbI and goldfish (Bossa et al. 1978).

Typically, in all fish haemoglobin chains so far studied, the amino terminal amino acid of α chains is acetylated (see Powers and Edmundson 1972; Bossa et al. 1978; Braunitzer and Rodewald 1980; Grujic-Injac et al. 1980).

Amphibians

Sullivan (1974b) has reviewed the structure and heterogeneity of amphibian haemoglobins. Haemoglobin heterogeneity has been observed in the blood of both salamanders (*Urodela*, e.g. *Necturus maculosus*; Weber et al. 1985) and frogs (*Anura*, e.g. *Rana catesbeiana*; Moss and Ingram 1968a; Tam et al. 1986). However, the polymorphism of adult haemoglobins is much more modest than in teleost fish. For example, *Rana catesbeiana* has two major haemoglobin components, which have similar β chains, but different α chains (Tam et al. 1986). The salamander *Necturus* has one major and one minor haemoglobin component, which have not been characterized in further detail (Weber et al. 1985). In the salamander, *Pleurodeles waltli*, adult haemoglobins appear to have three components, possibly resulting from combinations of two different α chains and two different β chains (Flavin et al. 1978a).

Tam et al. (1986) have sequenced the β chain of one component of adult *Rana catesbeiana* haemoglobin, and Chauvet and Acher (1972) a β chain of *Rana esculenta* haemoglobins. As compared to human globin chains, the β globin chain of *Rana catesbeiana* has 73 amino acid substitutions, and *Rana esculenta* 69 amino acid substitutions. The β chains of the frogs and man exhibit similar histidine contents. The first six amino terminal amino acids are missing from the β chain of adult *Rana*, whereby organic phosphate binding is reduced (see Sect. 10.3.4).

Reptiles

The structure and heterogeneity of reptilian haemoglobins has been reviewed by Sullivan (1974a). In the turtles (*Chelonia*), two major haemoglobin components are commonly seen, together with 0-3 minor haemoglobin components (e.g. Sullivan and Riggs 1967a). Of the *Squamata*, lizards often have only one electrophoretic haemoglobin component, although up to three components are found in some species (Guttman 1970a, b). In contrast, snakes appear to have multiple haemoglobin components (see e.g. Sullivan 1974a). The apparent multiplicity may result partly from the dissociation of tetrameric haemoglobins to dimers, as in the water snake *Liophis miliaris* (e.g. Matsuura et al. 1987) or polymerization due to disulphide bridge formation, as in some turtles (e.g. Sullivan and Riggs 1967b). In the sole surviving species of *Rynchocephalia*, studied by Abbasi et al. (1988), three different haemoglobins, consisting in altogether five different globin chains, were observed. In the *Crocodilia*, only one haemoglobin component has been found (Leclercq et al. 1981). Altogether, the data on reptilia indicate much smaller heterogeneity of haemoglobins of adult reptilia than of adult teleost fish.

Leclercq et al. (1981) have sequenced the α and β globin chains of the caiman *Caiman crocodylus*, the Nile crocodile *Crocodylus niloticus* and the Mississippi crocodile *Alligator mississippiensis*. The α chain of the Nile crocodile and the alligator differ from each other by 17 amino acids, the Nile crocodile and the caiman by 19 amino acids, and the caiman and alligator also by 19 amino acids. The corresponding differences between the β chains were 28, 31 and 38 amino acids. The crocodile α chains differ from the human α chain by 44–47 amino acids, and the β chains by 66–75 amino acids. Compared to the goldfish, the α chains of crocodiles differ by 60–70 amino acids, and the β chains by 80–85 amino acids. The histidine content of both α and β chain is greater than that of the corresponding human chains i.e. 11–12 in both chains. The number of sulfhydryl group (i.e. cysteine) residues is greater than in man. If the situation in other reptiles is similar, this may explain the susceptibility of reptilian haemoglobins to oxidation (i.e. polymerization due to disulphide bridge formation, e.g. Sullivan and Riggs 1967b, see also Sullivan 1974a). Incidentally, crocodiles are the first group with cysteine F9 in the β chain; the group is present in both the Nile crocodile and the alligator, whereas in the caiman, phenylalanine occupies this position (Leclercq et al. 1981).

Birds

The haemoglobin structure of avian haemoglobins has been reviewed by Schnek et al. (1985). In most cases two components, one major (Hb_A) and one minor (Hb_D), are present. It appears that in most cases the β chain is the same, whereas the α chains are different (α_A and α_D) in the different components.

At present, the adult globin chain sequences are available for ca. 20 species. Within species, the two α chains are markedly different, e.g. differing by 58 amino acids in the pheasant (*Phasianus colchinus*; Braunitzer and Godovac 1982). In contrast, variation within a single α chain between species is much smaller. The α_A chains of the Northern mallard (*Anas platyrhynchos*), the American flamingo (*Phoenicopterus ruber*), the golden eagle (*Aguila chrysaetos*) and the starling (*Sturnus vulgaris*) differ from the α_A chain of the pheasant by 18, 31, 18 and 24 amino acids respectively (see Godovac-Zimmermann and Braunitzer 1983, 1984; Oberthür and Braunitzer 1984; Oberthür et al. 1983; Braunitzer and Godovac 1982). Although this variation is relatively small, the variation within the β chains is even smaller; the β chains of the above species differ from the β chain of the pheasant by 4, 4, 8 and 11 amino acids respectively. The pheasant β chain differs by 47 amino acids from the human β chain, the α_A chain by 42 and the α_D chain by 58 amino acids from the human α chain.

Mammals

The haemoglobin components of (domestic) mammals have been reviewed by Garrick and Garrick (1983). As expected from the multiple globin genes, most mammals have multiple haemoglobin components. The camels and the donkey are exceptions, having only one α and β chain. A large number of mammalian globin chains have been sequenced and, on the basis of the differences in globin chain structure, evolution of haemoglobin in mammals has been estimated (e.g. Barnabas et al. 1971). Generally, within a single type of haemoglobin chains, the differences between globin chains of different mammals are relatively small. As an example, the α chains of the free-tailed bat (*Tadarida brasiliensis*; Kleinschmidt et al. 1987), the great Indian rhinoceros (*Rhinoceros unicornis*; Abbasi et al. 1987) and the lesser panda (*Ailurus fulgens*; Tagle et al. 1986) differ from the corresponding human chain by 18, 18 and 20 amino acids. The corresponding differences in the β chains are 23, 22 and 14 amino acids.

More than 500 human haemoglobin variants have been described. These have been reviewed, e.g. by Bunn and Forget (1986). Most of the globin chain variants have single amino acid substitution as compared to normal α or β chains, and can be used to probe the importance of single amino acid residues in the functional properties of haemoglobin. For example, modifications in the $\alpha_1\beta_2$ contact region destabilize haemoglobins 400–500 fold, whereas modifications outside the contact region only destabilize the molecule twofold (Ackers and Smith 1985). Modifications in the globin chain-haem contact generally lead to an unstable molecule (e.g. Bunn and Forget 1986).

3.5.2 Ontogenic Changes in Vertebrate Haemoglobins – Haemoglobin Switching

Changes in the haemoglobin pattern during ontogeny occur in most vertebrates. The larval (ammonochoete) and adult haemoglobins of lampreys have electrophoretically different mobilities (Potter and Brown 1975). Manwell's data on the skate, *Raja binoculata* (Manwell 1958b), and on the spiny dogfish, *Squalus acanthias* (Manwell 1958a), suggest that embryonic and adult haemoglobins are found in the elasmobranchs. Several teleost species have functionally different larval and adult haemoglobins. This situation is observed both in egg-laying teleosts, like rainbow trout (Iuchi 1973, 1985), and in viviparous teleosts, like *Embiotoca lateralis* (Ingermann and Terwilliger 1981b) and *Zoarces viviparus* (Weber and Hartvig 1984). Larval and adult haemoglobins differ from each other in many amphibians, e.g. *Pleurodeles waltli* (Flavin et al. 1978b), and *Rana catesbeiana* (Moss and Ingram 1968a; Maples et al. 1983). However, in the caelician, *Typhlonectes compressicauda,* the haemoglobins may be similar (Garlick et al. 1979). The multiple larval haemoglobins of, e.g. *Rana catesbeiana* and *Pleurodeles waltli* appear to have no chains in common with the adult haemoglobins (Moss and Ingram 1968a; Flavin et al. 1978b).

Watt et al. (1980) have sequenced the β chain of one haemoglobin component of *Rana catesbeiana* tadpole. The tadpole and adult β chains differ from each other by ca. 80 amino acids. The adult β chain differs from human β chain by ca. 75 amino acids, whereas the tadpole haemoglobin only differs from human β chain by 65 amino acids. These observations indicate that the genes coding for adult and tadpole β chains diverged early during evolution. Interestingly, the residues lining the organic phosphate binding site between the β chains are exactly the same in the tadpole of *Rana catesbeiana* as in man (Watt et al. 1980).

In the chicken, there is a marked shift in the proportions of the different globin chains during development (Groudine and Weitraub 1981; Haigh et al. 1982; Schalekamp et al. 1982; Fucci et al. 1983; Schalekamp and Van Goor 1984).

During ontogenesis of many mammals, two haemoglobin switches occur: the first, from embryonic to foetal (coinciding with the switch from nucleated to nonnucleated erythrocytes), and the second, from foetal to adult haemoglobin. Weber et al. (1987b) have sequenced the embryonic haemoglobin chains of the pig: the α-like embryonic chain differs from the adult chain by 65 amino acids, and the two β-like chains by 39 amino acids. The embryonic β-like chains have a very similar structure, differing only by 4 amino acids. In man, the embryonic α-like chain differs from the adult chain by 57 amino acid residues, whereas the β-like γ chains and ε chain only differ by 33 and 35 amino acids, respectively (see e.g. Bunn and Forget 1986).

Haemoglobin switching has been intensively studied in man, in some other mammals and in the chicken (e.g. Stalder et al. 1980; Haigh et al. 1982; Fucci et al. 1983; Mavilio et al. 1983; Groudine et al. 1983; Peschle et al. 1983; Garrick 1983; Schalekamp and Van Goor 1984). Some information is also available on amphibians (e.g. Moss and Ingram 1968a, b; Widmer et al. 1983; Maples et al.

1986) and teleost fish (e.g. Iuchi 1985). This section describes the suggested genetic and molecular mechanisms of haemoglobin switches. Oxygen transfer between maternal and foetal blood is described in Section 10.5.1.

The larval haemoglobins of rainbow trout appear to be associated with primitive erythrocytes, formed in the yolk sac and circulating in embryonic and newly hatched trout (see Iuchi 1985). Apparently the primitive erythrocytes do not produce adult haemoglobins, since they are not stained with a fluorescent antibody specific to adult haemoglobin (Iuchi and Yamamoto 1983). Also, the protein synthesis by primitive erythrocytes decreases markedly after 9 days from fertilization, indicating a decrease in globin synthesis, whereas marked protein synthesis is carried out by the definitive erythroid cells upon hatching (Yamamoto and Iuchi 1975). Thus, primitive and definitive erythrocytes produce different haemoglobins with no overlap in the production of the larval and adult haemoglobins (cf. Iuchi 1985).

In *Rana*, two sets of larval haemoglobin are formed. The first larval haemoglobin is expressed in erythrocytes produced by the intertubular regions of the mesonephric kidneys. The second larval haemoglobin is expressed in the red cell population produced by the liver (Broyles et al. 1981). During metamorphosis both larval erythrocyte types are replaced by a population of red cells that expresses only adult haemoglobins (Dorn and Broyles 1982). As in trout, each of the red cell populations produces a specific haemoglobin type only.

Thus, the haemoglobin switch in fish and amphibians fits the clonal model, in which stem cells are committed to the production of either larval or adult haemoglobins, and undergo clonal expansion and maturation (see e.g. Fucci et al. 1987). External factors affecting the haemoglobin switch in these vertebrate groups are largely unknown. However, thyroxine may be involved in the haemoglobin switch of the metamorphosizing frog (see Ingram 1985).

For the chicken, evidence supporting both the clonal model and the sequential gene activation model has been presented. According to the sequential gene activation model, red cells synthesize different haemoglobin types at different stages of maturation. In the sequential gene activation model, presented by Schalekamp et al. (1982) and Schalekamp and Van Goor (1984), the stem cells are initially committed to the erythroid line, whereby all globin genes are brought to a "permissive" condition and the genes are automatically activated in the ontogenetic order. Thereafter, appropriate environmental stimuli result in the active transcription of each of the genes. Primitive erythroid cells arise as an early cohort, have a short cell cycle (e.g. Lemez 1971) and transcribe all the globin chains, initially mostly embryonic ones but, with maturation, increasing amounts of adult chains. The definitive erythroid cells are many cell generations removed from the stem cells, and time-dependent changes in the chromatin structure may have occurred, so that only adult genes can be transcribed. In addition to the transcriptional control of globin gene expression, translational control may occur (see Schalekamp et al. 1982). In support of this model, primitive red cells express both the embryonic and adult α-globin chains (Schalekamp and Van Goor 1984). Correspondingly, both embryonic and adult α-globin genes are sensitive to DNAse I in the primitive erythrocytes (Stalder et al. 1980). The proportion of adult globins produced by primitive erythrocytes increases with the age of the cell (Schalekamp et al. 1982; Schalekamp and Van Goor 1984). The definitive erythroid cells do not express embryonic globin chains. Notably, the genes for embryonic α chains are fully methylated (Haigh et al. 1982), and the genes for embryonic β chains become resistant to DNAse I (Stalder et al. 1980).

In support of the clonal model, Fucci et al. (1987) have found that only embryonic haemoglobins are present in circulating primitive red cells throughout their development; the lack of adult haemoglobins is due to the lack of adult β-globin chain, synthesized only by the definitive erythrocytes. Furthermore, the definitive erythrocytes do not contain the major embryonic haemoglobins at any stage of development (Beaupain 1985; Fucci et al. 1987). Also, the differences in the methylation and DNAse I sensitivities in the primitive and definitive erythrocytes (see above) can be interpreted to mean that the erythroid lines have different programs of gene expression (Fucci et al. 1987).

Recently, Choi and Engel (1988) have provided evidence for a β-globin developmental stage selector element (SSE) within the adult β-globin promoter. Interaction between this element and the β-globin enhancer would make the developmental stage-specific regulation of ε-globin and β-globin synthesis possible. At present, it is not known how the SSE controls haemoglobin switching.

Apparently, the haemoglobin switching of nonmammalian vertebrates resembles the first haemoglobin switch of mammals from embryonic to foetal haemoglobins (see e.g. Peschle et al. 1983). Little is known about the factors which affect this switch in mammals. However, according to Peschle et al. (1985), the syntheses of ξ-globin and β-globin chains appear to be restricted to primitive and definitive erythrocytes, respectively. In contrast, Stamatoyannopoulos et al. (1987) have observed that both primitive and definitive erythroid cells express all the globin genes. Thereby, the switch from embryonic to foetal globin synthesis may be time-programmed, as with the perinatal haemoglobin switch which has been studied in more detail.

With regards to the perinatal haemoglobin switch, the sequential gene activation model is attractive. In the transcriptional wave hypothesis of Peschle et al. (1983) a wave of transcription would proceed along the globin gene cluster in the order embryonic-foetal-adult genes during cytogeny, both in foetal and adult life. Thereby, the foetal and adult genes could be sequentially transcribed during erythroid differentiation. The initial foetal type maturation of red cells would involve a peak in the γ chain production. During adult type maturation of red cells, the wave of transcription would have passed the γ-globin gene, causing a rapid decline of γ-chain synthesis. Some erythroid progenitors of the adult would start haemoglobin production less mature than normal cells, and be able to produce foetal haemoglobin. The proportion of these cells would increase in stress erythropoiesis, in which terminal cell divisions are skipped (see e.g. Peschle et al. 1983) and added divisions occur in the most primitive erythroid precursors (see Monette 1983). In this model, the effects of environmental factors on the haemoglobin switch could be taken into account, e.g. by supposing that they would be able to activate transcription.

According to Stamatoyannopoulos et al. (1983), progenitor erythroid cells (BFU-Es) would form a developmental sequence in which the foetal BFU-Es would be programmed to produce foetal haemoglobin. In the adult, the bone marrow would contain a mixture of precursor cells at different stages of maturation. Under normal erythropoiesis only the most mature precursors, programmed to produce adult haemoglobin, would undergo the final maturation to red cells. During stress, less mature BFU-Es, still capable of foetal haemoglobin production, would also undergo the final erythrocyte maturation.

Both these models are able to accommodate the experimental findings. The perinatal switch from foetal to adult haemoglobin appears to be time-programmed, and dependent on the stage of maturation of erythroid progenitors. It occurs after ca. 9-month gestation in man: in premature newborns after birth, and in postmature newborns before birth (see e.g. Peschle et al. 1983). Maturation of erythroid colonies grown in vitro is associated with a decrease in the ratio of foetal to adult haemoglobin produced (Papayannopoulou et al. 1979; Gianni et al. 1980; Comi et al. 1981). Other results indicate that environmental factors influence the perinatal haemoglobin switch. Foetal sheep haematopoietic cells produce adult haemoglobin when transplanted to an adult animal (Zanjani et al. 1979), and a factor in foetal sheep serum induces haemoglobin switching in human erythroid progenitors (Stamatoyannopoulos et al. 1983). However, the perinatal haemoglobin switch is at least partly irreversible or independent of the environment of the erythroid progenitors, since adult erythroid progenitors do not revert to foetal haemoglobin synthesis when transplanted into foetuses (Zanjani et al. 1981).

Chapter 4

Structure of Circulating Red Cells

4.1 General Anatomy of the Cell

The nonnucleated, under no-flow conditions, biconcave disc of most mammals is unique among red cells. Among mammals, the red cells of camels are ellipsoidal (see e.g. Cohen 1978a). The shape of lamprey red cells is similar to that of mammalian stomatocytes both in larvae and adults (see Potter et al. 1974; Sordyl, unpublished observation). The primitive erythrocytes of other vertebrates are round and flattened, with a central bulge caused by the nucleus. The definitive nucleated red cells are ellipsoidal. The size of erythrocytes varies markedly, as exemplified by the differences between the red cells of some amphibians with a volume of ca. 20 000 μm^3 and the red cells of goat with a volume of ca. 25 μm^3 (Fig. 4.1 gives dimensions of red cells for selected species from different vertebrate groups).

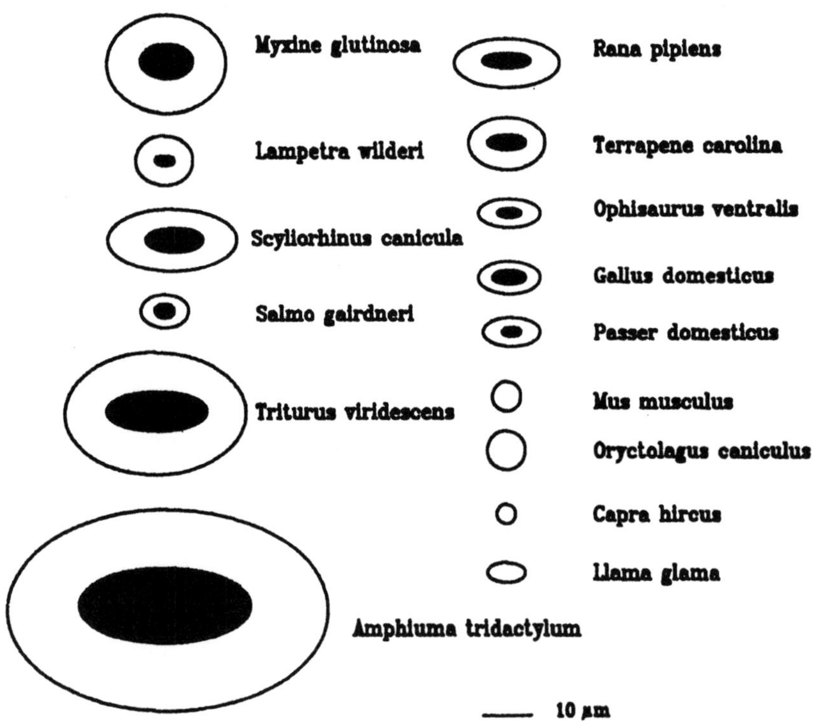

Fig. 4.1. The dimensions of some vertebrate erythrocytes. Data from Jordan (1938)

The amount of DNA in the nucleus of different vertebrate red cells varies markedly: the red cells of the parakeet (*Melopsittacus undulatus*) and the chicken (*Gallus domesticus*) contain ca. 40% of the amount of DNA found in human peripheral leukocytes, whereas the red cells of South American lungfish (*Lepidosiren paradoxa*) contain 35 times as much DNA as human peripheral leukocytes. The interspecies difference within vertebrate groups is greatest in fish (Atkin et al. 1965; Ohno and Atkin 1966).

Nucleated red cells also contain other cellular organelles, the number of which depends on the vertebrate group and the maturation stage of the erythrocyte. Micropinocytotic vesicles, which may be associated with haemoglobin synthesis, are seen in the red cells of cyclostomes and elasmobranchs (Mattison and Fänge 1977; Zapata 1980; Pulsford et al. 1982). Polyribosomes are present in immature fish erythrocytes, which produce large quantities of haemoglobin. The number of polyribosomes decreases upon maturation of the red cells, and the synthesis of haemoglobin slows down (e.g. Härdig 1978). Similarly, mitochondria and Golgi complexes are present in nucleated red cells, and their numbers decrease upon maturation of the cells (e.g. Brunner et al. 1975; Härdig 1978; Barni and Gerzeli 1985). Remnants of cytoplasmic organelles and nuclear material are found in vacuoles/cytolysosomes (e.g. Tooze and Davies 1965; Brunner et al. 1975; Zapata and Carrato 1981; Barni and Gerzeli 1985). In addition, marginal bands of microtubules are a characteristic feature of nucleated red cells (see Sect. 4.2.2.1).

4.2 The Red Cell Membrane

4.2.1 Membrane Lipids

Lipids make up 40% of total membrane weight in the human red cell. A little over 50% of the total lipids is phospholipids, more than 40% cholesterol, 2–3% free fatty acids and ca. 1% glycolipids (see e.g. Schwartz et al. 1985). The major phospholipids of the human red cell membrane are phosphatidylcholine, sphingomyelin, phosphatidylethanolamine (each 25–30% of total phospholipids) and phosphatidylserine (10–15% of total phospholipids). In addition, the membrane contains small amounts of lyso-phosphatidylcholine, phosphoinositides and phosphatidic acid. In pigeon red cells, phosphatidylcholine makes up 33%, phosphatidylethanolamine 30%, phosphatidylserine 8%, sphingomyelin 19% and phosphoinositides 4% of the total membrane phospholipids (Watts and Wheeler 1978).

One of the most characteristic features of the red cell lipid composition is the marked asymmetry of lipid species between the outer and inner leaflets of the cell membrane (e.g. Schwartz et al. 1984, 1985; Schlegel et al. 1985; Op den Kamp et al. 1985; Storch and Kleinfeld 1985). Glycolipids are found exclusively in the outer leaflet of plasma membrane (Storch and Kleinfeld 1985) and, of the phospholipids, 75% of phosphatidylcholine and 80% of sphingomyelin are in the outer leaflet. In contrast, 80% of phosphatidylethanolamine and 100% of phosphatidylserine is found in the inner leaflet (Schwartz et al. 1984). It appears that cholesterol is nearly equally distributed between the two leaflets and is able to move between the leaflets very rapidly: the half-time for transmembrane movement of cholesterol was shorter than 10 s at 37°C (Lange 1984). This half-time is much shorter than for, e.g. phosphatidylcholines for which, depending on the fatty acyl chains in the molecule, the half-times for transbilayer equilibration are 4–30 h (Op den Kamp et al. 1985). Thus, cholesterol may be the only constituent of the red cell membrane capable of transmembrane redistribution on a "circulatory" time scale.

Under physiological conditions, the red cell membranes are in a liquid-crystalline state. They are "fluid", i.e. the lateral mobility of membrane lipids is relatively high ($2–4 \times 10^{-9}$ cm s^{-1} at 25°C; Thompson and Axelrod 1980; Kapitza and Sackmann 1980), two orders of magnitude higher than the mobility of proteins embedded in the bilayer (Haest 1982). The fluidity of the cell membrane decreases (i.e. the acyl chain order of lipids increases) with increasing acyl chain length and with cholesterol enrichment, and increases with cis-unsaturation of the acyl chains (Cooper 1978; Storch and Kleinfeld 1985; Schwartz et al. 1985).

The way by which the membrane lipid asymmetry is created and maintained has been reviewed by Bishop and Bell (1988). The detailed mechanism is not quite clear. Membrane phospholipids are mainly produced by enzymes located at the cytoplasmic face of endoplasmic reticulum (e.g. Storch and Kleinfeld 1985). It is possible that the transmembrane movement of phospholipids produced is speeded up by flippases (Bishop and Bell 1985). These phospholipid transporters may produce phospholipid asymmetry in intracellular vesicles which are later inserted into the plasma membrane (Pfenniger and Johnson 1983). Similar flippases (ATP-dependent translocases) could also generate asymmetry in the plasma membrane: Seigneuret and Devaux (1984), and Daleke and Huestis (1985) have shown that a magnesium- and ATP-dependent protein selectively translocates aminophospholipids (phosphatidylethanolamine and phosphatidylserine) into the inner leaflet of the membrane. Middelkoop et al. (1988) have also demonstrated, using reversibly sickled red cells, that the asymmetric distribution of phosphatidylserine is maintained in part by ATP-dependent translocation. In addition, the membrane asymmetry depends on lipid-protein interactions. Phosphatidylserine interacts with membrane skeletal proteins, notably with spectrin (e.g. Haest 1982) and band 4.1 (Sato and Ohnishi 1983). Following oxidative cross-linking of spectrin molecules with diamide in intact human erythrocytes, the aminophospholipids became increasingly accessible to exogenously added phospholipase A_2. Similarly, in band 4.1-deficient red cells some of the phosphatidylserine appears in the outer leaflet of the cell membrane (Schwartz et al. 1985). In addition, the pH gradient across the cell membrane may contribute to the formation and maintenance of phospholipid asymmetry (see Hope and Cullis 1987).

Both the lipid composition of the membrane and the lipid asymmetry of the membrane are important for many cellular functions (for a review, see Deuticke and Haest 1987). For example, the sodium-potassium ATPase requires phosphatidylserine for full activation in human red cells (Roelofsen 1981). The activity of adenylate cyclase is regulated by the nature of its lipid environment (Houslay and Gordon 1983). The cholesterol content of the membrane affects protein function; the turnover number of the sugar transporter, reconstituted in dipalmitoylphosphatidylcholine membranes containing different amounts of cholesterol, varies markedly with variations in the cholesterol content of the membrane (see Connolly et al. 1985; Carruthers and Melchior 1986). Also, cholesterol suppresses the activity of the anion exchanger at cholesterol/phospholipid ratios higher than 0.3 (Grunze et al. 1980; Jackson and Morgan 1982). The lipid asymmetry and the association of lipids with membrane proteins may be important in the maintenance of cell shape and deformability (see Chap. 5). Furthermore, disturbed phospholipid asymmetry may promote coagulation: the presence of phosphatidylserine on the outer membrane leaflet appears to speed up coagulation markedly (see Schwartz et al. 1985). Loss of membrane lipid asymmetry also increases erythrocyte-endothelial cell interactions (Schlegel et al. 1985), and the presence of phosphatidylserine in the outer leaflet of the cell membrane increases the binding of red cells to macrophages and their phagocytosis (Schroit et al. 1985). Since senescent red cells accumulate calcium, which both decreases their deformability (Shiga et al. 1985) and disturbs the phospholipid asymmetry (see Schwartz et al. 1985), loss of phospholipid asymmetry may be an important factor in the recognition of senescent red cells and their removal from circulation.

The phosphoinositides, a minor class of phospholipids by molar percentage, only 3–4% of membrane phospholipids in human and bird erythrocytes (see Watts and Wheeler 1978; Ferrell and Huestis 1984), may be very important in controlling membrane (and cellular) functions. In several tissues, e.g. blood platelets and neutrophils, phosphatidylinositol-4,5-bisphosphate hydrolysis is sensitive to external stimuli. Both hydrolysis products, diacylglycerol and inositol trisphosphate, appear to function as intracellular second messengers (for reviews see Nishizuka 1984; Berridge and Irvine 1984; Majerus et al. 1985; Berridge 1987). Diacylglycerol operates within the plane of the membrane to activate protein kinase C, whereas inositol trisphosphate is released into the cytoplasm and functions as a second messenger for mobilizing intracellular calcium. It is not known for certain if hydrolysis of phosphatidylinositol-4,5-bisphosphate plays a role in intracellular signalling in any vertebrate red cells. However, recent studies by Cala (1986b) and McConnell and Goldstein (1988) suggest that the diacylglycerol-protein kinase C pathway may be involved in the activation of the volume regulatory ion transport pathways of *Amphiuma* and skate (*Raja erinacea*) red cells after osmotic disturbances. Phosphoinositide metabolism is also involved in the regulation of red cell shape in humans, either by affecting the membrane lipid bilayer balance (see Ferrell and Huestis 1984) or by affecting the binding of membrane skeletal protein band 4.1 to glycophorin (Anderson and Marchesi 1985).

The lipid composition of erythrocyte membrane, and changes in lipid composition in temperature adaptation, of poikilotherms are little known. Changes in temperature markedly affect acyl chain conformation, orientation and mobility (e.g. Tardieu et al. 1973; Meraldi and Slichter 1981). Thereby, a decrease in temperature leads to phase separations within the membrane, i.e. an increase in the number of rigid lipid domains dispersed in the fluid crystalline lipid phase and, ultimately, to phase changes from a fluid to gel state with corresponding disturbances in the lipid-associated membrane functions (see e.g. Hazel 1984). In temperature adaptation, however, the fluidity of most membranes is conserved via (1) a decrease in the proportion of saturated acyl chains in the membrane lipids, (2) an increase in the proportion of mono- or polyunsaturated acyl chains, (3) a decrease in the membrane phosphatidylcholine and sphingomyelin and (4) an increase in phosphatidylethanolamine contents (e.g. Cossins 1976, 1977; Hazel 1979). Homeoviscous adaptation of membranes and the mechanisms by which these adjustments may be achieved have been reviewed by Cossins (1983), Hazel (1984) and Cossins and Lee (1985). Bly et al. (1986) have shown that the proportion of saturated acyl chains decreases and polyunsaturated acyl chains increases in the erythrocyte membrane of the channel catfish, *Ictalurus punctatus*, during adaptation to decreased temperature. However, Bly and Clem (1988) were unable to show significant differences in the red cell membrane viscosity, as determined by the fluorescence depolarization method, of a teleost fish, *Ictalurus punctatus*, acclimated to 17° and 27°C. Thus, erythrocytes appear to follow the homeoviscous adaptation pattern poorly.

4.2.2 Structural and Membrane Proteins

Detailed information on the structure of membrane-associated proteins is available mainly on mammalian, especially human, red cells. The only nucleated red cells studied in some detail are avian red cells. Only fragmentary information is available on the structure of membrane proteins of reptilian, amphibian and fish red cells. From the available data it appears that the membrane skeletal proteins are conserved throughout the vertebrates (see also Dockham and Vidaver 1987), except for *Agnatha*, in which the membrane skeleton appears to be deficient (see Ellory et al. 1987). The variation in the integral membrane proteins of red cells within a given vertebrate group, and between different groups may be much greater. For example, Inaba and Maede (1988) recently isolated a transmembrane glycoprotein from the red cells of goat, constituting more than 10% of total membrane protein, which is also found in other ruminants, but is absent from the membrane of man, dog, rabbit and chicken.

4.2.2.1 Membrane Skeleton

The structure of the membrane skeleton of human red cells has been reviewed, e.g. by Branton et al. (1981); Pinder et al. (1981); Haest (1982); Marchesi (1984, 1985); Chasis and Shohet (1987) and Goodman et al. (1988). Table 4.1 gives the major proteins of the human red cell membrane. Figure 4.2 gives a schematic picture of the membrane skeleton and its interactions with integral membrane proteins. The membrane skeleton consists of spectrin, actin, tropomyosin, proteins 4.1, 4.2 and 4.9 and some myosin. The interactions between the membrane skeleton and the integral membrane proteins are mainly due to the

Table 4.1. The major membrane-associated proteins of the human red cell

Nomenclature on the basis of SDS-PAGE electrophoresis	Name of the protein	Molecular weight	Copies per cell
1	Spectrin, α subunit	240 000	200 000
2	Spectrin, β subunit	225 000	200 000
2.1	Ankyrin	210 000	100 000
2.2		190 000	
2.3		180 000	
2.6		165 000	
3	Anion exchanger	90–100 000	1 000 000
4.1a		80 000	100 000
4.1b		78 000	100 000
4.2		72 000	200 000
4.5	Glucose transporter	55 000	250 000
	Nucleoside transporter	45– 60 000	10 000
	Monocarboxylic acid transporter		
4.9		48 000	100 000
5	Actin	43 000	500 000
6	Glyceraldehyde-3-phosphate dehydrogenase	35 000	500 000
7		29 000	400 000
7a	Tropomyosin	29 000	60 000
7b	Tropomyosin	27 000	60 000
8		24 000	200 000
	Glycophorin α	43 000	600 000
	Glycophorin β	39 000	50 000
	Glycophorin γ	25 000	80 000

binding of band 4.1 to glycophorin, and the binding of ankyrin to spectrin and the anion exchanger. The interactions between the different components of the membrane skeleton, and between the membrane skeletal and integral membrane proteins are influenced by phosphorylation-dephosphorylation reactions of the proteins, by calmodulin, and by phosphatidylinositol phospholipids (for review, see Mische and Morrow 1988).

Spectrin is composed of α (MW 240 000) and β (MW 225 000) subunits, which self-associate to form heterodimers, tetramers and higher oligomers (e.g. Morrow and Marchesi 1981). Within the membrane skeleton, spectrin appears to exist mostly as tetramers, 100 000 tetramers per cell (see e.g. Chasis and Shohet 1987). Spectrin binds ankyrin, the "membrane attachment" protein, to a MW 50 000 polypeptide of the β chain, at a site 20 nm removed from the head of the spectrin molecule (Tyler et al. 1979, 1980; Morrow et al. 1980), and forms a complex with actin and band 4.1 at the end of the molecule not involved in self-association (e.g. Cohen et al. 1984; Cohen and Langley 1984). Furthermore, spectrin can bind 2.3-diphosphoglycerate, thus possibly affecting oxygen transport (Shaklai et al. 1978), and calmodulin (Sobue et al. 1981). Haemoglobin increases the spectrin dimer-dimer association, and increases the mechanical stability of the

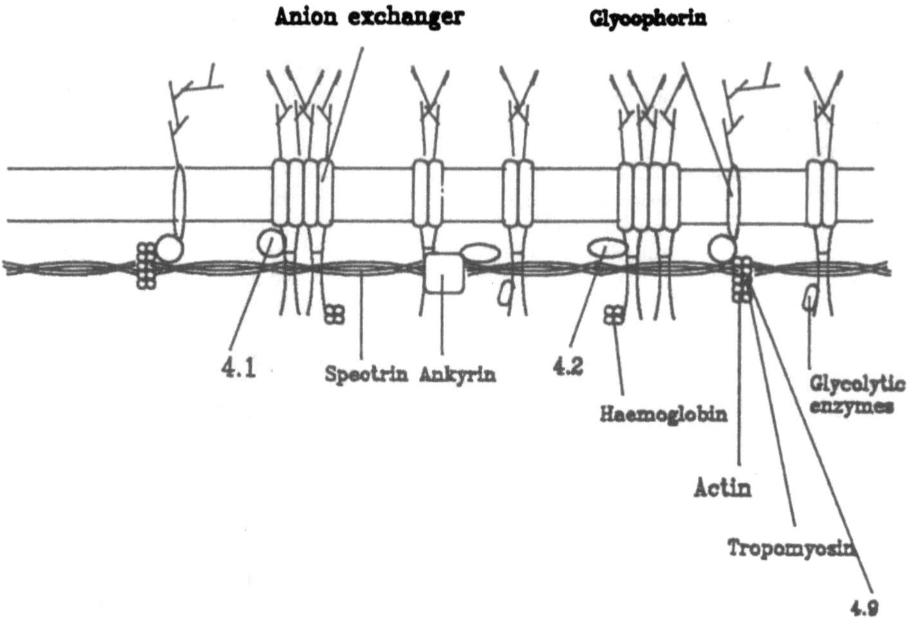

Fig. 4.2. A schematic picture of the membrane skeleton of mammalian erythrocytes and its interactions with the integral proteins

membrane (e.g. Liu and Palek 1984), whereas hemin weakens spectrin-protein 4.1 interactions and spectrin dimer-dimer associations.

Actin (MW 43 000), present in 500 000 copies in red cells, forms 15–20 unit polymers. The actin polymers may be stabilized by tropomyosin, which binds to the grooves between the helically coiled filaments (Fowler and Bennett 1984), and by actin-bundling protein (band 4.9; Siegel and Branton 1985). Each actin filament interacts with four spectrin tetramers. In the absence of band 4.1, the interactions are of low affinity (see e.g. Marchesi 1985).

Band 4.1, present in 200 000 copies (MWs 78 000 or 80 000) in human red cells, stabilizes the actin-spectrin network (see e.g. Cohen and Langley 1984), and links the skeletal proteins to the membrane (see e.g. Tyler et al. 1979, 1980). Band 4.1 binds to glycophorin (sialglycoproteins α, β, γ; glycoconnectin; see Chasis and Shohet 1987). The interaction between glycophorin and band 4.1 is very important for the deformability and mechanical stability of the cells. The mechanical stability of erythrocytes deficient in sialglycoproteins β and γ was only 50% of that in normal cells, and deformability only 40% of that in normal cells (Reid et al. 1987; membrane deformability and stability were measured using a continuously applied shear stress in the ektacytometer, see Chap. 5). Sialglycoprotein α (glycophorin A) deficiency does not appear to influence membrane deformability in normal situations (Reid et al. 1987). However, its association with protein 4.1 is regulated by polyphosphoinositides (see Anderson and Marchesi 1985) suggesting that phosphatidylinositol-4,5-bisphosphate metabolism may affect membrane-membrane skeletal protein interactions, thereby influencing the lateral mobility of membrane proteins. Band 4.1 is also capable of binding to the anion exchanger (band 3; capnophorin; see Pasternack et al. 1985).

The attachment of the anion exchanger to the membrane skeleton is mainly due to its interaction with ankyrin (goblin, syndein). Ankyrin (MW in human red cells 165 000–210 000) binds to the cytoplasmic domain of the anion exchanger with 1:1 stochiometry (Bennett and Stenbuck 1979, 1980). Only a fraction (ca. one-tenth) of the cell's anion exchanger molecules appears to bind ankyrin, possibly because only one conformation or oligomeric state of the anion exchanger can bind ankyrin (see e.g. Low 1986). Each ankyrin monomer also binds to a spectrin tetramer, thereby providing the membrane-membrane skeleton link.

Band 4.2 is present in ca. 200 000 copies in the human red cell membrane. It can bind both to ankyrin and to the cytoplasmic portion of band 3 (Korsgren and Cohen 1988). At present the function of band 4.2 is not known.

The membrane skeleton of nucleated erythrocytes has many elements similar to those in the membrane skeleton of mammalian erythrocytes. The presence of spectrin and actin (or proteins with similar molecular weights) has been demonstrated in such various groups as elasmobranchs (Cohen et al. 1982), amphibians (Centonze et al. 1986) and birds (e.g. Blikstad et al. 1983). However, the red cells of the hagfish. *Eptatretus stouti*, appear to be highly deficient in spectrin, and also in band 4.1 protein (see Ellory et al. 1987). In contrast, band 4.1 and ankyrin are present in the red cell membranes of birds (Chan 1977; Granger and Lazarides 1984).

In contrast to definitive mammalian red cells, definitive nucleated red cells have intermediary filaments composed of vimentin and synemin (Centonze et al. 1986) and a marginal band of microtubules (e.g. Barrett and Dawson 1974; Cohen 1978b; Zapata and Carrato 1981; Cohen et al. 1982). The microtubules are formed of tubulin, a protein with a molecular weight of 50–60 000 (Cohen et al. 1982). The β-tubulin variant of mature chicken erythrocytes is exclusively expressed in erythrocytes and thrombocytes, and appears in the erythroblasts at the same time as haemoglobin synthesis is initiated and the marginal band of microtubules starts to form (Murphy et al. 1986). In the earlier stages of erythroid differentiation, the microtubules of the cells are similar to those of other cell types, e.g. fibroblasts, i.e. centrosomal and radiating individually from the microtubule organizing centre to all domains of the cytoplasm (Murphy et al. 1986; Kim et al. 1987). However, although the tubulin variant of mature erythrocytes appears to be unique, the marginal bands of avian red cells can be reformed to detergent-extracted membrane skeletons by using calf brain tubulin, suggesting that the formation of marginal bands is partly dependent on the tubulin-membrane skeleton interactions (Swan and Solomon 1984). Notably, vimentin and synemin form contacts between the marginal band and the membrane skeleton, and between the marginal band and the nucleus (e.g. Centonze et al. 1986).

The primitive, nucleated red cells of the mouse also contain, in addition to a spectrin-actin network, microtubules which form loosely associated bundles close to the membrane, and intermediary filaments composed of vimentin (Koury et al. 1987).

4.2.2.2 *Functions of Membrane Skeleton*

The membrane skeleton stabilizes the red cell membrane. It increases shear deformability (see Chien 1987 and Chap. 5) and inhibits fragmentation and vesiculation of the membrane (see e.g. Haest 1982). The red cells of the spectrin-deficient house mouse (*Mus musculus*) show extensive vesicle formation in vivo (Palek and Liu 1981). Membrane skeleton also prevents cell fusion. The fusion of erythrocytes requires a formation of protein-free zones (e.g. Ahkong et al. 1975) which is prevented by an intact membrane skeleton, since the contact sites between integral membrane proteins and membrane skeleton appear to be distributed homogeneously over the whole membrane surface (see e.g. Haest 1982).

The membrane skeleton markedly influences the lateral mobility of the membrane proteins. For example, the diffusion constant of the anion exchanger in dimuristoylphosphatidylcholine membrane is ca. 1000 times greater than in erythrocyte membranes (Chang et al. 1981), and is markedly greater in spectrin-deficient mutant red cells than in normal red cells: in normal mouse red cells the diffusion constant of anion exchanger was 4.5×10^{-11} cm^2s^{-1} and in spectrin-deficient red cells 2.5×10^{-9} cm^2s^{-1} at 24°C (Sheetz et al. 1980). These mobilities are those measured for "free" anion exchanger; the ankyrin-bound anion exchanger is much more immobile with a diffusion constant less than 10^{-13} cm^2s^{-1} (e.g. Tsuji and Ohnishi 1986). Such a restriction, and release of restriction, of the lateral mobility of membrane proteins may be important for receptor function (see below and Axelrod 1983).

The marginal band of microtubules appears to be required for the formation of discoid shape in developing nucleated red cells (e.g. Barrett and Dawson 1974). However, "resting" mature red cells maintain their discoid symmetry even in the absence of the marginal band (Behnke 1970; Barrett and Dawson 1974; Cohen et al. 1982). The function of the marginal band in mature red cells may be to resist deformation and to return deformed cells rapidly back to the ellipsoidal shape: when subjected to mechanical stress, dogfish (*Mustelus canis*) erythrocytes without marginal bands often become folded and buckled, whereas those with marginal band remain flattened and ellipsoidal (Joseph-Silverstein and Cohen 1984). The intermediary filaments are probably required to keep the nucleus in its central position. The presence of intermediary filaments also restricts the cytoskeletal movement in nucleated red cells as compared to that in mammalian red cells (see Sect. 5.2.3.2).

4.2.2.3 *Receptors*

β-Adrenergic Receptor

The most intensively studied receptor-effector system in red cells is the β-adrenergic receptor (hormonally regulated adenylate cyclase; for detailed reviews, see e.g. Lefkowitz et al. 1982, 1983; Levitzki 1986, 1988; Benovic et al. 1988). Specific binding of β-adrenergic ligands to either intact red cells or red cell membranes have been observed in all vertebrate groups studied except *Cy-*

clostomata (Pajarinen and Nikinmaa, unpublished observation). The complete hormonally regulated adenylate cyclase system involves five functional units (see Fig. 4.3 and, e.g. Levitzki 1984, 1986; Lefkowitz et al. 1985):

1. The stimulatory receptor (β-adrenergic receptor). β-adrenergic receptors are glycoproteins, with a molecular weight of 58 000–62 000 in mammalian and frog systems; in bird red cells the molecular weight of the receptor appears to be 40 000–50 000 (see Lefkowitz et al. 1983). The structure of the β-adrenergic receptor resembles that of the porcine muscarinic acetylcholine receptor and that

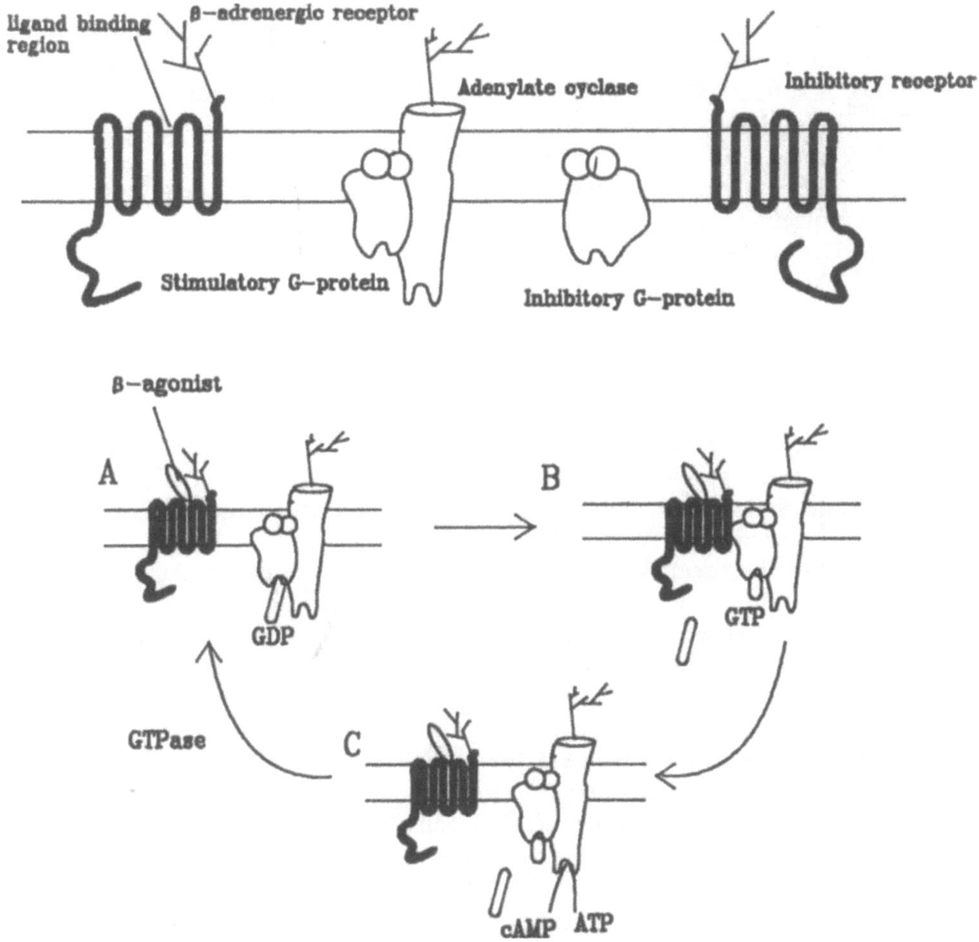

Fig. 4.3. A schematic picture of the β-adrenergic receptor-effector system (hormonally regulated adenylate cyclase), and its activation by β-adrenergic agonists. After binding of β-adrenergic agonist to the receptor (*A*), the receptor moves laterally to interact with the Gs protein-adenylate cyclase complex (*B*). The guanosine diphosphate (*GDP*) is replaced by guanosine triphosphate (*GTP*), and adenylate cyclase is activated (*C*), carrying out the hydrolysis of adenosine triphosphate (*ATP*) to cyclic adenosine monophosphate (*cAMP*). ATP hydrolysis continues as long as GTP is bound to the Gs protein. The GTPase activity of the Gs protein then hydrolyzes GTP to GDP. See text for further details

of rhodopsin (e.g. Nathans and Hogness 1983; Dixon et al. 1987). Each of these molecules have seven membrane-spanning hydrophobic helices, which are the most homologous parts of the different molecules, and four hydrophilic extra- and intracellular portions. The different molecules vary most in their C-terminal regions and the putative third intracellular segments (see Dixon et al. 1987). At present, the binding site of the β-adrenergic ligands on the receptor molecules is not accurately known. However, the results of Dixon et al. (1987) and Strader et al. (1987) indicate that there is no large hydrophilic domain associated with β-adrenergic ligand binding, but that ligands bind to the transmembrane region of the receptor.

2. The stimulatory guanyl nucleotide regulatory protein (Gs protein) binds guanosine triphosphate, and is tightly associated with the catalytic unit of the system. Gs protein is a triheteromer: the α subunit has a molecular weight of 43 000, β subunit 35 000 and γ subunit 5 000–10 000 (Sternweis et al. 1981; Codina et al. 1984). The α subunit binds guanine nucleotides and has GTPase activity (see e.g. Stryer and Bourne 1986).

3. The catalytic unit (adenylate cyclase) is a transmembrane glycoprotein with a molecular weight of ca. 150 000 (Pfeuffer et al. 1985).

4. The inhibitory receptor.

5. The inhibitory guanyl nucleotide regulatory protein (Gi protein), which is very similar to the Gs protein: the β and γ subunits are identical to those of Gs protein, and the α subunit has a molecular weight of 40–41 000 (Codina et al. 1984; Hildebrandt et al. 1984).

The mechanism of adenylate cyclase activation, and desensitization of the β-adrenergic receptor has been reviewed, e.g. by Lefkowitz et al. (1983); Levitzki (1986) and Sibley et al. (1987). Figure 4.3 gives a schematic model for the β-adrenergic activation of adenylate cyclase.

Initially, the hormone is bound to the receptor. Thereafter, the hormone-receptor complex interacts with the Gs protein-catalytic unit complex. This interaction probably requires lateral diffusion of the hormone-receptor complex into collisions with the Gs protein-catalytic unit complex (see e.g. Tolkowsky and Levitzki 1978; Axelrod 1983). The lateral mobility of the β-adrenergic receptor may be limited by the lipid environment of the molecule or receptor-membrane skeletal protein interactions (see e.g. Axelrod 1983). Cherksey et al. (1980) suggest that agonists (but not antagonists) induce a release of the receptor from the membrane skeletal constraints. Thereafter the receptor would be free to move laterally to collide with the Gs protein-catalytic unit complex.

Each hormone-receptor complex can activate numerous catalytic units via transient collisions with the Gs protein-catalytic unit complex (e.g. Levitzki 1986). Activation occurs provided that guanosine triphosphate (GTP) is bound to the α subunit of the Gs protein. ATP is converted to cyclic AMP at the catalytic unit as long as GTP remains bound to the Gs protein. However, the GTPase activity of the Gs protein rapidly hydrolyses GTP to GDP, and inactivates the complex. GDP is released before the next activation cycle (for further details see Levitzki 1986).

In addition to the Gs protein, the β-adrenergic receptor interacts with the Gi protein. Surprisingly, this interaction appears to enhance the hormonal accumulation of cyclic AMP, possibly by decreasing the basal adenylate cyclase activity in the absence of agonists (Cerione et al. 1985).

Desensitization is the tendency of biological responses to wane over time despite the continuous presence of a stimulus of constant intensity. Homologous desensitization is agonist-specific, i.e. only the response to the desensitizing hormone is reduced, whereas in heterologous desensitization the responses to different agonists operating through distinct receptors but via a common effector system are reduced. In the hormonally regulated adenylate cyclase system, desensitization of β-adrenergic receptor involves the phosphorylation of the receptor (see e.g. Sibley et al. 1985; Stadel et al. 1983). In heterologous desensitization, the receptor is phosphorylated by the cAMP-dependent protein kinase, whereas homologous desensitization appears to involve a specific β-adrenergic receptor kinase which has recently been purified (Benovic et al. 1987). For a review of desensitization through protein phosphorylation see Sibley et al. (1987).

Insulin Receptor

Binding of insulin to red cell membranes was first described by Haugaard et al. (1954). Ginsberg et al. (1977) demonstrated specific insulin binding to turkey red cells: the properties of insulin receptors appeared to be similar to those of, e.g. human hepatocytes. Human erythrocyte insulin receptors were characterized in 1978 by Gambhir et al. (1978), again showing functional similarities to those of other cell types. The apparent number of insulin receptors per cell is 3000 for turkey red cells (Ginsberg et al. 1977) and 2000 for human red cells (Gambhir et al. 1978).

It is likely that the insulin receptors from various sources share a similar structure. The receptor appears to be a complex of two α subunits (MW 130 000) and two β subunits (MW 95 000), linked together by disulphide bonds (for reviews, see e.g. Czech et al. 1981; Lewis and Czech 1988). The insulin receptor is an insulin-activated protein kinase, in which the β subunit is the substrate for phosphorylation. Effects of insulin on mammalian red cells include the inhibition of spectrin phosphorylation (Hesketh 1986), a decrease in the microviscosity of the membrane (Dutta-Roy et al. 1985) and stimulation of the sodium pump (Baldini et al. 1986). Stimulation of sodium pump may be the reason for the increased red cell potassium and decreased red cell sodium concentration, observed in rainbow trout red cells treated with insulin (Houston 1988). The physiological importance of the effects of insulin on red cells is not known at present.

Other Receptors

The binding of insulin to human erythrocyte membrane is increased by prostaglandin E_1 or prostacyclin (Ray et al. 1986). Dutta-Roy and Sinha (1985) have shown that there are specific prostaglandin binding sites on the red cell membrane. In the frog, *Rana pipiens*, incubation of the red cells with prostaglandin E_1 in the presence of a phosphodiesterase inhibitor caused an accumulation of cyclic AMP (Rudolph and Greengard 1980). However, the physiological significance of prostaglandin receptors on red cells is not clear at the moment. Similarly, the importance of high-affinity L-triiodothyronine binding to rat red cell membranes (Botta et al. 1983) is not clear.

4.2.2.4 Transport Proteins

The Anion Exchanger.

The anion exchanger (capnophorin, band 3 protein) is the major protein of the human erythrocyte membrane, making up 25–30% of total membrane protein. The estimated number of band 3 molecules in intact human red cells is ca. 1 000 000. In SDS-PAGE (sodium dodecyl sulphate-polyacrylamide gel) electrophoresis, the anion exchanger forms a diffuse band within the molecular weight range of 90 000–100 000 (e.g. Fairbanks et al. 1971; Yu and Steck 1975).

The presence of band 3 in other vertebrates has been deduced mainly on the basis of a protein band comigrating with human band 3 in SDS gel electrophoresis, and on the basis of stilbene disulphonate-inhibitable facilitation of anion exchange. Based on these criteria, the anion exchanger appears to be present in large numbers in the red cells of all vertebrates apart from *Agnatha*. Ellory et al. (1987) observed very little DIDS-inhibitable anion exchange activity in the red cells of the hagfish, *Eptatretus stouti*. DIDS did not affect the very slow chloride efflux across the red cell membrane of the lamprey, *Lampetra fluviatilis* (Nikinmaa and Railo 1987). The numbers of anion exchanger may be also small in primitive, nucleated erythrocytes of mammals; the amount of band 3 in the SDS gel electrophoresis of primitive murine erythrocytes appeared to be negligible as compared to definitive erythrocytes (Koury et al. 1987).

The following discussion initially outlines the structure and the major structure-function relationships of mammalian band 3, and thereafter gives some of the fragmentary information on the structure-function relationships of band 3 in other vertebrate groups.

Band 3 is a glycoprotein, and consists of two structural and functional domains (Fig. 4.4). The membrane spanning 55 kDa domain is involved in the transport of anions, and possibly also water and cations (see e.g. Solomon et al. 1983). The cytoplasmic, hydrophilic 43 kDa domain is the attachment site for ankyrin (see Sect. 4.2.2.1) and cytoplasmic proteins (see Passow 1986; Low 1986). In addition, band 3 contains antigenic determinants for blood group specificity (e.g. Childs et al. 1978) and possibly senescent cell recognition (Kay et al. 1983; Low et al. 1985).

In the native erythrocyte membrane, band 3 protein exists at least as dimers, but probably also as tetramers (see e.g. Jennings 1985; Jennings and Nicknish 1985). Band 3 molecules copolymerize with hemichromes (denatured haemoglobin; see Waugh and Low 1985; Waugh et al. 1987). The anion transport, however, requires only the monomeric form of the molecule (see Jennings 1985).

The membrane-spanning, mostly hydrophobic domain crosses the membrane at least eight, but most likely 12–13 times (Kopito and Lodish 1985;

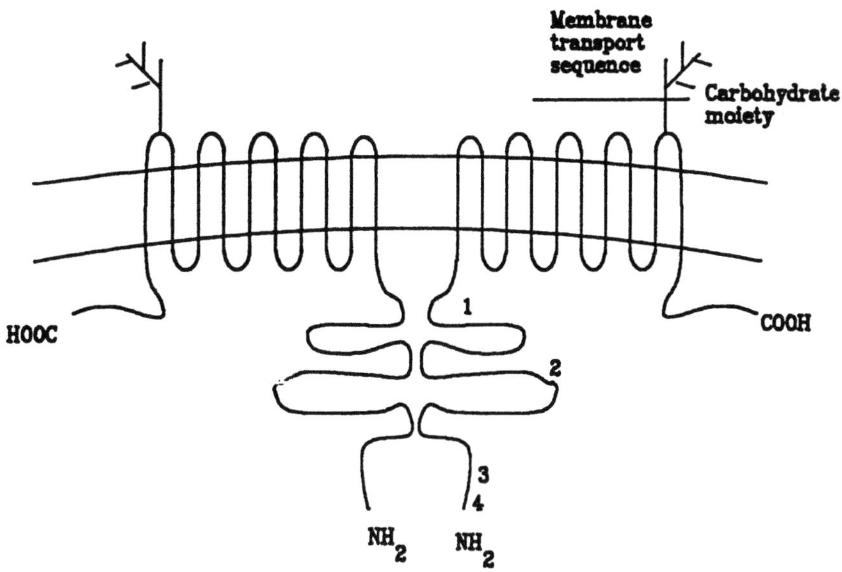

Fig. 4.4. The major features of the structure of human band 3 and the approximate sites of its interactions with *1* ankyrin, *2* band 4.2, *3* glycolytic enzymes and *4* haemoglobin. For further details, see text and Low (1986)

Jennings et al. 1986a). Although the primary structure of mouse band 3 is now known (see Kopito and Lodish 1985), the portions of the membrane spanning domain, which are involved in the translocation of anions across the membrane either directly ("transfer sites") or by allosterically modifying the activity of the transfer sites ("modifier sites"), are still little known. Transport of anions may involve an aqueous pore which is interrupted by permeability barrier(s), probably relatively close to the surface (see Passow 1986 for a detailed account on the models of anion transport). Jay and Cantley (1986) discuss in detail the possible amino acid residues involved in the transport of anions. The "extracellular loops" of the membrane-spanning parts of the molecule are predominantly positively charged, and initiate the selection for anions. The actual transport site appears to be a positively charged "funnel". Three to four arginine and three lysine residues would act as an "anion selector" near both the cytoplasmic and the exterior entrance of the hydrophilic funnel. Deeper in the cavity, negatively charged aspartate and glutamate residues could be involved in the actual gating mechanism, controlling the passage of anions.

The cytoplasmic domain of band 3 can be cleaved from the membrane-spanning domain without the loss of function in either. Thus, these two portions of the molecule are relatively independent (see Low 1986). The cytoplasmic domain appears to exist as a dimer in most conditions. It is elongated, possibly with a length of 25 nm (see Low 1986). The elongated form of the cytoplasmic domain may be important for generating the surface area for the numerous interactions between cytoplasmic proteins and band 3. The most important of these interactions are depicted in Fig. 4.4. In addition, an endogenous tyrosine kinase can phosphorylate band 3. The major target for phosphorylation appears to be a tyrosine residue close to the N-terminus of the molecule (Dekowski et al. 1983).

The N-terminus of the human band 3 molecule is highly acidic, containing 18 glutamic or aspartic acid residues within the first 33 amino acid residues (Kaul et al. 1983). It is the binding site for haemoglobin, hemichromes and glycolytic enzymes (for a review see Low 1986). Tyrosine phosphorylation of band 3 inhibits the binding of most of these proteins (Low et al. 1987). There is enough of band 3 to bind practically all of the cell's glycolytic enzymes. Thereby, band 3 could form a glycolytic enzyme complex. The results of Chetrite and Cassoly (1985) suggest that the T-state of haemoglobin may bind to band 3 inside the intact human red cell. However, at present the physiological significance of the binding of glycolytic enzymes and haemoglobin to band 3 is not known. The copolymerization of band 3 with hemichromes may generate a site for autologous antibody binding, and provide a mechanism for selective removal of denatured haemoglobin from the cell (Low et al. 1985; Waugh et al. 1987; see also Sect. 2.4).

The ankyrin binding site of band 3 is removed from the extreme N-terminus of the molecule. Apparently, not all the band 3 molecules are capable of binding ankyrin (see Low 1986). The differences in ankyrin binding appear not to reside in the primary structure of band 3, but may be a result of differences in conformation or oligomeric state of the protein in the cell membrane (see Low 1986). Interestingly, despite the fact that the binding of ankyrin to band 3 plays a major role in membrane skeleton-membrane interactions, the regulation of the binding has been little studied. For example, it is not known if treatments which alter cell shape affect band 3-ankyrin interactions.

Band 3 can also bind band 4.1 (see Sect. 4.2.2.1). The relative importance of this binding depends on the phosphorylation state of phosphatidylinositol in the membrane, which regulates band 4.1-glycophorin association (see Anderson and Marchesi 1985). In addition, band 4.2 interacts with band 3.

The molecular weight of band 3 varies in mammals: mouse, pig, camel and llama band 3 all have a higher molecular weight than human band 3 (see Ralston 1975; Whitfield et al. 1983; Khodadad and Weinstein 1983). The N-terminus of the cytoplasmic domain of murine band 3 is considerably less acidic than in man (Kopito and Lodish 1985). Notably, band 3 of another rodent, the rat, does not bind glyceraldehyde-3-phosphate dehydrogenase (Ballas et al. 1985). The anion exchanger of the chicken also has a less acidic N-terminus than human band 3, and does not bind glyceraldehyde-3-phosphate dehydrogenase (Jay 1983). Furthermore, antibodies to the cytoplasmic domain of the human band 3 do not crossreact with the chicken band 3 (Low 1986), indicating that considerable differences in structure must exist. The nature of these differences, and their effect on the function of the cytoplasmic domain are not known. There is no information at present on the structural and functional properties of the cytoplasmic domain of band 3 in reptiles, amphibians and fish. Furthermore, the structural basis for the differences between the anion transport function of band 3 in a poikilotherm, *Salmo gairdneri,* and mammals is not known. Romano and Passow (1984) have shown that the temperature dependence of anion exchange in rainbow trout is much smaller than in mammals. In addition, DIDS-inhibitable sulphate transport increases with increasing pH from 6.5 to 7.5, in contrast to that observed in mammals.

Glucose Transporter

The human erythrocyte glucose transporter is a glycoprotein in the 4.5 region of SDS-PAGE electrophoresis, with a molecular weight of 55 000, constituting approximately 5% of total membrane protein (e.g. Allard and Lienhard 1985). The estimated number of glucose carrier molecules in the human red cell membrane is 250 000. The glucose transporter is a glycoprotein consisting of three major domains: (1) 12 α-helical membrane spanning regions, (2) a large, highly charged cytoplasmic part, and (3) an exterior domain with the carbohydrate moiety. The α helices may be arranged into a water-filled channel. The structure of the human glucose transporter has recently been reviewed by Walmsley (1988). In the native red cell membrane the protein appears to exist either as di- or as tetramer (Wheeler and Hinkle 1985).

Avian erythrocytes also show specific and saturable facilitated diffusion of glucose and related monosaccharides across the cell membrane (Wood and Morgan 1969). Although the structure of this transporter has not been studied, the cross reaction between antibodies raised against the human erythrocyte glucose transporter and chicken embryo fibroblasts (Salter et al. 1982) suggests that glucose transporters may generally share similar structures.

Nucleoside and Monocarboxylate Transporters

The red cell nucleoside and monocarboxylate (lactate) transporters also occupy the 4.5 region in the SDS-PAGE electrophoresis (see Whitfield et al. 1985). The human red cell nucleoside transporter has an apparent molecular weight similar to that of the glucose transporter (45–60 000), but the apparent number of transport sites is an order of magnitude smaller, 10 000 (Young and Jarvis 1983). The monocarboxylate transporter appears to be a 43 000 kDa protein in rabbit red cells, which have much more pronounced lactate transport than human red cells (Jennings and Adams-Lackey 1982). However, even in man a very weak band at a molecular weight of 43 000 is observed (Jennings and Adams-Lackey 1982).

Active Ion Pumps

The properties of the sodium pump (Na,K-ATPase) have been reviewed, e.g. by Glynn and Karlish (1975); Cavieres (1977); Wallick et al. (1979); Schuurmans-Stekhoven and Bonting (1981); Cantley (1981) and Kaplan (1985). Sodium pump actively transports sodium out of the cell and potassium into the cell in the ratio of three Na^+ ions for two K^+ ions. In human red cells the pump is present in only 100–1000 copies (Drickamer 1975; Cavieres 1977). Thus, although most of the functional information on the pump has been obtained using mammalian red cells, the structural properties of the enzyme have mainly been studied on other cell types. The sodium pump has been purified at least from mammalian kidney (e.g. Jorgensen 1974), shark rectal gland (Hokin et al. 1973), eel electroplax (Perrone et al. 1975) and duck salt gland (Hopkins et al.

1976). Since the purified sodium pump is similar in all cases, it is likely that its structure is strongly conserved, and similar also in red cells. The molecule contains two subunits: the α subunit (MW 100 000 kDa) contains the ATP binding site and the phosphorylation site of the molecule, whereas the β subunit (MW 50 000 kDa) may be the sodium ionophore of the molecule (see e.g. Shuurmans-Stekhoven and Bonting 1981). In the membrane, the molecule appears to be a $\alpha_2\beta_2$ tetramer (Peters et al. 1981).

The function of sodium pump requires a phospholipid environment (e.g. Wallick et al. 1979), possibly for proper subunit interactions of the pump (e.g. Schuurmans-Stekhoven and Bonting 1981). Although the exact requirement for specific phospholipids is not known, Roelofsen and Van Deenen (1973) have reported that phosphatidylserine is essential for the activation of sodium pump in human red cells. On the other hand, DePont et al. (1978) did not observe any specific phospholipid requirements for the activation of sodium pump from rabbit kidney outer medulla.

The calcium pump is present in human red cells in ca. 400 copies (Drickamer 1975; for reviews on the structure and properties of the calcium pump, see e.g. Carafoli and Zurini 1982; Al-Jobore et al. 1984; Schatzmann 1986). Its presence in nucleated red cells has also been demonstrated (Ting et al. 1979; Asai et al. 1976). The pump extrudes calcium from the cell, probably in exchange for protons (Smallwood et al. 1983), and thus maintains the very low intracellular calcium concentration. The molecular weight of the protein is ca. 140 000, and in the native membrane it behaves as a dimer (Cavieres 1984). The activity of the pump is regulated by calmodulin (Jarrett and Penniston 1977). Carafoli and Zurini (1982) have presented a model for the functional architecture of the calcium pump, based on the proteolytic fragmentation of the erythrocytic molecule. A MW 90 000 polypeptide contains the active site and the calmodulin binding site of the molecule, and is capable of transporting calcium in a calmodulin-sensitive process (Zurini et al. 1984). Similar to the sodium pump, the activity of the calcium pump depends on the phospholipid environment. Maximal activation is achieved in the presence of acidic phospholipids (phosphatidylserine, phosphatidylinositols and phosphatidic acid; see Al-Jobore et al. 1984).

Both the sodium and calcium pumps appear to be closely associated with the anion exchange pathway (Fossel and Solomon 1981; Waisman et al. 1981), as indicated, e.g. by the fact that inhibitors of anion exchange inhibit the calcium pump. A possible functional significance of this, provided that band 3 binds glycolytic enzymes (e.g. glyceraldehyde-3-phosphate dehydrogenase, i.e. band 6) in vivo, is that glycolysis continuously provides fuel (ATP) for the function of the pumps. At least the sodium pump uses "membrane-bound" ATP as a fuel (Shoemaker and Hoffman 1985).

In addition to information concerning the sodium and calcium pumps, Drickamer (1975) presented evidence for an erythrocytic magnesium pump. However, I am not aware of structural information on this molecule.

There is very little information on the structural basis of "novel" transport pathways for cations, e.g. the sodium/proton exchanger of amphibian and fish red cells (e.g. Cala 1983b; Baroin et al. 1984a; Nikinmaa and Huestis 1984b;

Cossins and Richardson 1985). Huot et al. (1989) have, however, purified a renal amiloride-binding protein with a molecular weight of 25 000, which has the properties of the sodium/proton exchanger. Solomon et al. (1983) have suggested that cation movements (as well as anion and water movements) through the red cell membrane would generally take place via an aqueous pore, formed between the two monomers of dimeric band 3. Although anions can exchange through monomeric band 3 (e.g. Passow 1986), the above suggestion for the movement of cations is attractive, since it would explain the close relationship between cation and chloride movements, observed in catecholamine-stimulated avian red cells (for a review see e.g. Palfrey and Greengard 1981) and, to a lesser extent, in amphibian (e.g. Cala 1983b) and fish (e.g. Motais and Garcia-Romeu 1987) red cells. Cytoskeletal elements have also been implicated in the adrenergic stimulation of cation transport: phosphorylation of goblin (ankyrin; syndein, MW 180 000–240 000) has been observed in both turkey (Rudolph et al. 1978) and frog (Rudolph and Greengard 1980) erythrocytes. Notably, ankyrin is associated with band 3. The calcium-induced increase in red cell potassium permeability (Gardos-effect; Gardos 1959) may involve a 23 000 kDa membrane protein (Plishker et al. 1986).

Chapter 5

Red Cells in Circulation: Factors Affecting Red Cell Shape and Deformability

5.1 Blood Viscosity

Vascular architecture and blood viscosity (including the capability of red cells to deform to enter capillaries) are the main determinants of the resistance to blood flow in circulation. The apparent viscosity of blood in a given vessel is determined by plasma viscosity, the number of cells in unit volume of blood, and the aggregation and deformability of the cells, especially the red cells, since they are by far the most numerous cell type in blood. The major part of the apparent viscosity of blood can be attributed to the red cells. For example, in rainbow trout, at 15° and at a shear rate of 90 s^{-1} in a cone-plate viscometer, the plasma viscosity was ca. 2 cP and that of blood ca. 6 cP (Fletcher and Haedrich 1987). With a decrease in the shear rate, the relative importance of the red cells in determining blood viscosity increases further.

The viscosity of blood can be measured using, e.g. capillary and rotational viscometers. In capillary viscometers blood is allowed to flow through a known length (L) of a capillary (with a known radius R) at a given pressure head (Δ P), and the volume flow rate (Q) is measured. The apparent viscosity of blood (η app, which supposes that blood flow in the tube is laminar) can then be calculated from Poiseuille's law (see e.g. Bayliss 1962):

$$\eta \text{ app} = (\pi \times \Delta P \times R^4)/(8 \times Q \times L). \qquad (5.1)$$

The apparent viscosity of blood in microvessels can be similarly obtained from the pressure gradients, vascular architecture and volume flow rate (see Chien 1985). The principle of operation of rotational viscometers is given using coaxial cylinder viscometer as an example (Fig. 5.1). In this type of viscometer, blood is placed in the space between the cylinders, and the outer cylinder is rotated at a constant speed. Because of the viscosity of the fluid in the space between the cylinders, the inner cylinder will also start to rotate, unless prevented from doing so by an external constraint. If the inner cylinder is held stationary, the torque produced on the cylinder is proportional to the shear stress. The shear rate (γ) can be calculated from the rotation speed and the geometry of the outer cylinder, and the shear stress (τ) from the torque produced with a use of geometry-conversion factor, determined by viscosity-standard oils. The apparent viscosity of blood is then the ratio τ/γ (see e.g. Chien et al. 1971).

In laminar flow, shear rate is the velocity gradient between two fluid layers (units s^{-1}). Shear stress is the magnitude of the force which has to be applied per unit area to produce a given rate of shear (units dyne cm^{-2} or N m^{-2}). Viscosity of the fluid is the ratio between shear stress and shear rate (units poise = the shear

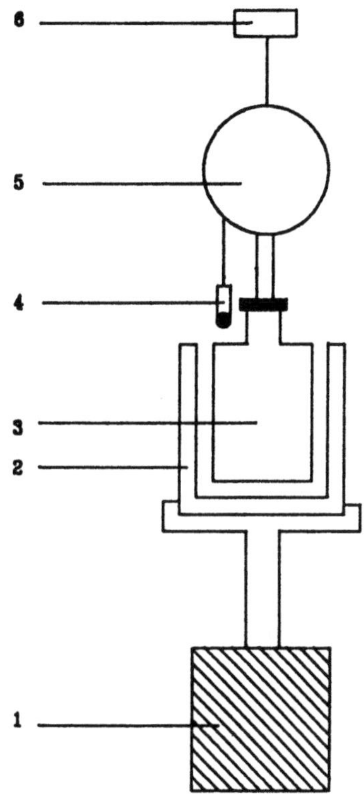

Fig. 5.1. The principle of operation of a rotational viscometer. Blood is placed in the space between the outer (*2*) and inner (*3*) cylinder. The outer cylinder is rotated at a constant speed by a motor (*1*). Owing to the viscosity of blood the inner cylinder tends to start to rotate. The slightest movement of the inner cylinder is detected by a photodetector (*4*). This movement is converted into an electrical signal and used to activate a torque motor (*5*), which returns the inner cylinder to its initial position. The voltage required to keep the inner cylinder in the initial position is the displayed measuring signal (*6*)

stress of 1 dyne cm^{-2} produces 1 s^{-1} shear rate, or N m^{-2} s^{-1}). The shear rate (γ) on the blood vessel wall can be roughly calculated from the relation

$$\gamma = 8 \times \text{flow velocity of blood/vessel diameter} \qquad (5.2)$$

(which assumes Newtonian behaviour, i.e. laminar flow with a parabolic flow velocity distribution in the vessel). See e.g. Bayliss (1962) for further details.

Figure 5.2 gives apparent viscosities for the blood of selected vertebrates at two shear rates as a function of temperature. Regardless of the vertebrate group, several features are similar.

Blood is a non-Newtonian fluid, i.e. its viscosity, measured in a viscometer, decreases with increasing shear rate. The major factors causing the shear-dependence of blood viscosity are red cell aggregation at low shear rates and red cell deformability and orientation to flow at high shear rates (see e.g. Chien 1970). Red cell aggregation, deformation and orientation are likely to occur in the circulation, since the shear rates on blood vessel walls vary from 0 to 2500 s^{-1} (e.g. Graham and Fletcher 1983; Stoltz and Donner 1987). In blood flowing through tubes, the shear dependence of apparent viscosity of blood is also influenced by the thickness of plasma layer in the tube periphery. Reinke et al. (1986) measured the apparent viscosity of human blood in 29–94 μm diameter tubes, and found that the relative viscosity of blood remained unchanged throughout the range of shear rates (0.7–100 s^{-1}) studied. They suggested that while the red cell ag-

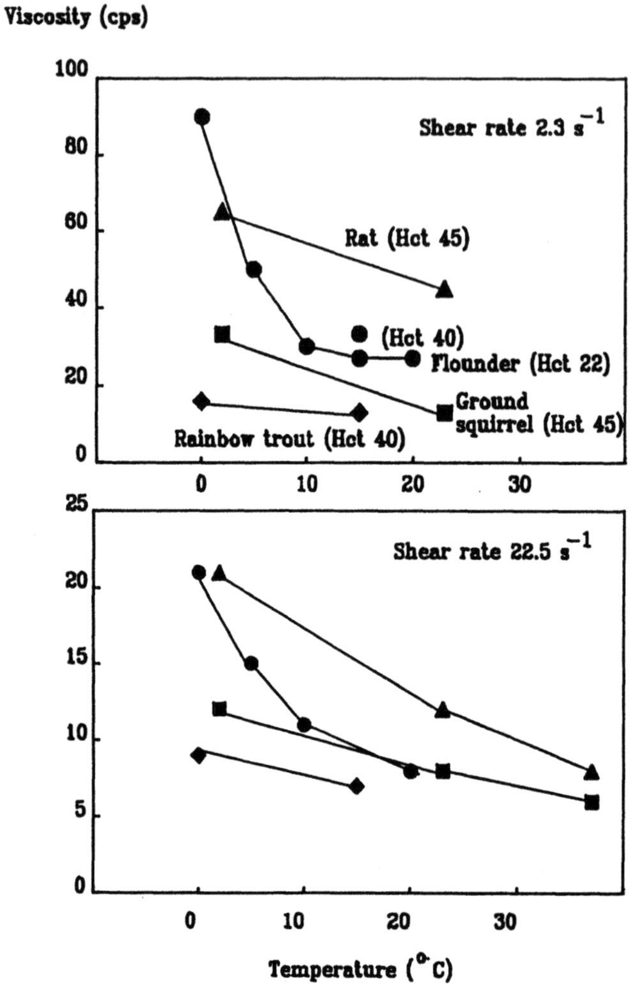

Fig. 5.2. The blood viscosity of some vertebrates as a function of temperature at two shear rates. Data on the ground squirrel (*Spermophilus tridecemlineatus*) and rat from Maclean (1981), on the winter flounder (*Pseudopleuronectes americanus*) from Graham and Fletcher (1983), and on the rainbow trout (*Salmo gairdneri*) from Fletcher and Haedrich (1987)

gregation increases the local viscosity in the center of the tube at low shear rates, this effect is counteracted by an increase in the thickness of the marginal, cell-depleted layer of fluid, leading to a decrease in the local viscosity at the peripheral parts of the tube.

At low shear rates, the viscosity of blood of poikilothermic vertebrates (with nucleated red cells) and hibernating mammals is often much lower than that of homeothermic mammals. This may be related to the presence of a putative plasma factor preventing red cell aggregation at low shear rates in hibernators (Maclean 1981), and the lesser tendency of nucleated red cells to aggregate than mammalian red cells (Chien et al. 1971). At high shear rates (at similar temperatures and haematocrit values) the blood viscosity of mammals and vertebrates with nucleated red cells appears to be similar in spite of the greater resistance of nucleated red cells to deformation (see Sect. 5.2.3.2). This contrasts with the observations on red cell suspensions in Ringer solutions, in which nucleated red cells (of Peking duck, *Anas platyrhynchos*) have much higher relative viscosity than human red cells at a given haematocrit value (Gaehtgens et al. 1981b).

At a constant shear rate, blood viscosity increases with decreasing temperature (Fig. 5.2). The relationship between temperature and blood viscosity appears to be biphasic. The increase in viscosity is more marked at temperatures below 22°C than above in human blood (Rampling and Whittingstall 1987). In addition to the general dependence of fluid viscosity on temperature, the increase is, to some extent, due to red cell aggregation, which occurs at increased shear rates when temperature is decreased (Neumann et al. 1987). Notably, in hibernators and in fish experiencing large fluctuations in body temperature, the dependence of blood viscosity on temperature is much less at low shear rates than in, e.g. laboratory rats (see Fig. 5.2 and Maclean 1981; Fletcher and Haedrich 1987).

The apparent viscosity of blood decreases with decreasing tube diameter, until the diameter decreases enough that the red cells must deform in order to enter the tube, whereafter viscosity increases (see e.g. Skalak and Chien 1981; Gaehtgens et al. 1981a,b).

5.2 Deformability of Red Cells

5.2.1 Basic Principles

Effective deformation of red cells is required if the red cells are to traverse the capillary bed, since capillaries often have a smaller diameter than the red cells. Also, as stated in the preceding section, the deformation of red cells and their orientation to flow are important determinants of blood viscosity. The deformation of red cells is their ability to change shape in response to deforming forces. The major factors affecting red cell deformability are the geometry of the cell (mainly the surface-to-volume ratio, but also the shape of the cell), the intracellular viscosity, the rheological properties of the membrane and the nature of the deforming force (see e.g. Chien 1987). The surface-to-volume ratio can be affected by osmotic swelling or shrinking of the cell; the cell shape by various treatments outlined in Section 5.2.4.3; and the intracellular viscosity mainly by changes in the cellular haemoglobin concentration, e.g. as a result of osmotic swelling or shrinking. The major types of membrane deformation are (1) an elongation or "shear" of the membrane without a change in the surface area and without bending, (2) an increase in the surface area and (3) a bending of the membrane. These deformations are characterized by shear modulus, area expansion modulus and bending modulus. A larger modulus indicates a greater resistance to that particular form of deformation (Hochmuth and Waugh 1987).

5.2.2 Methods for Measuring Red Cell Deformability

The deformability of red cells can be measured in three different types of situations: (1) deformation of red cells in bulk flow, (2) deformation of red cells attached to the bottom of a flow channel and (3) deformation of cells in narrow capillaries (see Chien 1977).

5.2.2.1 Deformation of Red Cells in Bulk Flow

The rotational cone-plate viscometer is commonly used in the determinations of red cell deformability (e.g. Fischer et al. 1978a). When it is used for red cell deformability measurements, the transparent cone-plate viscometer is mounted on an interference-contrast microscope, and red cells are either photographed or filmed with a video camera. The shear rate (in s^{-1}) is practically uniform in the conical gap, and equals the difference in the peripheral speed of the cone and the plate. The shear stress (in dynes cm^{-2}) can be obtained from the shear rate by multiplying it by medium viscosity. The elongation index (E′) of the red blood cells, which gives an index of deformability, can be calculated from the projected lengths (L′) and widths (B) of the photographic images (Fischer et al. 1978a):

$$E' = (L' - B)/(L' + B). \qquad (5.3)$$

In ektacytometry (e.g. Bessis and Mohandas 1975), blood suspension is placed in the gap of a rotational viscometer crossed by a laser beam, a constant shear rate applied and the light scattering spectra produced by the red cells determined. The diffraction rings obtained are characteristic of the shape and orientation of the cells studied (for a detailed explanation, see Stoltz et al. 1981).

5.2.2.2 Deformation of Cells in a Flow Channel

In the flow channel method, the red cells are allowed to attach themselves to the glass bottom of a narrow channel, whereafter cell-free suspension is pumped through the channel at a constant rate. The degree of cell elongation at different flow rates can be determined by microphotometry, and the shear stress (τ) on the channel floor calculated from the pressure drop between the inlet and the outlet of the channel and the channel geometry (see Chien 1977):

$$\tau = \Delta P \times \text{channel width}/(2 \times \text{outlet-inlet distance}). \qquad (5.4)$$

5.2.2.3 Deformation of Cells in Narrow Channels

In the micropipette technique, a micropipette is placed on a red cell and a constant negative pressure applied, whereby a portion of the red cell membrane is aspirated in the pipette (e.g. Waugh and Evans 1976; Crandall et al. 1978;

Chabanel et al. 1987). The membrane extensional rigidity, which reflects the steady-state resistance to deformation, can be estimated from the relationship between the stress applied (applied negative pressure × internal radius of the micropipette) and the strain induced (maximum length of the aspirated portion of the red cell membrane in the micropipette/internal radius of the micropipette; see e.g. Chabanel et al. 1987). From the time course of cell entry during aspiration or cell recovery during release, the time constant for the viscoelastic response of the cell membrane and the surface extensional viscosity can be estimated (see Hochmuth et al. 1979).

In the filtration method, red cell suspensions are passed through filters with narrow pore sizes. Polycarbonate filters are most commonly used, since their pore size is rather uniform and the channels are straight (Reinhart et al. 1984; Reinhart and Chien 1985). Two major experimental approaches have been adopted: either the cells are sucked (pressed) through the filter at a constant negative (positive) pressure and the transit time for a volume of blood (or an individual cell) determined (e.g. Reid et al. 1976; Hanss 1983; Kikuchi et al. 1983; Acquaye et al. 1987) or the cells are filtered through the polycarbonate sieve at a constant flow, and the pressure generated by the red cells passing through the filter determined (e.g. Usami et al. 1975; Leblond and Coulombe 1979; Reinhart et al. 1984). In the former case, the initial flow rate can be taken as the index of deformability; in the latter, the initial pressure rise.

5.2.2.4 Aspects of Deformability Given by the Different Methods

The different methods of cellular deformability measurement reflect different aspects of deformability. Reinhart and Chien (1985) have investigated the relative roles of cell geometry and cellular viscosity in the filtration test by manipulating cell volume and mean cellular haemoglobin concentration via osmolality changes. The results show that red cell filtratability through pores close to the limiting size for red cell filtration is mainly determined by the surface/volume ratio (i.e. the geometry) of the cell. In filtration through larger pores, the viscous properties (membrane and cytoplasmic viscosity) become increasingly important. However, accurate estimations of the different determinants of deformability are difficult with this method.

Micropipette aspiration can be used to determine different aspects of membrane rheology (see Hochmuth and Waugh 1987) – neither cell geometry nor cytoplasmic viscosity influence the results. Similarly, the flow channel technique is insensitive to changes in cell geometry or internal viscosity, and thereby suited to studies of the rheological properties of the cell membrane (see Chien 1977). Both the micropipette and flow channel techniques make it possible to study the rheological properties of single cells. In rotational viscometers, the factors contributing to the cell deformation depend on the shear rate. At high rates of shear the internal fluid viscosity is an important determinant of the deformation. At low shear rates the membrane properties and cell geometry become important.

5.2.3 Deformation of Red Cells in Circulation

5.2.3.1 Mammalian Red Cells

The resting or "minimum energy" shape of mammalian erythrocytes is a biconcave disc. However, whenever mammalian red cells are under shear stress, their shape changes. In the cone-plate viscometer, red cells are deformed to ellipsoids at high shear rates, and are oriented in the direction of flow (e.g. Fischer et al. 1978a; Fischer 1978). It is likely that this kind of deformation and orientation also occurs in the circulation, at least on the arterial side, since the shear rates in arterial circulation ($> 700\ s^{-1}$) are high. As shown by Chien (1970), the deformation of red cells decreases the apparent viscosity of blood throughout the range of physiological shear rates. In the cells under shear, the membrane moves around cytoplasm in a tank-treadlike motion. Such membrane motion has been elegantly demonstrated by Fischer et al. (1978b), using small latex particles as probes for membrane movement.

During capillary entrance the red cells are markedly deformed, and flow edge on in a slipper-like shape in the capillaries (Bagge and Brånemark 1981).

Red cell aggregation, which occurs at low shear rates, appears to result from adsorption of macromolecules on the red cell membrane and formation of macromolecular bridges between two red cells. Even the formation of the aggregates may require that the red cell is easily deformed, since this makes a closer contact between adjacent cells possible and may thereby facilitate the bridge formation (see Chien and Jan 1973).

The major determinant of the shear modulus in red cells is the cytoskeletal network (Chien 1985). When spectrin is crosslinked with diamide (Fischer et al. 1978a; Chasis and Mohandas 1986) or with malonyldialdehyde (Pfafferott et al. 1982), the elongation of red cells in shear stress is reduced. Wheat germ agglutinin, which binds to membrane sialglycoproteins, decreases red cell deformability, as measured by ektacytometry (Chasis et al. 1985), and rigidifies the membrane in micropipette aspiration (Smith and Hochmuth 1982; Evans and Leung 1984). Further, red cells deficient in sialglycoproteins β or γ had reduced membrane deformability (Reid et al. 1987), measured in an ektacytometer. Chasis and Mohandas (1986) have presented a model for cytoskeletal deformation during shear elongation of human red cells at constant surface area (Fig. 5.3).

5.2.3.2 Nucleated Red Cells

Nucleated red cells are generally more resistant to shear-induced shape changes than mammalian red cells. The shear modulus of the membrane of nucleated erythrocytes is 5–15 times greater than that of mammals (Waugh and Evans 1976). This may be because the cytoskeletal system of nucleated erythrocytes is much more complex than that of the mammalian erythrocytes (see Sect. 4.2.2.1). The cytoskeletal "movement", depicted in Fig. 5.3, may be much more difficult in nucleated red cells than in mammalian red cells, since the spectrin-actin network of nucleated red cells is further restrained by the intermediary filaments

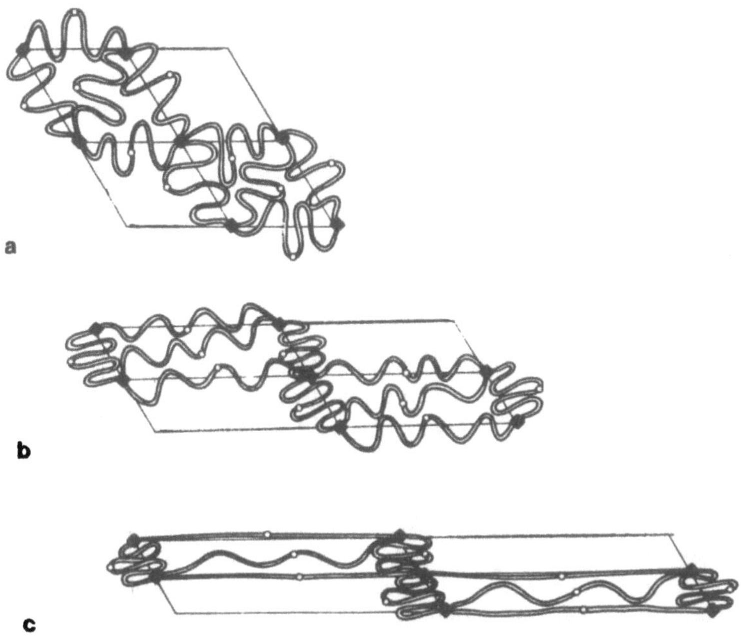

Fig. 5.3 a-c. A model of reversible deformation of erythrocyte membrane. Reversible deformation occurs with a change in geometric shape but at a constant surface area. *a* The nondeformed membrane. With increased shear stress, the membrane becomes increasingly extended (*b* and *c*). Further extension of the membrane beyond that shown in *c* would result in an increase in surface area and in the breaking of the junction points. This is the stage at which membrane fragmentation occurs. ♦, Protein 4.1, actin and spectrin association points; ○, spectrin-spectrin association points; linear coils, spectrin dimers. Reproduced from J. A. Chasis and N. Mohandas (1986), the Journal of Cell Biology 103:343–350 by copyright permission of the Rockefeller University Press

and marginal band of microtubules. As a result, the deformation of nucleated red cells in shear stress is much smaller than the deformation of mammalian red cells (Gaehtgens et al. 1981a,b). Also, the complexity of the cytoskeleton may explain the lack of tank-treading in bird and amphibian red cells (Schmid-Schönbein and Gaehtgens 1981).

Because of the limited deformation of nucleated red cells in shear stress, the decrease in blood viscosity, observed with increasing shear rates in rotational viscometers (Chien et al. 1971; Fletcher and Haedrich 1987; Graham and Fletcher 1983), must be largely due to a better alignment of the red cells in the direction of flow at high shear rates than at low ones (see Fischer 1978; Gaethgens et al. 1981a,b). Koyama (1985) has shown that frog red cells are oriented with their long axis in the direction of flow in the microvasculature of frog lungs in vivo, even at low wall shear rates (16 s^{-1}). Thus, the data on native blood suggest that the lack of orientation to flow of frog red cells suspended in a dextran solution, even at high shear rates (Fischer 1978), is not observed in circulation.

Although the resistance of nucleated red cells to shear deformation is greater than that of mammalian red cells, they are markedly deformed when entering the capillaries. Notably, frog and *Amphiuma* red cells can pass through filters with a pore diameter as small as one-quarter of the minor oval diameter of the cell (the value for man is ca. one-third; Chien et al. 1971). It appears that the marginal band of microtubules and its associations with other cytoskeletal elements are involved in the recovery of normal cell shape after deformation, since elasmobranch red cells without the marginal band of microtubules often remain wrinkled and deformed after passing through narrow capillaries (Joseph-Silverstein and Cohen 1984).

5.2.3.3 Physiological Effects of Reduced Red Cell Deformability

Decreased deformability of the red cells leads to their becoming trapped in the spleen (e.g. Groom 1987) and other tissues, like the sternum, the lung, the liver and the femur bone (Simchon et al. 1987). It is probable that in most tissues the red cells become mechanically trapped in the capillaries, whereby the microcirculatory blood flow in the tissue is drastically reduced (e.g. Simchon et al. 1987). However, it is not likely that mechanical trapping explains the trapping of the red cells in the spleen, since the reticular meshwork of the spleen resembles a Millipore filter with a pore size of 1.5 – 3 x the diameter of the red blood cell. More likely, the red cells are retained by adherence to reticular fibers and macrophages (see Groom 1987).

5.2.4 Factors Affecting Red Cell Deformability

5.2.4.1 Osmotic Perturbations and Cell Volume Changes

Red cell shrinkage in hypertonic medium leads to an increase in the surface-to-volume ratio of the red cell, whereby red cell deformability would be expected to increase. Simultaneously, however, the internal viscosity, which is a function of internal haemoglobin concentration (Herrmann and Müller 1986) increases, leading to a decrease in deformability. In hypotonic medium, the opposite changes occur. Clark et al. (1983) have studied the elongation of red cells at different osmolalities by ektacytometry. The elongation of red cells in shear stress is minimal at low osmotic concentrations, indicating the importance of surface-to-volume ratio in determining deformability. With increasing osmolality, elongation at constant shear stress is facilitated, until at above-physiological osmolalities, elongation is restricted because of the increase in internal viscosity. Reinhart and Chien (1985) have investigated the net effect of osmolality-induced changes in the surface-to-volume ratio, and internal viscosity, on the passage of human red cells through narrow capillaries. The minimum resistance of red cell filtration through polycarbonate sieves with pore diameter (2.6 μm) close to the critical size for red cell passage was achieved at 400 mOsm osmolality. In this case, the surface-to-volume ratio affected red cell filtration more than the changes in internal viscosity at physiological osmolalities. Resistance to filtration through larger pores (6.9 μm), in constrast, decreased to 200 mOsm osmolality, indicating that decreasing internal viscosity was a more important determinant of filtratability than cell geometry. These observations suggest that an increase in cell volume may either inhibit or facilitate red cell passage through microvasculature, depending on the vascular architecture. Hughes and Kikuchi (1984), working on rainbow trout red cells, observed that the swollen red cells of hypoxic fish filtered through 8 μm Nuclepore filters faster than control cells. This may be largely due to the fact that the resistance to filtration through large pores depends mainly on the intracellular viscosity.

5.2.4.2 Temperature

The red cells become more deformable with increasing temperature. The shear modulus of human red cells decreases 30–40% with an increase in temperature from 6° to 37°C, and the membrane surface viscosity decreases sixfold (Waugh and Evans 1979; Hochmuth et al. 1980). Kikuchi et al. (1985) and Hughes et al. (1982b) have studied the filtratability of fish red cells through 5 μm polycarbonate sieves at different temperatures, and found that the pore passage time increases markedly with decreasing temperature. In addition to the deformation of individual red cells, this may reflect increased red cell aggregation at reduced temperatures (see Sect. 5.2.3).

5.2.4.3 Red Cell Shape, and Shape-Transforming Agents

The normal resting biconcave shape of mammalian red cells can be perturbed by various treatments. Nakao et al. (1960) observed that ATP-depleted red cells became echinocytes (see Fig. 5.4 for the nomenclature of major mammalian red cell shapes). Later, echinocyte formation was observed in cells loaded with calcium (e.g. Palek et al. 1974). Further, incubation of cells with a variety of amphipathic drugs causes the red cells to become either echinocytes or stomatocytes (Sheetz and Singer 1974). On the basis of their findings, Sheetz and Singer (1974) proposed a bilayer-couple mechanism for drug-induced shape changes of mammalian erythrocytes. According to the model, echinocytes are formed if the amphipathic drug is preferentially equilibrated in the outer lipid monolayer of the red cell membrane, expanding it relative to the inner monolayer. Stomatocytes are formed if the drug equilibrates into the inner monolayer, which then expands relative to the outer monolayer. The bilayer-couple hypothesis can explain most of the experimental data obtained concerning shape transformation of red cells with amphipathic drugs or lipid incorporation into the red cell membrane (see Ferrell et al. 1985). Isomaa et al. (1987) have recently suggested that amphiphiles may also form intrabilayer nonbilayer phases, which affect the transbilayer redistribution of both the intercalated amphiphiles and bilayer lipids. Further, Kuypers et al. (1984) and Christiansson et al. (1985) have shown that the shape changes can depend on the lipid molecular shape: if phosphatidylcholines with acyl chains different from the acyl chains of native phosphatidylcholine are introduced on the red cell membrane, they cause the membrane to bend either inwards or outwards, depending on the shape of the introduced phosphatidylcholine species.

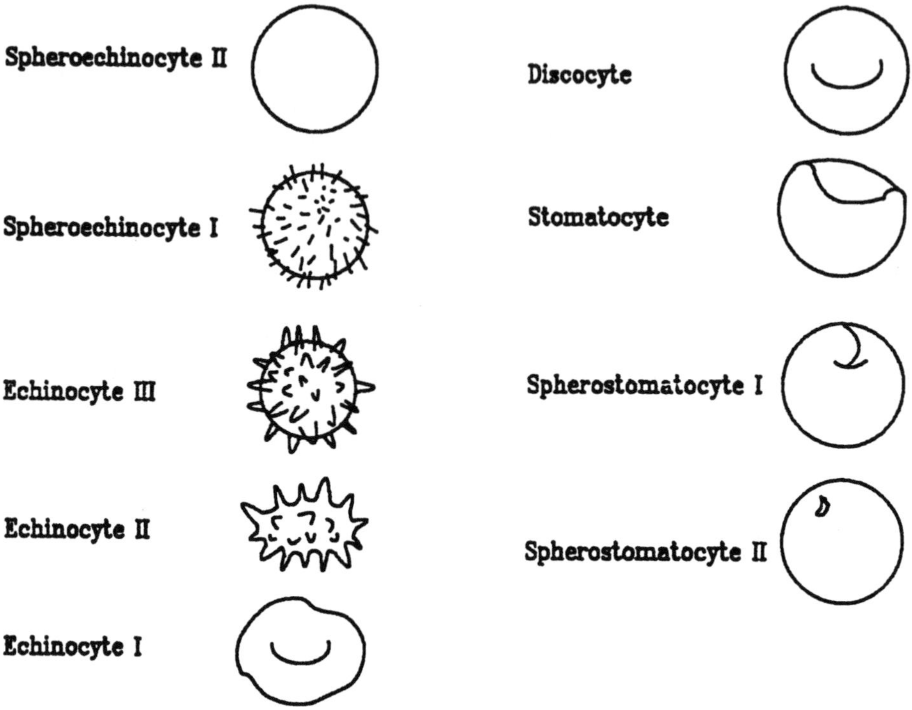

Fig. 5.4. Nomenclature of the major mammalian red cell shapes according to Bessis (1973)

The bilayer-couple hypothesis has also been used to explain the changes in red cell shape after metabolic ATP depletion and calcium loading. Ferrell and Huestis (1984) observed that the time courses of changes in membrane phosphatidylinositol-4,5-bisphosphate concentration and metabolic crenation coincided. During crenation, phosphatidylinositol-4,5-bisphosphate is broken down to phosphatidylinositol, which has a smaller, less highly charged head group. As a consequence, and if phosphatidylinositol does not move to the outer monolayer, shrinkage of inner monolayer relative to outer monolayer would occur and cause echinocytosis. Ferrell and Huestis calculated the shrinkage of the inner monolayer expected, as well as the volume of echinocytic spicules, and reached good agreement. Also, in cells crenated as a result of calcium loading, changes in the lipid composition and shape changes follow similar time courses (see Allan and Thomas 1981). Vanadate is an effective echinocytosis-inducing compound. Changes in the membrane lipid phosphorylation and shape changes coincide also in vanadate-treated cells (Backman 1986). Giraud et al. (1984) observed that stomatocytic transformation of cholesterol-depleted cells was correlated with a decrease in the turnover rate of phosphatidylinositol bisphosphate.

Although the effects of lipid incorporation and amphipathic drugs can largely be explained by the bilayer-couple hypothesis, with modifications, membrane skeleton may also be involved in the shape changes. If membrane skeletal proteins are crosslinked with diamide (which rigidifies the membrane), both echinocytic and stomatocytic changes induced by a variety of treatments are inhibited (Haest et al. 1980). Lovrien and Anderson (1980) have shown that wheat germ agglutinin, which binds to glycophorin, prevents echinocyte-discocyte transformation. Wheat germ agglutinin also markedly rigidifies the membrane (Smith and Hochmuth 1982; Chasis et al. 1985). In addition, an alternative explanation for the polyphosphoinositide effect on the red cell shape may involve changes in the cytoskeleton-glycophorin and band 3 interaction. In the presence of phosphatidylinositol-4,5-bisphosphate (PIP_2), band 4.1 binds to glycophorin with a high affinity, whereas in the absence of PIP_2 the binding affinity between the two proteins is weak, and band 4.1 may bind to band 3 (Anderson and Marchesi 1985). Such changes in the membrane skeleton-integral protein interactions could lead to shape changes. Jinbu et al. (1984) have observed that ankyrin (which binds membrane skeleton to band 3) appears to be required for shape changes. The surface charge of the red cell membrane, located on the sialic acid residues of the glycocalix, may also influence the red cell shape. Treatment of human red cells with enzymes that reduce the surface charge exerts a "stomatocytogenic" effect on the cells (Grebe et al. 1988). Thus, as stated by Elgsaeter et al. (1986), neither the protein gel nor the lipid bilayer alone, but the two together, are responsible for erythrocytic shape.

In bird red cells, which are more resistant to shear deformation than mammalian red cells, neither dimyristoyl phosphatidylcholine nor indomethacin treatment caused changes in the gross morphology of the cells (Nikinmaa and Huestis 1984a), although they are effective echinocytic agents in human red cells (Ott et al. 1981; Fujii et al. 1979). The only effects observed were small irregularities in the membrane surface. In contrast, both calcium treatment (Allan et al. 1982; Thomas et al. 1983; Nikinmaa and Huestis 1984a) and ATP depletion (Nikinmaa and Huestis 1984a) caused extensive wrinkling, shedding of vesicles and spherication of bird red cells. These shape changes almost certainly involve rearrangements of the cytoskeleton. Thomas et al. (1983) observed that the wrinkling, spherication and shedding of vesicles from calcium-treated cells was associated with a breakdown of several membrane-associated proteins, including spectrin, goblin (ankyrin) and microtubule-associated-proteins.

Treatments which affect red cell shape also influence cellular deformability. Spherostomatocytes, produced by chlorpromazine treatment, were less filtratable than normal cells, most likely because, at a constant volume, the surface area of the cells decreased. In contrast, filtration of spheroechinocytes, produced by sodium salicylate treatment, proceeded more easily than that of control cells, because, at a constant volume, the surface area of the cells increased (Reinhart and Chien 1986). ATP-depleted echinocytes are swollen, whereby the surface area-volume ratio of the cells is reduced, and leads to decreased deformability of the cell (see Chien 1987). In contrast, echinocytes produced by calcium loading are dehydrated, whereby their internal viscosity increases markedly and limits deformability (Mohandas et al. 1981; Clark et al. 1981). It appears that the membrane viscoelastic properties are little affected by treatments that change cell shape. Meiselman et al. (1978) reported that ATP depletion did not affect mechanical properties of the membrane when micropipette and flow channel techniques were used. Baker (1981) obtained similar results using 0.6 μm diameter Nuclepore filters for studying the membrane deformability of ATP-depleted cells. However, Chabanel et al. (1987) found that chlorpromazine-induced stomatocytes and sodium salicylate-induced echinocytes showed increased resistance to extension in micropipette studies. In conclusion, the effects of cellular geometry and internal viscosity appear to be the predominant factors affecting red blood cell deformability of shape-transformed cells, although effects on membrane viscoelastic properties may also occur.

5.2.4.4 Red Cell Age

During aging, red cell volume decreases, whereby cellular haemoglobin concentration increases (see e.g. Nash and Meiselman 1983) with little change in the surface-to-volume ratio (see e.g. Sutera et al. 1985). The increase in internal viscosity decreases cellular deformability. Although the membrane shear modulus appears to be independent of red cell age (Nash and Wyard 1981), the viscosity of red cell membrane increases with aging, as shown by the observation that shape recovery after micropipette aspiration slows down in old cells (Nash and Meiselman 1983). In addition, old cells are less deformed in shear stress, and some of them lose the tank-treading ability (Sutera et al. 1985). In contrast, cell age has very little effect on the passage of human red cells through 5 μm diameter pores (Micheli et al. 1987). Sordyl (unpublished data) observed that the young red cells of rainbow trout filtered more easily than old cells.

5.2.4.5 Other Factors

At unphysiologically low extracellular pH values (< 6.5), the deformability of red cells in micropipette aspiration decreases markedly (Crandall et al. 1978). Similarly, the shear elastic modulus of red cell membrane, studied using the micropipette aspiration method, decreases when extracellular glucose concentration increases above 5 g l^{-1} (Traykow and Jain 1987). Kikuchi and Koyama (1984) have shown that an increase in the total protein concentration of plasma, within the physiological range, increases the pore passage time of human red cells. Dutta-Roy et al. (1985) have observed that physiological concentrations of insulin affect the filtration rate of human red cells. These observations show that the deformability of red cells may be influenced by several external factors, and suggest that the control of red cell deformability may contribute significantly to the flow properties of red cells in circulation.

Chapter 6

Energy Metabolism and Regulation of Organic Phosphate Concentrations

6.1 Energy Consumption

The major energy consuming processes in the mature mammalian erythrocyte, incapable of protein or significant de novo lipid synthesis (e.g. Bishop 1964), are reviewed by Rapoport (1986). The ion pumps which maintain the internal ion concentrations consume a little over one-quarter of the total energy consumption. The sodium pump accounts for 25% of the total energy consumption of the human erythrocyte, and the calcium pump ca. 1%. Phosphorylation-dephosphorylation reactions of the red cells account for another 25% of the total energy consumption. Energetically the most important of these reactions (ca. 20% of the total energy consumption) are the phosphorylation-dephosphorylation reactions of phosphoinositides, which take part in the maintenance of cell shape (see Sect. 5.2.4.3), affect the membrane protein/cytoskeleton interactions (see Sect. 4.2.2.1) and may play a role in intracellular signalling (see Sect. 4.2.1). Spectrin phosphorylation (and spectrin-dependent ATPase) appears to play a role in the restoration of the smooth biconcave shape after shear deformation (Kodicek et al. 1987). Phosphorylation of the anion exchanger to a tyrosine residue in the cytoplasmic portion of the molecule by an endogenous tyrosine kinase influences the binding of glycolytic enzymes to the molecule (Boivin et al. 1986; Low et al. 1987). In addition, energy is required for the generation of reducing equivalents for the maintenance of glutathione in the reduced state (see e.g. Rose 1971). The energy required is supplied almost exclusively by glycolysis at a rate of 1–2 mmol glucose utilized (2–4 mmol ATP produced) l^{-1} cells h^{-1} (see e.g. Schweiger 1962). In resting conditions, a small proportion (ca. 10%) of the glucose utilized is channeled through the pentose phosphate pathway (hexose monophosphate shunt). The small oxygen consumption of mature mammalian erythrocytes (0.3 mmol O_2 l^{-1} cells h^{-1} at 37 °C) is explained by the two oxidative steps of the pentose phosphate pathway (Schweiger 1962).

Mammalian reticulocytes are metabolically much more active than mature erythrocytes, with an ATP production of 100–200 mmol l^{-1} cells h^{-1} (see Rapoport 1986). A large part of the energy consumption is due to the synthesis of proteins, mainly globin (ca. 25% of total ATP consumption in a medium containing amino acids and glucose), but also catalase, carbonic anhydrase, 2,3-diphosphoglycerate synthase, band 4.1 and lipoxygenase. Ion pumps account for ca. 20% of the total energy consumption, and proteolysis for ca. 15%. The relative proportions of membrane phosphorylation-dephosphorylation reactions, of cell membrane movements etc. are not known (see Rapoport 1986). In contrast to mature erythrocytes, about 45% of glucose utilization proceeds via the Krebs cycle, and about the same percentage yields lactate (see Siems et al. 1982). Since the complete oxidation of glucose via the Krebs cycle yields approximately ten times more ATP than the production of lactate from glucose, more than 90% of the total ATP production is produced by aerobic metabolism. The oxygen consumption of reticulocytes ranges from 15 to 50 mmol O_2 l^{-1} cells h^{-1}.

With regard to energy metabolism, nucleated red cells resemble reticulocytes rather than mature mammalian erythrocytes. Thus, the presence of functional Krebs cycle has been demonstrated (see e.g. Schweiger 1962), and the oxygen consumption of some fish erythrocytes at 20°C approaches that of mammalian reticulocytes at 37°C (10 mmol l^{-1} cells h^{-1}; Bushnell et al. 1985), although generally the values are somewhat lower, ranging from 2 to 10 mmol l^{-1} cells h^{-1} (Hunter and Hunter 1957; Schweiger 1962; Rapoport 1986; Ferguson and Boutilier 1988). Similarly, the ATP-consuming processes of nucleated red cells are largely similar to those of reticulocytes: immature circulating erythrocytes produce haemoglobin (see Sect. 3.1), adaptation of red cell membranes to different temperatures in poikilothermic animals possibly requires a more versatile lipid metabolism than in mammalian erythrocytes (see Sect. 4.2.1), and concentrative ion and amino acid transport mechanisms (based on the sodium gradient) are used both in volume regulation and in the control of intracellular pH (see Chap. 8). In addition, ATP is consumed in the proteolytic reactions and in the breakdown of lipids, which take place in circulating nucleated red cells during the breakdown of cytoplasmic organelles upon maturation (see Sect. 4.1).

6.2 Transport of Substrates into the Red Cells

6.2.1 Amino Acid Transport

Amino acids play several different roles in the metabolism of red cells. First, they, especially alanine, glutamine, glutamate and aspartate, are used as substrates of energy metabolism in the reticulocyte (see e.g. Rapoport et al. 1971; Rapoport 1986), and possibly in nucleated erythrocytes (see Sect. 6.5). Second, cysteine, glycine and glutamate are the building blocks of glutathione (e.g. Srivastava 1971). Third, amino acids are required for globin synthesis as well as for nucleotide synthesis, which is carried out in reticulocytes (see Rapoport 1986) and possibly nucleated erythrocytes. In addition, taurine is an important component of the volume regulation in fish erythrocytes (Fugelli and Thoroed 1986; Fincham et al. 1987), and effective transport of amino acids out of the cells may be important in the proteolytic reactions. Since the functions of amino acids are more varied in reticulocytes and in nucleated erythrocytes than in mammalian erythrocytes, it is not surprising that their amino acid transport systems are more versatile than those of mature mammalian erythrocytes.

The amino acid transport systems are defined operationally, e.g. by competitive transport and inhibition by different amino acids or by substrate analogues, and by the use of genetic variants of red cells with transport properties different from "normal" red cells (see Young and Ellory 1977). In addition to the amino acids mentioned below, the different transport systems may transport other amino acids, although with a lower affinity. The transport of amino acids into the red cell has been studied almost exclusively in mammals and birds.

As a broad generalization, the sodium-dependent, concentrative amino acid transport pathways disappear upon maturation of mammalian red cells, but are retained in the nucleated bird erythrocytes. Also, there are marked differences in the amino acid permeabilities between different mammalian red cells. Ruminant red cells are characterized by a very small amino acid permeability as compared to human, rabbit and rat red blood cells (see Young and Ellory 1977).

Several operationally defined sodium-dependent amino acid transport systems have been described for the red cells.

1. *System Gly* has been intensively studied in nucleated pigeon erythrocytes (e.g. Vidaver and Shepherd 1968). The transport system is also present in guinea pig reticulocytes (Fincham et al. 1984) and human erythrocytes (Ellory et al. 1982), but is absent in mature rabbit, sheep or guinea pig erythrocytes (Fincham et al. 1984). In addition to sodium, the system appears to require chloride for activity, and the transport of glycine is inhibited by sarcosine. In addition to system Gly, rabbit reticulocytes appear to have another sodium-dependent glycine transport system (see Young et al. 1980).

2. *System A* is present at least in developing mammalian erythroid cells (Winter and Christensen 1965; Vadgama et al. 1987). The transport activity diminishes in maturation, and the route is not present in mature mammalian red cells. The transport route favours alanine, but also transports threonine, serine, phenylalanine and leucine.

3. *System ASC* has been defined for developing mammalian erythroid cells (Vadgama et al. 1987), nucleated pigeon erythrocytes (Eavenson and Christensen 1967; Vadgama and Christensen 1985) and human red blood cells (Ellory et al. 1981b). The transport route is absent in most mature mammalian red cells studied. The system transports medium-sized neutral amino acids with a preference for alanine, serine and cysteine. Proline is also transported to a considerable extent.

4. *System β* is present at least in the nucleated red cells of the pigeon (Eavenson and Christensen 1967) and the eel (Fincham et al. 1987). The system transports β-amino acids, mainly taurine and β-alanine, and is responsible for the very high (up to 30 mmol l^{-1} red cell water) concentration of taurine in the red blood cells of euryhaline fish (Fincham et al. 1987).

5. In addition, dog and cat red cells appear to have a selective, sodium-dependent transport system for aspartate and glutamate (see e.g. Ellory et al. 1981a).

The sodium-independent facilitated diffusion pathways appear to be more conserved throughout erythroid cells than the sodium-dependent systems.

1. *System* Ly^+ (y^+) appears to be present in most red cells, and transports cationic amino acids, e.g. lysine and arginine. Sheep and horse erythrocytes lack this transport route (see Young et al. 1980; Fincham et al. 1988).

2. *System L* transports zwitterionic large amino acids e.g. leucine, isoleucine, tyrosine, tryptophan and phenylalanine in both mammalian (e.g.

Young et al. 1980) and nucleated (e.g. Eavenson and Christensen 1967; Vadgama and Christensen 1985) red cells. This transport system may be heterogenic (see Vadgama and Christensen 1985).

3. *System T* preferentially transports amino acids with a ring in the side chain (Rosenberg et al. 1980; Young et al. 1980; Vadgama and Christensen 1985). The transport system is present in human red cells (e.g. Rosenberg et al. 1980), but absent from sheep and pigeon red cells (Young et al. 1980; Vadgama and Christensen 1985).

4. *System asc* transports mainly amino acids with three to four carbons (alanine, serine and cysteine), and resembles system ASC except for being sodium-independent. This transport system is present in mammalian reticulocytes (Vadgama et al. 1987), the red cells of sheep and horses (e.g. Young et al. 1975; Fincham and Young 1983), nucleated pigeon red cells (Vadgama and Christensen 1985) and even in the red cells of the Pacific hagfish, *Eptatretus stouti* (Fincham et al. 1986). In addition to transporting neutral amino acids, this transport system is the only pathway for mediated transport of cationic amino acids in horse erythrocytes. Horse red cells have genetic variants both with high-affinity (asc_1) and low-affinity (asc_2) transport. In addition, both sheep and horse red cells have genetic variants lacking this transport system (e.g. Fincham et al. 1988).

Also, the anion exchanger may transport glycine, serine and cysteine in most red cells studied (see e.g. Young 1983).

6.2.2 Monosaccharide Transport

The mechanism of monosaccharide transport into human red cells has been reviewed, e.g. by Naftalin and Holman (1977), Widdas (1980, 1988) and Jones and Nickson (1981). Comparative aspects of sugar transport into erythrocytes have been reviewed by Bolis (1973) and Ingermann et al. (1985a).

The transport of sugars into most red cells is a facilitated diffusion process, i.e. it shows saturation kinetics, stereospecificity and can be inhibited by competitive inhibitors like phlorizin and phloretin (see Bolis 1973). The glucose transporter also generally accepts other sugars: in addition to D-glucose, e.g. D-mannose and D-galactose are transported (whereas D-sorbitol, D-fructose and L-glucose are not; see Ingermann et al. 1985a).

The glucose permeability of mature human red cells is much higher than most other mammals. The apparent number of transport sites (which are in the Zone 4.5 in the red cell membrane proteins, see Sect. 4.2.2.4) is 124 000–190 000 (Lowe and Walmsley 1987). The permeability of glucose through the red cell membrane is 100 000–1 000 000 times greater than for lipid bilayers. Jacquez (1984) has suggested that, owing to the high permeability for glucose, red cells may participate in the transport of glucose to the brain.

Generally, glucose transport across mammalian reticulocyte membrane is much faster than across the membrane of mature red cells. In addition, glucose transport is faster in foetal than in adult red cells (see Kim 1983; Ingermann et al. 1985a; Rapoport 1986). An extreme among mammals are pig erythrocytes. In the reticulocytes and foetal red cells of the pig, a facilitated glucose transport pathway is present, but the red cells of an adult pig lack the glucose transporter (see Kim 1985; Woffendin and Plagemann 1987).

The transport of glucose across the red cell membrane of chicken embryos and adult chicken proceeds at a similar speed (Ingermann et al. 1985b). The half-time for equilibration across the membrane is very slow, 2–3 h (Ingermann et al. 1985b). However, as in mammals, the transport is carrier-mediated diffusion. The transport of glucose across chicken red cells is stimulated by inhibitors of aerobic metabolism (Wood and Morgan 1969) or by anoxia as such (Whitfield and Morgan 1973). Avian sugar transport is also stimulated by exogenous ATP (Whitfield and Morgan 1973), catecholamines (Whitfield et al. 1974) and calcium (Bihler et al. 1982a,b). The high concentrations of catecholamines required, and the apparent lack of involvement of either α or β receptors in the stimulation of glucose transport, suggest that the catecholamine effect may be quite unspecific.

The red cells of several fish, including brown trout (*Salmo trutta*; Bolis et al. 1971), armored catfish (*Pterygoplichthys*) and piracuru (*Arapaima gigas*; Kim and Isaacks 1978) are essentially impermeable to glucose. In other fish, e.g. electric eel (*Electrophorus electrocus*) and arawana (*Osteoglossum bicirrhosum*; Kim and Isaacks 1978) glucose crosses the membrane by simple diffusion. Carrier-mediated diffusion of glucose has been observed in both embryonic and adult seaperch (*Embiotoca lateralis*; Ingermann et al. 1985a). However, even in this case the half-time for equilibration of glucose across the membrane is 6–7 h. Ribose generally crosses the membrane of fish red cells much faster than hexose sugars (see Kim and Isaacks 1978), and also permeates the glucose-impermeable red cells (see also Bolis 1973).

Hagfish (*Eptatretus stouti*) erythrocytes take up glucose from the medium very rapidly. The half-time for the equilibration of glucose in the carrier-mediated diffusion is ca. 20 s (Ingermann et al. 1984). The glucose transporter has a substrate specificity similar to human red cells, and transport is inhibited by phloretin and phlorizin, as in human erythrocytes (see Ingermann et al. 1985a).

6.2.3 Transport of Nucleosides, Nucleic Acid Bases and Nucleotides

The transport of nucleosides, nucleic acid bases and nucleotides across the membranes of animal cells has been reviewed, e.g. by Plagemann and Wohlhueter (1980) and Plagemann et al. (1988).

Nucleosides enter mammalian red cells by facilitated diffusion (e.g. Oliver and Paterson 1971; Plagemann and Wohlhueter 1980; Templeton and Chilson 1981). A single transport system, located in the zone 4.5 of membrane proteins in the human erythrocytes (see Sect. 4.2.2.4) and with a broad specificity, appears to transport nucleosides into the red cells. Transport is inhibited by nitrobenzylthioinosine (NBTI) and dipyridamole (see e.g. Berlin and Oliver 1975). Two forms of nucleoside transport can be distinguished on the basis of sensitivity to NBTI inhibition. One form (NBTI-sensitive) is strongly inhibited by nanomolar

NBTI concentrations, the other (NBTI-resistant) is inhibited only by micromolar concentrations of NBTI. Although most cell types express both nucleoside transport forms in varying proportions, human red cells only express the NBTI-sensitive transport (see e.g. Woffendin and Plagemann 1987). Red cells of several ruminants (see e.g. Young 1983), and cats and dogs (see e.g. Duhm 1974), appear to lack a functional nucleoside carrier.

Free nucleic acid bases are not transported via the nucleoside carrier in erythrocytes. Dipyridamole, which effectively inhibits nucleoside transport, does not inhibit the transport of hypoxanthine across the human red cell membrane, and uridine does not inhibit hypoxanthine transport (Plagemann et al. 1987). Plagemann et al. (1987) suggested that independent carriers may transport hypoxanthine, uracil and adenine across the membrane of human red cells. No carrier is present for cytosine (Plagemann et al. 1987). Later studies by Domin et al. (1988) suggest that all the purine nucleobases are transported by a single carrier, distinct from the nucleoside carrier. The transport is also weakly inhibited by pyrimidine nucleobases, thymine and uracil. This possibly indicates that these bases also cross the membrane via the same carrier. Cytosine does not interact with the transporter. Templeton and Chilson (1981) have found both a high-affinity and a low-affinity transport site for adenine in rabbit red cells. They suggest that the high-affinity adenine transporter of rabbit red cells may also transport other purine bases.

Although the red cell membrane is generally impermeable to nucleotides, apparently active extrusion of cyclic AMP has been demonstrated in avian red cells (e.g. Heasley and Brunton 1985; Heasley et al. 1985). This extrusion can be inhibited, e.g. by prostaglandins.

6.3 Glycolysis

Red cell metabolism and glycolysis has recently been reviewed, e.g. by Agar and Board (1983). Table 6.1 gives glycolytic enzyme activities in some vertebrates.

In human and most mammalian red cells, glucose transport into the red cell is not rate limiting for glycolysis. However, the glycolytic pathway of pig erythrocytes is inoperative, owing to the impermeability of the cells to glucose. When the cells are rendered permeable to glucose (and other hydrophilic solutes) by treating them with amphotericin B, increasing the amount of glucose in the medium increases the rate of lactate accumulation (Kim and McManus 1971a). In nucleated red cells the transport of glucose into the cells may be commonly rate limiting for glycolysis. Bird red cells, which have slow glucose transport (e.g. Ingermann et al. 1985b), do not consume detectable amounts of glucose (Rosa et al. 1983). Extracellular glucose does not influence the respiration of the glucose-impermeable red cells of brown trout (Bolis et al. 1971), and even in the glucose-permeable red cells of lungfish (*Lepidosiren paradoxa*), ATP levels are not maintained after extracellular supplementation with glucose (Kim and Isaacks 1978). Similarly, Bachand and Leray (1975) suggest that, although all the

Table 6.1. The activities of glycolytic enzymes in various vertebrate red cells[a]

Enzyme	*Pleurodeles waltlii* (Amphibia) 1.	*Perca flavescens* (Osteiichthyes) 2.	Chicken (Aves) 3.	*Eudyptula minor* (Aves) 4.	Rabbit (Mammalia) 5.	Goat (Mammalia) 6.
Hexokinase	1.9	0.08	1.9	0.6	0.3	0.8
Phosphoglucose isomerase	81.5	28.3	7.5	24.9	69.9	61.0
Phosphofructokinase	–	1.4	5	2.3	1.5	1.2
Glyceraldehyde phosphate dehydrogenase	14.5	27.3	40	221.5	106	48.2
Phosphoglycerate kinase	30.9	79.0	37.5	698	54	38.2
Pyruvate kinase	19.6	13.2	39	22.3	1.3	1.4
Lactate dehydrogenase	–	233.5	56	630.6	114	25.7

[a] Enzyme activities in $\mu mol\ min^{-1}\ g^{-1}$ Hb, – activity not measured. Sources 1. Audit et al. (1976), 2. Bachand and Leray (1975), 3. Rosa et al. (1983), 4. Nicol et al. (1988), 5. Agar and Smith (1974), 6. Agar (1979).

enzymes of the glycolytic pathway are found in the red cells of yellow perch (*Perca flavescens*), glycolysis is limited by the slow transport of glucose and consecutive low intracellular concentration of the substrate.

The nucleated red cells of the hagfish are exceptions to the rule – glucose permeability cannot be the rate-limiting step for glycolysis with a 20-s half-time for equilibration (Ingermann et al. 1984). Furthermore, the relatively high concentrations of 2,3-diphosphoglycerate in the red cells of the lamprey (*Entosphenus tridentatus*) and several amphibians (see Table 6.3 and Bartlett 1980) suggest that the red cells of these animals metabolize glucose via glycolytic pathway.

In addition to glucose, red cells are able to use both mannose (Kaloyianni-Dimitriades and Beis 1984) and galactose (e.g. Ng 1971) as substrates for glycolysis. However, owing to their low concentration in plasma and slower permeation into the red cells than the permeation of glucose, they are not likely to be physiologically important substrates. Trioses like dihydroxyacetone and glyceraldehyde may also enter the glycolysis of erythrocytes.

In the red cells with high glucose permeability, the low activity of hexokinase is rate limiting for glycolysis (see e.g. Fujii and Miwa 1986). The hexokinase activity decreases with red cell age, and appears to be responsible for the decrease in cellular ATP concentration: the velocity of hexokinase reaction is similar to the glycolytic flux (Fornaini et al. 1985). In physiological situations the enzyme is markedly inhibited by glucose-6-phosphate. When the concentration of glucose-6-phosphate is decreased with methylene blue, glucose utilization is enhanced (see Rose 1971). Additional inhibitors are MgADP, 2,3-diphosphoglycerate (2,3-DPG) and glucose-1,6-phosphate (e.g. Seider and Kim 1979).

The marked inhibition of hexokinase by glucose-6-phosphate means that the activity of phosphofructokinase may also regulate the glycolytic flux by relieving the inhibition from hexokinase (see Rose 1971). Phosphofructokinase is markedly inhibited by a decrease in pH, by MgATP, by 3-phosphoglyceric acid, phosphoenolpyruvate and 2,3-DPG (see Rose 1971). In addition, inositol pentaphosphate effectively inhibits phosphofructokinase (Rosa et al. 1983).

Pyruvate kinase is the third enzyme in the glycolytic pathway catalyzing an irreversible reaction. However, the maximal activity of pyruvate kinase in mammalian red cells is about 50 times greater than the activity needed to account for the rate of lactate production. Thus, even with the product inhibition by ATP, the enzyme can maintain the steady-state level of phosphoenolpyruvate below saturation (see Rose 1971). One notable difference between mammalian and nucleated red cells is that the relative activity of pyruvate kinase is very much higher than either phosphofructokinase or hexokinase in the nucleated red cells (see Table 6.1). The significance of this difference is not known as yet.

The most characteristic feature of mammalian red cells is the high intracellular concentration of 2,3-DPG. In addition to being present in mammalian red cells, 2,3-DPG is found in significant amounts only in the red cells of lampreys, amphibia and the embryos of turtles and birds (see Sect. 6.7.1). 2,3-DPG is produced from 1,3-diphosphoglycerate (1,3-DPG) and degradated to 3-phosphoglycerate (3-PG) by a bifunctional 2,3-diphosphoglycerate synthase/phosphatase enzyme (for a review see Sasaki et al. 1983). Of the overall glycolytic flux, 15–25% goes via the 2,3-DPG pathway. The substrates for 2,3-DPG synthase are 3-PG, which is present in a concentration 100-fold higher than the K_m value, and 1,3-DPG, the concentration of which is only one-tenth of the K_m value. Thus, the synthase is severely inhibited by the lack of this substrate. Also, the enzyme is inhibited by the endproduct 2,3-DPG. As a result, any treatment that increases the concentration of 1,3-DPG will increase the 2,3-DPG synthase activity and, thereby, the 2,3-DPG concentration. The phosphatase, on the other hand, is always saturated with 2,3-DPG, thereby being the rate-limiting step in the flux via this pathway. The phosphatase activity is inhibited by 3-PG.

The energy production by glycolysis may be closely coupled to the energy consumption by the sodium pump. Fossel and Solomon (1978) have shown that the activity of the glycolytic enzyme complex is affected by ouabain, an inhibitor of the sodium pump. Interestingly, both the glycolytic enzymes and the sodium pump may be in close contact with the anion exchange protein of the red cell membrane (see Sect. 4.2.2.4).

6.4 Pentose Phosphate Pathway and Glutathione Metabolism

6.4.1 Pentose Phosphate Pathway

The pentose phosphate pathway (see Fig. 6.1 and, e.g. a review by Dische 1964) produces NADPH from NADP, whereby oxidized glutathione (GSSG) can be reduced to glutathione (GSH) by glutathione reductase. Also, the pentose phosphate pathway provides ribose-5-phosphate for the production of nucleotides (see e.g. Marks 1964). Furthermore, nucleosides and 5-sugars can be

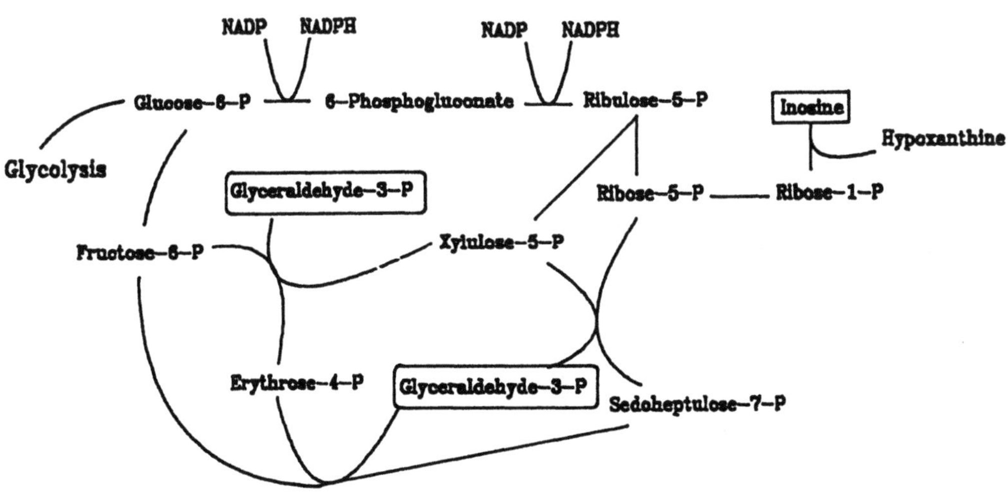

Fig. 6.1. The pentose phosphate pathway in erythrocytes. The *rectangle* indicates the point where purine nucleosides enter the pathway. *Rounded rectangles* indicate the points where the glycolytic pathway is re-entered

used as substrates of pentose phosphate pathway, and thereby used in the energy metabolism of the red cells.

In unstressed situations, 5–10% of the total glucose consumption of human red cells is channeled via the pentose phosphate pathway (e.g. Gaetani et al. 1974). The flux is limited by the first enzyme of the cycle, glucose-6-phosphate dehydrogenase, which, owing to inhibition by high NADPH/NADP ratio, operates at less than 1% of its capacity (see e.g. Thorburn and Kuchel 1985). In oxidative stresses (e.g. methylene blue treatment), the NADPH/NADP ratio decreases and the inhibition of glucose-6-phosphate dehydrogenase is relieved. In this case 90–100% of the total glucose consumption may be channeled via the pentose phosphate pathway, and the activity of hexokinase becomes rate limiting (e.g. Thorburn and Kuchel 1985). Similarly, in the erythrocytes of the marsupial, *Didelphis virginiana,* ca. 7% of the glucose flux passes through the pentose phosphate pathway in unstressed conditions. The percentage increases to 75% of the total flux in the presence of methylene blue (Bethlenfalvay et al. 1984). There are large variations in the activities of enzymes of the pentose phosphate pathway in mammals (see Table 6.2). With regards to nucleated erythrocytes, estimates on the importance of the pentose phosphate pathway in the utilization of glucose have not been made. However, the enzymes of the pentose phosphate pathway have been found in all species studied so far.

Glucose impermeable red cells or red cells with low glucose permeability may use the pentose phosphate pathway as an energy-producing pathway. Notably, fish red cells impermeable to glucose are permeable to ribose, and lactic acid is formed in piracuru red cells after incubation with ribose (Kim and Isaacks 1978). Glucose impermeable pig red cells are permeable to both ribose, deoxyribose and nucleosides (Kim and McManus 1971b; Jarvis et al. 1980).

To enter the pentose phosphate pathway, ribose must be phosphorylated by ATP in a reaction catalyzed by ribokinase (see e.g. Kim and McManus 1971b). Thereafter, some ribose-5-phosphate must be converted to xylulose-5-phosphate by the action of phosphopentose isomerase and phosphopentose epimerase. Inosine (and other purine nucleosides) can enter the pentose phosphate pathway after reacting with inorganic phosphate in a reaction catalyzed by

Table 6.2. Activities of the oxidative enzymes of the pentose phosphate pathway, glucose-6-phosphate dehydrogenase (G-6-PDH) and 6-phosphogluconate dehydrogenase (6-PGDH), and glutathione reductase (GR) in some mammals

Species[a]	G-6-PDH	G-PGDH	GR
Sheep (1)	0.78	–	2.55
Goat (2)	0.75	1.30	5.77
Rabbit (3)	10.4	0.88	2.00
Rat (4)	19.4	8.0	1.27
Camel (4)	15	3.5	1.3
Man (4)	8	6.7	6.2

[a] Enzyme activities in μmol min^{-1} g^{-1} Hb, – activity not measured. Sources: 1. Agar and Smith (1972), 2. Agar and Smith (1973/74), 3. Agar (1979), 4. Suzuki et al. (1984).

purine nucleoside phosphatase, in which hypoxanthine and ribose-1-phosphate are formed (see e.g. Duhm 1974). Ribose-1-phosphate is converted to ribose-5-phosphate by the action of phosphoribomutase.

Kim et al. (1980) and Jarvis et al. (1980) have shown that inosine and other purine nucleosides can support ATP production in pig erythrocytes. Kim et al. (1980) have further shown that the nucleosides liberated from the liver cell are the physiological substrate for pig red cells: hepatectomy of an adult pig results in the loss of red cell ATP and, in vitro, net synthesis of ATP by pig red cells can be achieved by exposing them to isolated liver cells.

6.4.2. Glutathione Metabolism

Glutathione is a tripeptide of glutamic acid, cysteine and glycine, with a highly reactive sulfhydryl group. Its metabolism in erythrocytes has been reviewed, e.g. by Srivastava (1971) and Board and Agar (1983). The concentration of glutathione in red cells is high, 1–3 mM in both mammalian and nucleated erythrocytes (Srivastava 1971; Buckley 1982; Härdig and Höglund 1983; Suzuki et al. 1984; Fisher et al. 1986). Glutathione is continuously synthetized in the red cells in two steps (see e.g. Srivastava 1971):

$$\text{L-glutamic acid} + \text{L-cysteine} + \text{ATP} \rightleftharpoons \text{L-}\gamma\text{-glutamyl-cysteine} + \text{ADP} + P_i \text{ and} \quad (6.1)$$

$$\text{L-}\gamma\text{-glutamyl-cysteine} + \text{glycine} + \text{ATP} \rightleftharpoons \text{glutathione} + \text{ADP} + P_i. \quad (6.2)$$

These reactions are catalyzed by γ-glutamyl-cysteine synthetase and glutathione synthetase respectively.

The production of glutathione may be limited either by the transport of the required amino acids or by enzyme deficiencies. Bannai and Tateishi (1986) have reviewed the role of membrane transport in glutathione metabolism. The transport of cysteine (cystine) may be critical for glutathione production, since intracellular cysteine concentration is lower than the apparent K_m value of γ-glutamyl-cysteine synthetase for cysteine. Ohtsuga et al. (1988) have recently obtained data on human red cells, which suggest that cystine transport for glutathione synthesis is induced when the red cells are exposed to oxidative stresses. In addition, the production is controlled by the feed-back competition of glutathione with the glutamate binding site of γ-glutamyl-cysteine synthetase. This mechanism limits the formation of γ-glutamyl-cysteine and, consequently, glutathione (see Board and Agar 1983).

In sheep, both mutants lacking the cysteine transport, and mutants with low γ-glutamyl-cysteine synthetase activity are common (see e.g. Fisher et al. 1986). Thus, the importance of glutathione for red cell function, and the different steps in its production can be evaluated using these cells. In the transport-deficient sheep erythrocytes, glutathione levels decrease with red cell age. Simultaneously, the concentration of denatured haemoglobin is markedly elevated (Fisher et al. 1986).

Glutathione protects haemoglobin (and other easily oxidized proteins and lipids) from oxidation, because its sulfhydryl group is highly reactive. GSH (reduced glutathione) reacts with H_2O_2 to form GSSG (oxidized glutathione) and water in a reaction catalyzed by glutathione peroxidase. GSSG is reduced back to GSH in the presence of NADPH in a reaction catalyzed by glutathione reductase. The NADPH can be regenerated by pentose phosphate pathway (see above and Srivastava 1971).

GSSG can also form mixed disulphides with haemoglobin. The formation of these may be an intermediate step in Heinz body (an aggregate of denatured haemoglobin molecules) formation. In the presence of glutathione reductase, the mixed disulphide can be cleaved to form haemoglobin and glutathione (see Srivastava 1971).

In mammalian red cells the turnover rate of glutathione is 2 to 10 days (e.g. Board and Agar 1983). This is due to the rapid transport of oxidized glutathione out of the cell (see Beutler 1983). GSSG is transported out of the cell against a concentration gradient (see Srivastava 1971; Bannai and Tateishi 1986). Thus, in addition to the oxidation/reduction cycle, the de novo synthesis of reduced glutathione and the excretion of oxidized glutathione play an important role in the defence against oxidative stresses.

Although the glutathione metabolism in nucleated red cells has been little studied, the considerable concentrations of glutathione in fish red cells (e.g. Buckley 1982; Härdig and Höglund 1983) and the presence of glutathione reductase in the erythrocytes of the fish, yellow perch (Bachand and Leray 1975) and the amphibian, *Rana ridibunda* (Kaloyieanni-Dimitriades and Beis 1984) suggest that glutathione metabolism in nucleated erythrocytes is similar to that of mammalian erythrocytes.

6.5 Krebs Cycle and Oxidative Phosphorylation

Rapoport (1986) has reviewed the respiration of mammalian reticulocytes. Schweiger (1962) reviewed the early information on the role of Krebs cycle and oxidative phosphorylation in the metabolism of nucleated erythrocytes. By then, all the Krebs cycle enzymes had been demonstrated in bird red cells, as had been the role of oxidative phosphorylation in ATP generation: the oxygen consumption of bird red cells is almost completely inhibited by the use of cyanide. Since 1962, relatively little additional information on the aerobic energy metabolism of nucleated red cells has been gathered. The presence of mitochondria, considerable oxygen consumption and a functional pyruvate dehydrogenase complex have been demonstrated in fish erythrocytes (see Sect. 4.1; Eddy 1977; Salama 1986; Bushnell et al. 1985; Ferguson and Boutilier 1988). Greaney and Powers (1978) have shown that the ATP production of killifish (*Fundulus heteroclitus*) red cells is inhibited by cyanide. Lane (1984) has demonstrated that in rainbow trout the effectiveness of cyanide in decreasing cellular organic phosphate concentrations decreases with the age of the cell, whereas the effect of iodoacetate persists or increases. Iodoacetate inhibits glycolysis at the level of glyceraldehyde 3-phosphate dehydrogenase (e.g. Lehninger 1975). These observations suggest that the role of oxidative phosphorylation in the energy production decreases with maturation of nucleated erythrocytes, and that anaerobic energy production (either glycolysis or pentose phosphate pathway) assumes a greater role. This is in keeping with the decreasing number of cellular organelles with maturation of nucleated erythrocytes (see Sect. 4.1).

In addition to glucose, substrates of the pentose phosphate pathway and glycolytic intermediates, the Krebs cycle can directly utilize amino acids, especially glutamine, glutamate, aspartate and alanine, and fatty acids after β-oxidation. Alanine, aspartate, glutamate and glutamine can reach 80% relative carbon dioxide production (carbon dioxide produced from the amino acid accounts for 80% of the oxygen consumption of the cell) in reticulocytes. They are thus very effectively used in the Krebs cycle (see Rapoport 1986). Although their role in the metabolism of nucleated red cells has not been studied, they are potential substrates for energy metabolism, since the concentrations of aspartate, glutamate and alanine in the red cells of carp exceed those of plasma (Dabrowski 1982). Short-chain saturated fatty acids (e.g. octanoate) are also good substrates of aerobic energy production in mammalian reticulocytes (Rapoport 1986). Since fatty acids easily diffuse across the red cell membrane, and since their concentration in the plasma of the nucleated erythrocytes of fish is relatively high (0.5–1.5 mM; L. Forsman, personal communication), their contribution to the energy metabolism of nucleated red cells may be significant. Of the glycolytic intermediates, the concentration of pyruvate and lactate appear very high in the nucleated erythrocytes of the fish *Perca flavescens* (Leray and Bachand 1975). Lactate appears to be passively distributed (distribution ratio is approximately the same as that for chloride), whereas the pyruvate concentration of the red cells is much higher than that of plasma (Leray and Bachand 1975). This may indicate that pyruvate is actively taken up into the red cells.

In conclusion, many of the substrates of Krebs cycle appear to be taken concentratively into nucleated red cells. Since these cells generally have a low permeability to glucose, and since the activity of hexokinase and phosphofructokinase are quite low in relation to pyruvate kinase, for example, these red cells may bypass the early part of glycolysis almost completely, and use either pentose phosphate pathway or Krebs cycle directly as the major pathway for energy metabolism.

6.6 Organic Phosphate Metabolism

6.6.1 Nucleotide Metabolism

Mature mammalian erythrocytes are incapable of de novo biosynthesis of nucleotides (see e.g. Schweiger 1962; Lerner and Lowy 1974; Rapoport 1986). In contrast, reticulocytes incorporate radioactive glycine and formate into both adenine and guanine nucleotides. It is not known whether nucleated red cells carry out de novo synthesis of nucleotides. However, since both free purine and pyrimidine bases and nucleosides permeate the red cells easily (see Sect. 6.2.3), the red cells can use exogenous nucleosides/free bases in nucleotide metabolism. The most likely source for red cell nucleosides/free nucleic acid bases is the liver. In pig erythrocytes, inosine from the liver can support ATP production (see Sect. 6.4.1). The perfusion of rabbit liver with ^{3}H hypoxanthine and a suspension of human red cells resulted in the labelling of the red cell adenine nucleotides (Lerner and Lowy 1974). Since human red blood cells cannot convert hypoxanthine to adenine, the conversion must have taken place in the liver. This finding shows that the liver can be the source of adenine/adenosine utilized in the formation of red cell adenine nucleotides. Apparently, the permeation of nucleosides/free nucleic acid bases is not the limiting step in the overall production of nucleotides (see Plagemann and Wohlhueter 1980).

Figure 6.2 outlines the major pathways of adenosine and guanosine phosphate production from the free bases and nucleotides. Essentially, all the adenine and guanine taken up into rabbit red cells is converted into ATP and GTP respectively (Hershko et al. 1967). The incorporation of purines is stimulated by inorganic phosphates. Inorganic phosphate stimulates phosphoribose-pyrophosphate synthetase, the activity of which determines the intracellular availability of phosphoribose pyrophosphate. The conversion of free nucleic

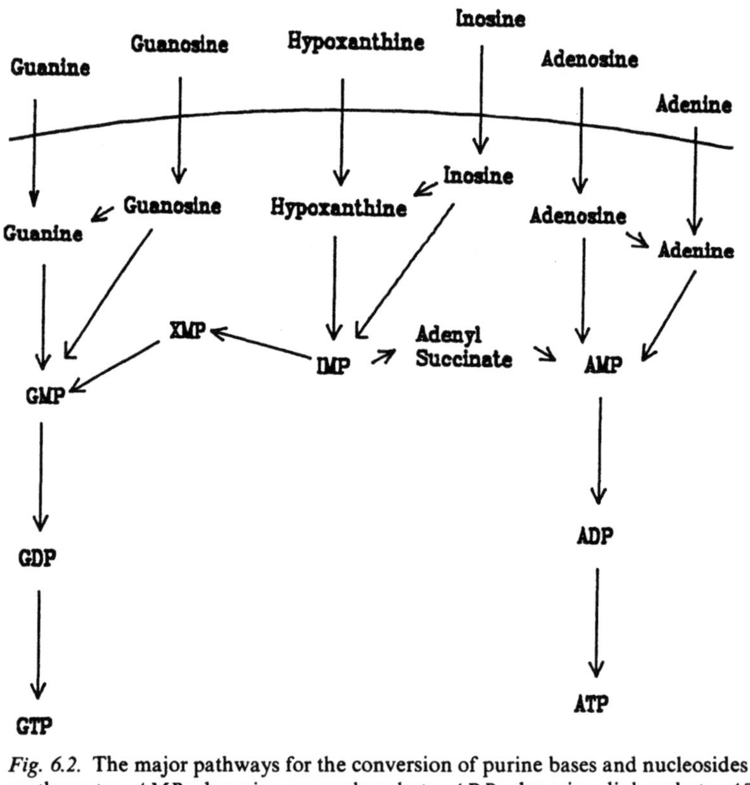

Fig. 6.2. The major pathways for the conversion of purine bases and nucleosides to nucleotides in erythrocytes. *AMP* adenosine monophosphate, *ADP* adenosine diphosphate, *ATP* adenosine triphosphate, *IMP* inosine monophosphate, *XMP* xanthosine monophosphate, *GMP* guanosine monophosphate, *GDP* guanosine diphosphate, *GTP* guanosine triphosphate, – red cell membrane. See text for details

acid bases to nucleotides may be limited by the availability of this substrate (see Salerno et al. 1987). Notably, the dependence of both the incorporation of purines to purine nucleotides and the phosphoribose pyrophosphate formation on the inorganic phosphate concentration are similar (Hershko et al. 1967). In addition, the activities of the specific (adenine or guanine) phosphoribosyl transferases or the activity of adenosine kinase may limit nucleotide synthesis from free nucleic acid bases or nucleosides. Both AMP and GMP can also be formed from inosine monophosphate in most erythrocytes, although human red cells are incapable of this conversion (see e.g. Hershko et al. 1967; Bartlett 1978).

Many fish erythrocytes have a high concentration of GTP. At present, the metabolic reason for this is not known. The activity of guanine phosphoribosyl transferase in fish red cells is not known, nor are the activities of inosinate dehydrogenase and guanylate synthetase, which convert inosine monophosphate to guanosine monophosphate. However, Parks et al. (1973) observed that the activity ratio GMP kinase/AMP kinase was much higher in the eel (*Anguilla rostrata*), which has a very high red cell GTP concentration, than in the other species (*Myxine glutinosa, Squalus acanthias, Phoca vitulina* and man) studied. This suggests that, in the eel, the guanine nucleotide synthetic pathway may be very active.

The major catabolic pathways of purine nucleotides are depicted in Fig. 6.3. The catabolism proceeds mainly via inosine monophosphate, inosine and hypoxanthine. As pointed out by Hershko et al. (1967), inosine monophosphate is not phosphorylated further to di- and triphosphates, and is thereby continuously exposed to the hydrolytic action of 5′-nucleotidase which initiates the catabolic pathway to the freely diffusible hypoxanthine. Most of the adenosine monophosphate and guanosine monophosphate catabolized are first converted to inosine monophosphate by the action of AMP deaminase and GMP reductase respectively. The catabolic pathways may be important in nucleated erythrocytes in which nucleic acid catabolism is still expected to occur.

6.6.2 Inositol Pentaphosphate Metabolism

Two pathways are available for inositol pentaphosphate synthesis: first, the conversion of glucose-6-phosphate to inositol-1-phosphate by myoinositol-1-phosphate synthetase and second, direct phosphorylation of myoinositol by a kinase. After the initial phosphorylation, subsequent phosphate groups can easily be added. Direct phosphorylation of myoinositol appears to be the pathway for inositol pentaphosphate production in bird erythrocytes (Isaacks et al. 1982). Isaacks et al. (1982) were unable to demonstrate any radioactivity in inositol pentaphosphate when labelled glucose or inosine was used as a substrate, whereas exogenous tritium-labelled inositol was transported into the red cell and phosphorylated to yield large amounts of tritium-labelled inositol pentaphosphate. Once formed, inositol pentaphosphate appears not to be catalyzed in chicken red cells. Isaacks et al. (1982) did not observe any change in the inositol pentaphosphate concentration during a 72 h incubation of chicken red cells, whereas the ATP concentration decreased to ca. one-third of the original value.

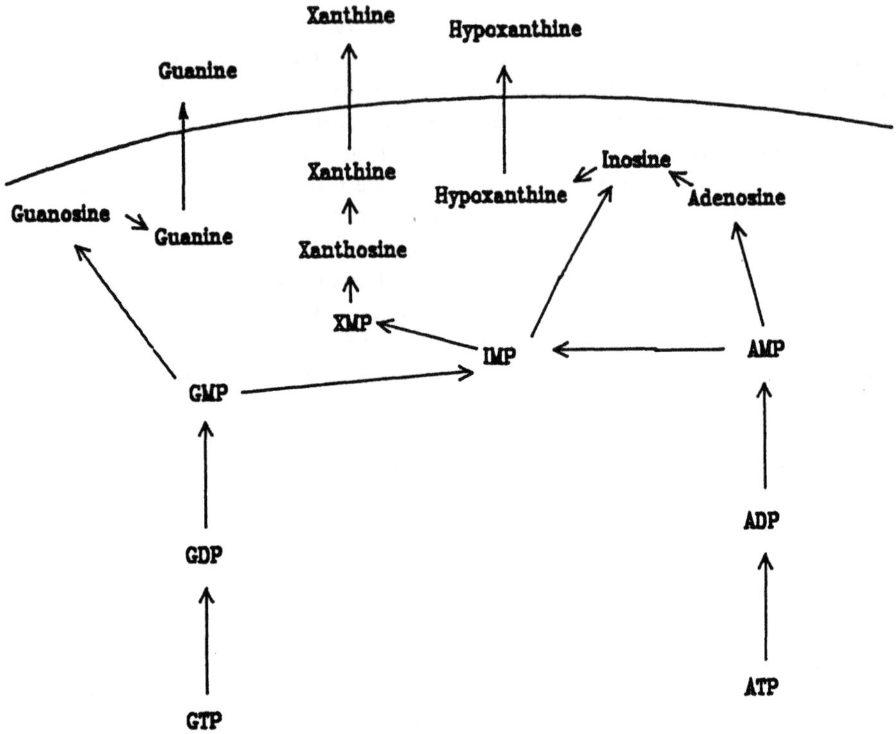

Fig. 6.3. The major pathways for the catabolism of purine nucleotides. See text for details

6.7 Cellular Control of Red Cell Organic Phosphate Concentrations

6.7.1 Organic Phosphate Concentrations

The characteristic feature of red cells is the high concentrations of selected organic phosphates within the cell. The major organic phosphates found are 2,3-diphosphoglycerate (2,3-DPG), inositol pentaphosphate (IPP), adenosine triphosphate (ATP) and guanosine triphosphate (GTP). The major organic phosphates in the red cells of different vertebrates are given in Table 6.3.

From the available data, several general conclusions can be made:

1. Regardless of the vertebrate group, the total organic phosphate concentration of red cells is low in species in which haemoglobin-oxygen affinity is not affected by organic phosphates. This is true for cyclostomes (Bartlett 1982a; Nikinmaa and Weber 1984), crocodiles (see Bartlett 1980) and some ruminants (see Bartlett 1980; Isaacks and Harkness 1983).

2. Mammalian red cells are characterized by a high concentration of 2,3-diphosphoglycerate. In addition, significant amounts of 2,3-DPG are found

Table 6.3. The major organic phosphates of the red cells of selected vertebrates (in mmol l^{-1} red blood cells)

Species		2,3-DPG	ATP	GTP	IPP	other	Source
Agnatha							
Eptatretus stouti	Hagfish		0.6	0.1		2.8	Bartlett 1982a
Entosphenus tridentatus	Lamprey	0.3	1.7	0.1		0.8	Bartlett 1982a
Elasmobranchii							
Squalus acanthias	Spiny dogfish		1.1	0.2		0.2	Bartlett 1978
Triakis semifasciata	Leopard shark		4.6	4.4			Bartlett 1980
Osteiichthyes							
Lepidosiren paradoxa	American lungf.		1.6	3.1		7.1	Isaacks et al. 1978b
Protopterus amphibius	African lungf.		2.2	3.9			Johansen et al. 1976
Electrophorus electricus	Electric eel		0.7	2.8		1.9	Isaacks et al. 1978b
Arapaima gigas	Piracuru		2.1	2.6	1.9	2.6	Isaacks et al. 1978b
Anguilla anguilla	European eel		2.3	8.8			Weber et al. 1976a
Anguilla rostrata	American eel		3.8	16.8			Geoghegan and Poluhowich 1974
Cyprinus carpio	Carp		6.1	4.0			Lykkeboe and Weber 1978
			8.4	2.7			Albers et al. 1983
Tinca tinca	Tench		5.6	3.8			Jensen et al. 1983
Salmo gairdneri	Rainbow trout		3.8	0.4			Tetens and Lykkeboe 1981
Coregonus lavaretus	Whitefish		4.1	0.7			Salama and Nikinmaa unpubl.
Perca fluviatilis	Perch		2.6	2.0			Salama and Nikinmaa unpubl.
Amphibia							
Xenopus laevis(adult)		3.7	1.0	0.2			Bridges et al. 1985
Rana catesbeiana	Bullfrog						
(tadp.)		2.0	2.6		0.07		Bartlett 1976
(adult)		1.7	3.7		0.1		Bartlett 1976
Ambystoma tigrium(gill)		5.5	2.7	0.3			Bartlett 1976

Table 6.3. (Continued)

Species		2,3-DPG	ATP	GTP	IPP	other	Source
Reptilia							
Chelonia							
Chelonia mydas	Green turtle						
(45-d embryo)		7	2.0				Bartlett 1976
(15-month)		0.3	4.0		0.3		Bartlett 1976
(44-d embryo)		5.5	3.5			5.0	Isaacks et al. 1978a
(50-d embryo)		4.7	2.9			3.7	Isaacks et al. 1978a
(14-d turtle)		4.1	6.1			5.6	Isaacks et al. 1978a
(60-d turtle)		1.3	3.55		0.11	3.6	Isaacks et al. 1978a
(1-yr turtle)		0.50	3.47		0.30	1.9	Isaacks et al. 1978a
(mature)		0.3	2.5		0.4	1.7	Isaacks et al. 1978a
Caretta caretta	Loggerhead turtle						
(60-d embryo)		7.4	2.7			2.8	Isaacks et al. 1978a
(mature)		0.4	3.3		0.4	2.0	Isaacks et al. 1978a
Crocodilia							
Alligator mississip.	Alligator		1.3				Bartlett 1976
Squamata							
Iguana iguana	Green iguana		7.8	0.3			Bartlett 1980
Boa constrictor			8.2	1.7			Bartlett 1980
Aves							
Passer domesticus	House sparrow		7.7		3.1		Maginniss 1985
Dromaius novahollandiae	Emu						
(24-d embryo)		15.5	1.1			3.9	Bartlett 1982c
(adult)		5.1	0.2		1.6	1.3	Bartlett 1982c
Meleagris callopavo	Turkey						
(23-d embryo)		4.8	2.7		0.6	2.6	Isaacks et al. 1976
(adult)			1.0		4.1	3.0	Isaacks et al. 1976

Anas domesticus	Duck				
(23-d embryo)		3.5	1.4	0.4	Bartlett and Borgese 1976
(6-d duckling)		0.16	3.7	2	Bartlett and Borgese 1976
(49-d duck)		0	2.4	2.6	Bartlett and Borgese 1976
Mammalia					
Bos taurus	Cow				
(140-d foetus)		2.5			Zinkl and Kaneko 1973
(60-d calf)		0.01			Zinkl and Kaneko 1973
Sus scrofa	Pig				
(100-d foetus)		2.4			Baumann et al. 1973
(5-d piglet)		10.7			Baumann et al. 1973
(adult)		8.5			Baumann et al. 1973
Felix domesticus	Cat				
(60-d foetus)		2.33			Dhindsa and Metcalfe 1974
(adult)		2.5			Dhindsa and Metcalfe 1974
Oryctolagus caniculus	Rabbit				
(19-d foetus)		2.00	2.25		Jelkmann and Bauer 1977
(22-d foetus)		0.8	2.3		Jelkmann and Bauer 1977
(25-d foetus)		0.4	2.0		Jelkmann and Bauer 1977
(30-d foetus)		0.2	2.1		Jelkmann and Bauer 1977
(11-d rabbit)		4.7	3.3		Jelkmann and Bauer 1977
(adult)		9.5	1.2		Jelkmann and Bauer 1977

in the red cells of lamprey (Bartlett 1982a), amphibians (Bartlett 1980) and, for a short period of time, in the embryonic and newly hatched birds and turtles (e.g. Isaacks and Harkness 1980).

3. Birds especially, but turtles as well are characterized by significant changes in the pattern of nucleotides in the red cells during development (e.g. Isaacks and Harkness 1980). Depending on the age of the embryo/animal, ATP, 2,3-DPG or IPP concentrations are high (see Sect. 6.7.2).

4. Reptiles have three distinct organic phosphate patterns. IPP and 2,3-DPG are present in addition to ATP in the red cells of turtles during ontogeny (Bartlett 1976; Isaacks et al. 1978a). In lizards, ATP is by far the most important organic phosphate (e.g. Bartlett 1976). The total organic phosphate pool of crocodilia is very small (e.g. Bartlett 1980).

5. Fish red cells contain ATP and variable quantities of GTP. In addition, the red cells of some species contain IPP, 2,3-DPG, uridosine triphosphate and inositol diphosphate (see e.g. Weber 1982).

6.7.2 Ontogenetic and Environmental Factors Influencing the Organic Phosphate Concentrations

6.7.2.1 Ontogenetic Changes

Throughout the vertebrate group, variable ontogenic changes in the organic phosphate concentrations of red cells are observed. The red cells of the foetus of the viviparous fish, *Embiotoca lateralis*, have a lower NTP concentration than the red cells of adult fish (Ingermann and Terwilliger 1981a). In contrast, foetal and adult red cells of another viviparous fish, *Zoarces viviparus*, have similar NTP levels (Hartvig and Weber 1984). In the amphibian, *Typhlonectes compressicauda*, the erythrocytes of the adult contain three times as much ATP, the major organic phosphate, as do the foetal erythrocytes (Garlick et al. 1979).

In early bird embryos, ATP is the predominant organic phosphate. Shortly before hatching, the ATP concentration decreases markedly, and 2,3-DPG appears transiently. After hatching, 2,3-DPG concentration rapidly decreases and ATP concentration increases. Thereafter ATP is gradually replaced by IPP (see Isaacks and Harkness 1980). The appearance and disappearance of 2,3-DPG are associated with the appearance and disappearance of 2,3-DPG synthetase activity (Ruiz-Ruano et al. 1984). Ingermann et al. (1983) observed that the transient rise of 2,3-DPG concentration in chicken red cells could be prevented by incubating the embryos in 70% oxygen. This also attenuated the decrease in ATP concentration. In contrast, hypoxia accentuated the decrease in ATP concentration and increased the production of 2,3-DPG (Baumann et al. 1983; Ingermann et al. 1983). The results suggest that the relative hypoxia of the late chicken embryo causes a shift in the energy metabolism of the red cells from aerobic (favouring the formation of ATP) towards glycolytic (favouring the formation of 2,3-DPG) pathway. However, the exact regulatory enzymatic steps affected are not known as yet.

In mammals, the red cell 2,3-DPG concentration increases markedly in pigs, dogs and rabbits from the foetus and neonatal animal to the adult; it increases slightly in horses, decreases in cows, goats and sheep, and does not change significantly in cats (see e.g. Isaacks and Harkness 1983). The low concentration of 2,3-DPG in the foetuses of rodents is due to the highly active foetal pyruvate kinase. After birth, the foetal pyruvate kinase is gradually replaced by the adult enzyme, whereby the 2,3-DPG concentration increases (Jelkmann and Bauer 1978, 1980a, b; Franzke and Jelkmann 1982).

6.7.2.2 Hormonal Effects

Mairlbäurl and Humpeler (1981) observed that catecholamines caused the accumulation of cyclic AMP in rat red cells, whereafter red cell pyruvate kinase was inhibited and phosphofructokinase activated. As a result of these changes, the 2,3-DPG concentration increased and the ATP concentration decreased. Nikinmaa (1983b) observed that catecholamines caused a drop in the red cell ATP concentration of trout. Also in this case pyruvate kinase activity appeared to decrease. However, it is also possible that the decrease in ATP concentration is a result of increased ATP consumption by the sodium pump, which is activated in catecholamine-treated red cells because of the marked sodium influx (see Sect. 8.6.1). Milligan and Wood (1987) have since confirmed the effect of catecholamines on the red cell ATP concentration of trout. In contrast, catecholamines did not affect the ATP concentration of starry flounder (*Platichtys stellatus*) erythrocytes (Milligan and Wood 1987).

6.7.2.3 Oxygen Availability

In man, the red cell 2,3-DPG concentration of high altitude dwellers appears to be higher than that of people living at sea level (for a review see e.g. Baumann et al. 1987). Similarly, if normoxic mammals are subjected to moderate environmental hypoxia, the red cell 2,3-DPG concentration increases (e.g. Lenfant et al. 1972; Baumann et al. 1987). This is probably due to the hyperventilation-induced increase in the plasma and red cell pH, since it can be prevented by treating the subjects with acetazolamide prior to hypoxic exposure, thereby preventing the plasma pH rise (e.g. Lenfant et al. 1972). The increased pH activates phosphofructokinase (see e.g. Trivedi and Danforth 1966). As a result, 1,3-diphosphoglycerate concentration would tend to increase, leading to an increase in the activity of 2,3-DPG synthetase, and increased 2,3-DPG levels in the red cell. In contrast, some mammalian species, e.g. *Manis pendactyla*, adapted to hypoxic burrow conditions, have lower red cell DPG concentrations than similar sized, non-burrowing mammals (Weber et al. 1986).

The red cell IPP concentrations of adult highland and lowland geese are similar (Petchow et al. 1977), as are the red cell 2,3-DPG concentrations of highland and lowland goose embryos (Snyder et al. 1982). Hypoxia does not appear to affect the red cell ATP concentration (Pionetti and Bouverot 1977) in adult birds. These observations suggest that red cell organic phosphates do not play a role in the interspecies adaptation to hypoxia, or in the adaptation to hypoxia of adult birds. However, hyperoxia inhibits the transient increase of 2,3-DPG concentration in embryonic chickens, and hypoxia causes an elevation of the 2,3-DPG concentration from the normal levels (Ingermann et al. 1983).

Hypoxic exposure causes a decrease in the red cell NTP concentration of fish (e.g. Wood and Johansen 1972, 1973b; Wood et al. 1975; Greaney and Powers 1978; Soivio et al. 1980; Tetens and Lykkeboe 1981; Nikinmaa and Soivio 1982). Whenever GTP is present, its concentration decreases faster than the concentration of ATP, both in vivo (e.g. Lykkeboe and Weber 1978; Jensen and Weber 1985c) and in vitro (e.g. Jensen and Weber 1985c). Early studies suggested that the decrease in red cell NTP concentration would be a relatively slow process, taking 1–3 weeks of hypoxia exposure to be completed (Greaney and Powers 1978; Soivio et al. 1980). However, later studies have indicated a much faster time course: acute exposure of rainbow trout to deep hypoxia (P_{O_2} = 35 mmHg) induced a significant reduction of red cell NTP concentration within 1 h (Tetens and Lykkeboe 1985), and in hypoxic-hypercapnic tench red cell NTP concentrations reached a new steady-state level within a 24 h exposure period (P_{O_2} = 28 mmHg; P_{CO_2} = 7.5 mmHg; Jensen and Weber 1985c).

Greaney and Powers (1978) suggested that the hypoxia-induced decrease in the red cell ATP concentration of fish would be a direct consequence of decreased oxidative phosphorylation, resulting from inadequate oxygen supply. This is unlikely, however, since in other cell types, e.g. liver cells, oxidative phosphorylation is not affected before the oxygen tension decreases below 5 mmHg (DeGroot and Noll 1987). The red cell NTP concentration in vivo decreases at much higher blood oxygen tensions (e.g. 35–40 mmHg in trout; Soivio et al. 1980). Furthermore, the red cell ATP concentration of *Fundulus heteroclitus* (Greaney and Powers 1978) and trout (Tetens and Lykkeboe 1981) in vitro decreased only in complete anoxia.

Tetens and Lykkeboe (1981) suggested that a humoral factor would induce the decrease in the ATP concentration. Catecholamines are one possible candidate, since they have been shown to affect red cell ATP concentration, and since their concentration increases in hypoxia (e.g. Tetens et al. 1988). However, Tetens (1987) observed that whereas the total nucleotide pool of rainbow trout red cells decreased during hypoxia exposure in vivo, in vitro incubation with saturating concentrations of noradrenaline did not affect the total nucleotide pool, but caused a drop in the ATP concentration and an increase in the ADP concentration (see Fig. 6.4). Salama and Nikinmaa (1988) have shown that catecholamines do not affect the red cell GTP concentration of carp in vitro, whereas the GTP concentrations are preferentially decreased in hypoxic conditions. Thus, the nature of the humoral factor affecting the NTP concentrations and the mechanism of its action are not known.

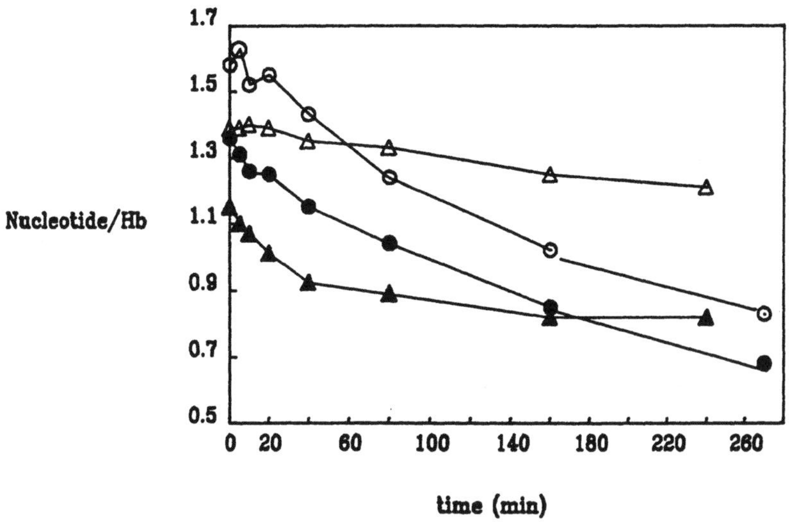

Fig. 6.4. Effect of hypoxia (oxygen tension 30 mmHg; ○, ●) and adrenergic stimulation (noradrenaline concentration 10^{-5} M △, ▲) on the ATP/Hb (●, ▲) and (ATP+ADP+GTP)/Hb ratio of rainbow trout red cells (○, △). Data are from Tetens (1987)

6.7.2.4 Exercise

In man, physical training increases the red cell 2,3-DPG concentration (e.g. Mairbäurl et al. 1983; Lijnen et al. 1988; Hespel et al. 1988). This is due to a higher proportion of young erythrocytes in circulation, with a high intracellular concentration of 2,3-DPG (Mairbäurl et al. 1983).

Exhausting exercise (chasing) causes a decrease in the red cell NTP concentration of striped bass, *Morone saxatilis* (Nikinmaa et al. 1984), trout (Milligan and Wood 1987), and the antarctic fish, *Dissostichus mawsoni* (Qvist et al. 1977), but does not affect the red cell NTP concentration in either tench, *Tinca tinca*, (Jensen et al. 1983) or starry flounder (Milligan and Wood 1987). It is likely that the effect is due to adrenergic stimulation of the red cells, since the decrease of NTP concentration can be prevented by treating the animals with propranolol, a β-adrenergic antagonist, before the exercise (Nikinmaa et al. 1984), and since chasing does not affect the red cell NTP concentrations of the species with catecholamine-unresponsive red cells.

6.7.2.5 Temperature and Metabolic Rate

The 2,3-DPG concentration of the red cells of hibernating mammals decreases during hibernation. This has been observed in the ground squirrel *Citellus tridecemlineatus* (Musacchia and Volkert 1971), in the hamster *Mesocricetus auratus* (Tempel and Musacchia 1975), in the hedgehog *Erinaceus europaeus* (Kramm et al. 1975; Tähti et al. 1981) and in the woodchuck *Marmota monax*

(Harkness et al. 1974). The decrease in 2,3-DPG concentration may be related to the relative acidification of blood in hibernating animals (blood and red blood cell pH remains nearly constant despite a 30 °C decrease in body temperature, see e.g. Tähti 1978; Tähti et al. 1981; Malan 1985), which may inhibit phosphofructokinase and decrease the production of 2.3-DPG.

Aestivating lungfish (*Protopterus amphibus*), which, like hibernating mammals, have a markedly reduced metabolic rate, show a pronounced decrease in the red cell GTP concentration as compared to active specimens (Johansen et al. 1976).

In poikilothermic vertebrates, temperature affects the red cell organic phosphate concentrations. The effect appears to depend on the temperature range studied. Nikinmaa et al. (1980) found that the red cell ATP concentration of rainbow trout (*Salmo gairdneri*) increased when temperature increased from 10° to 15 °C. Dobson and Baldwin (1982) observed a similar increase between the temperatures 10° and 20 °C in the blackfish *Gadopsis marmoratus*. Tetens et al. (1984) acclimated the Antarctic fish *Pagothenia borchgrevinki* to an increased temperature (temperature was increased from −1.5 °C to +4.5 °C), and found that the ATP/haemoglobin molar ratio increased. Laursen et al. (1985) observed that in the temperature range 2°–17 °C the red cell GTP concentration of eel (*Anguilla anguilla*) blood significantly increased.

At higher temperatures, the red cell organic phosphate concentrations appear to decrease. Nikinmaa et al. (1980) observed that when the environmental temperature was increased from 15° to 18 °C, the red cell ATP concentration of rainbow trout significantly decreased. Houston and Koss (1984) showed that the NTP/Hb ratio of rainbow trout decreased from 14° to 24 °C. Laursen et al. (1985) similarly observed that at high acclimation temperature (29 °C) the GTP concentration of eel red cells decreased as compared to 17 °C acclimated animals. Wood et al. (1978) reported that the ATP concentration of the blood of the tortoise *Malacochersus tornieri* decreased in thermal acclimation: 20 °C-acclimated animals had higher red cell ATP concentrations than 35 °C-acclimated animals. In contrast, an increase in temperature from 5° to 15 °C did not affect red cell ATP concentration of the elasmobranch. *Cephaloscyllium isabella* (Tetens and Wells 1984).

Chapter 7

Major Ion Transporting Pathways

The ion distribution across red cell membrane (see Table 7.1 for ion concentrations in the plasma and red cells of various vertebrates) is a result of ion transport across the membrane. Ion transport across the red cell membrane is also responsible for the generation of membrane potential, maintenance and regulation of cell volume, and control of intracellular pH. In this chapter the properties of the major pathways for ion transport are examined. Furthermore, the factors involved in the generation and maintenance of membrane potential are discussed. Control of cell volume and pH are the subject of Chapter 8.

Membrane transport pathways can be classified in four categories (see e.g. Ellory and Hall 1987):

1. Basal permeability or electrodiffusive leak, operationally defined as the residual flux after all known transport pathways have been inhibited.
2. Facilitated transport via a "channel" or "carrier". This type of transport is non-concentrative, and often characterized by large fluxes. An example of a channel in red cells is the calcium-activated potassium channel (see below and Lew and Ferreira 1978), and examples of carriers are the anion exchanger, glucose, nucleoside and several amino acid carriers.
3. Secondarily active transport uses an actively maintained ion (often sodium) gradient to transport an ion (or a molecule) against an electrochemical gradient. The actual transport pathways involved in this type of transport are passive. Thus, the difference between carrier-mediated passive transport and secondarily active transport is that the latter is coupled to an electrochemical gradient for one (or more) of the transported ions.
4. Primarily active transport (ion pumps) is driven by the hydrolysis of ATP, and generates the characteristic disequilibrium ratios for sodium, potassium and calcium across the red cell membrane.

Whenever charge is transported across the membrane by a pump, the term "electrogenic transport" is used (e.g. $3Na^+/2K^+$ pump). If a secondarily active transport is charge-carrying (as e.g. Na^+/Ca^{2+} exchange), the term "rheogenic transport" is often used (see e.g. Heinz 1981). In electrosilent cotransport (symport) systems, the charges of cations and anions transported in the same direction balance each other, and in electrosilent countertransport (antiport, exchange) the charge of ions transported in the opposite direction are the same.

Information about ion transport can be obtained from flux studies, the use of ion substitutions and specific transport inhibitors. In addition, although it is not always possible for red cells, the flux data should be coupled to information on the electrical phenomena associated with the transport.

Table 7.1. Red cell and plasma ion concentrations of selected vertebrates (in mmol l^{-1} water)

Species	Na		Cl		K		Ca		Mg		Source[a]
	plasma	rbc	plasma	rbc	plasma	rbc	plasma	rbc	plasma	rbc	
Man	155	19	112	78	5.0	136	–	–	–	–	1.
Man	135–147	6	95–113	78	3.3–4.8	109–114	2.1–2.8	0.001	0.7–1.1	1.5–2.4	2.,3.
Cat	158	142	112	84	4.6	8	–	–	–	–	1.
Dog (adult)	153	135	112	87	4.8	10	–	–	–	–	1.
Dog (l-d-o)	–	–	–	–	4.2	21.4	–	–	–	–	4.
Cattle (HK)	–	15	–	–	–	70	–	–	–	–	5.
Cattle (LK)	–	87	–	–	–	7	–	–	–	–	5.
Cattle (LK, newborn)	153	29	90	–	8	92	2.9	–	1.0	–	2.
Cattle	136–168	28–102	83–118	85	4.3–10	24–85	2.1–4.8	–	0.6–1.7	0.2–1.3	2.
Sheep (HK)	–	10–43	–	–	–	60–88	–	–	–	–	5.
Sheep (LK)	–	79–121	–	–	–	8–26	–	–	–	–	5.
Sheep	151–160	98	116	78	4.8–5.9	46	2.9	–	0.9–1.1	0.5–1.3	2.
Chicken	154	–	120	–	3.4–6.5	106–143	3.3–5.9	–	1.2	4.1	2.,6.
Turkey	–	6	–	–	–	102	–	–	–	–	7.
Turtle (23°C) (*Chrysemus picta bellii*)	127	15.5	91.9	31.6	4.3	136.3	2.2	0.3	1.5	11.2	8.
Carp (*Cyprinus carpio* 20°C)	135	18	118	30	4.0	107	–	–	–	–	9.
Atlantic salmon (*Samo salar* 10°C)	140	38	138	80	4.4	145	–	–	0.8	7.2	10.
Lamprey (*L. fluviatilis* 2°C)	111	28	116	73	3.6	73	2	–	0.7	3.4	11.

[a] Sources: 1. Bernstein (1954), 2. Altman and Dittmer (1961), 3. Yoshida et al. (1986), 4. Miles and Lee (1972), 5. Ellory and Tucker (1983), 6. Arad et al. (1983), 7. Gardner et al. (1975), 8. Maginniss and Hitzig (1987), 9. Fuchs and Albers (1988), 10. Virtanen et al. (1988), 11. Nikinmaa and Weber (1984)

The flux of an ion is its rate of movement across the cell membrane, and depends on the concentration difference across the membrane, the permeability of the membrane and time. Detailed accounts on flux measurements have been written e.g. Ellory (1982). Fluxes are commonly determined using radioactive isotopes (e.g. $^{22}Na^+$, $^{35}SO_4^{2-}$, $^{36}Cl^-$, $^{86}Rb^+$), although often, after osmotic disturbances, for instance, and in adrenergically stimulated cells, net fluxes of ions are determined. The tracer fluxes include the contribution of both conductive transport and fluxes via electrosilent transport pathways. Thus, without additional electrical information, the nature of ion transport cannot be determined.

Ion substitutions are commonly used to elucidate the ion requirements of a transport pathway. Sodium is commonly replaced by choline, and the specific requirement for chloride ions in transport pathways can be investigated by substituting other permeable anions for chloride. If the coupling between cation and anion transport is electrical, substitution of thiocyanate or nitrate for chloride should speed up transport, since the conductive permeancy of these two anions is greater than that for chloride (see e.g. Parker 1983b). If, on the other hand, transport requires chloride as one of the transported ions in the cotransporter, these substitutions should markedly diminish transport.

Several transport inhibitors are available for studies of ion transport (see Table 7.2). Although they can give useful information on the nature of ion transport pathways, "blind" reliance on the results obtained can be misleading. For example, amiloride, which is used in red cell studies to block the sodium/proton exchanger, is also an inhibitor of sodium channels (e.g. Sariban-Sohraby and Benos 1986) at much lower concentrations than are required to inhibit the sodium/proton exchanger. Furthermore, amiloride also acts as a protonophore (Dubinsky and Frizzell 1983), and binds to adrenergic receptors on the red cell membranes at concentrations similar to those required to inhibit the sodium/proton exchanger (Mahe et al. 1985). Similarly, furosemide, which is used to inhibit sodium/potassium/chloride cotransport, also inhibits the anion exchange pathway at concentrations used in red cell studies (see e.g. Nikinmaa et al. 1987b).

7.1 Electrodiffusive Leak

Transport via electrodiffusive leak should follow Fick's law of diffusion and be nonsaturable. Further, with regards to ions, it should respond both to chemical and electrical gradients, i.e. be conductive (see e.g. Macey 1980). Although, theoretically, electrodiffusive leak is the basal permeability of cell membrane, operationally it can be defined only as the residual flux after all known transport pathways have been inhibited. Thus, when "new" transport pathways are found, the flux via "electrodiffusive leak" decreases.

The basal cation permeability of human red cells increases with increasing pressure (Hall and Ellory 1986). The passive permeability of potassium generally decreases with decreasing temperature (Hall and Willis 1984; Ellory and Hall

Table 7.2. Transport inhibitors commonly used in studies of ion transport

	Inhibitor	I_{50}	Remarks
Gardos channel	Oligomycin	2–3 μg l^{-1}	Also inhibits sodium pump
	Quinine, Quinidine	5–100 μM	Also inhibits Ca/Na exchange
Anion exchange	DIDS (4,4-diisothiocyano-stilbene-2,2-disulphonic acid)	0.1–5 μM	Disulphonic stilbene derivatives are effective inhibitors of anion exchange, DIDS being the most potent one (50% inhibition of transport with DIDS at < 1 μM concentration). Initially (< 1 s) reversible binding, later (hours) irreversible covalent binding.
	SITS (4-acetamido-4-isothiocyano-2,2-disulphonic stilbene)		
	DBDS (4,4-dibenzoamide stilbene-2,2-disulphonic acid)		
	ANS (1,8-anilinonapthalene sulfonic acid)	< 30 μM	
	DNFB (1-fluoro-2,4-dinitrobenzene)		Increases cation permeability
	Eosin 5-maleimide		Irreversible inhibition
	NAP-taurine (N-(4-azido-2-nitro-phenyl)-2-aminoethylsulfonate)	20 μM	Competitive inhibitor of anion transport. Inhibition at the external surface more effective than at internal surface. Photoactive anion
	Furosemide	1 mM	
	Niflumic acid		
Na/K/2 Cl cotransport	Bumetanide	0.1–0.5 μM	
	Furosemide	10 μM	Also inhibits anion exchange
K/Cl cotransport	Bumetanide	10 μM	
	Furosemide	1 mM	
Na/H exchange	Amiloride	10–100 μM	Also inhibits sodium channels (I_{50} 0.3 μM) Na/Ca exchange, sodium pump, adenylate cyclase, adrenergic receptors. The potency of amiloride derivatives ranges from I_{50} < 0.1 μM to 1 mM.
Na/Ca exchange	Quinidine		Also inhibits Gardos channel
	Amiloride	1 mM	
Sodium pump	Quabain	< 1 μM	I_{50} in man, marked species variation in sensitivity
	Digoxin	< 1 μM	
	Digitoxin	< 0.1 μM	
	Other cardiac glycosides		
	Vanadate	4 μM	
	Oligomycin		
Calcium pump	Vanadate	15 μM	
	Quercetin	4–6 μM	
	Trifluoperazine		Echinocytic agent
	Cardiac glycosides		Inhibitors generally the same as for sodium pump

Sources: Sarkadi and Gardos (1985); Cabantchik and Rothstein (1972); Fortes (1977); Passow (1986); Ellory and Hall (1988); Palfrey et al. (1980); Kleyman and Cragoe (1988); Glynn (1985); Makonnen and Est (1987); Schatzmann (1982).

1987). In guinea-pigs, the passive permeability of potassium increases markedly with decreasing temperature in the presence of calcium. This is probably due to the opening of the calcium-activated potassium channel (Gardos channel, see Sect. 7.2) at low temperatures because of the failure of the calcium pump (Hall and Willis 1984). In man, the passive potassium permeability increases at temperatures below 12°C, even in the absence of calcium (see e.g. Ellory and Hall 1987), suggesting that a yet unknown potassium transporting pathway can be unmasked in primate red cells at low temperatures.

The passive conductive fluxes of chloride probably represent two modes of transport. The conductive flux of chloride is partially inhibited by DIDS, an inhibitor of the anion exchange pathway (Knauf et al. 1977; see also Passow 1986). Thus, it is likely that this portion of conductive flux represents diffusion (or slippage) via the anion exchange pathway. However, 30–40% of the conductive transport is insensitive to DIDS, possibly representing the true electrodiffusive leak. The apparent total permeability of lamprey (*Lampetra fluviatilis*) red cells to chloride is ca. 10^{-9} cm s^{-1} at 20°C (Nikinmaa and Railo 1987), a value close to the chloride permeability of lipid bilayers at the same temperature (Tosteson 1972). Since the transport is also insensitive to DIDS, the movements of chloride ions across lamprey red cell membrane at steady state may represent a true electrodiffusive leak.

7.2 Ion Channels

The calcium-activated potassium channel (Gardos channel) is the best known ion channel in red cells (see reviews by Lew and Ferreira 1978; Sarkadi and Gardos 1985). In human red cells, elevation of intracellular calcium concentration either by the ionophore A23187, by ATP depletion with iodoacetamide, or by treating the cells with fluoride, triose reductone, lead salts or propranolol (all in the presence of extracellular calcium) causes a rapid increase in the potassium permeability of human red cells and consecutive cell shrinkage. Alvarez and Garcia-Sancho (1987) have estimated the number of calcium-dependent potassium channels to be ca. 150. The calcium-dependent potassium loss can be inhibited by carboxymethylation of a MW 23 000 cytoplasmic protein (Plishker et al. 1986), and by antibodies raised against the protein. Calcium increases the association of this protein to the cell membrane. The calcium-activated potassium channel is inhibited by quinine and quinidine, and oligomycin. The channel may also be blocked by carbocyanine dyes, used to probe membrane potential (for a more detailed account on the inhibitors and activators of the calcium-activated potassium channel see Lew and Ferreira 1978; Sarkadi and Gardos 1985). Hall and Willis (1984) have studied the temperature sensitivity of the calcium-activated potassium channel in ground squirrel and guinea pig red cells. They found that in both species the A23187-induced 42-potassium accumulation was insensitive to temperature at temperatures above 15°C, but that the rate constant decreased by a factor of 2 between 0° and 15°C.

Intracellular calcium also activates a potassium channel in the nucleated *Amphiuma* red cells (Lassen et al. 1976). In the nucleated red cells of the chicken, and the teleost (*Ctenopharyngodon idella*), calcium loading markedly increases both the potassium and sodium permeabilities (Marino et al. 1981). Thus, the cell volume is little affected by calcium loading in a medium containing both sodium and potassium.

Although the calcium-activated potassium channel is present in many red cells, there are also cells which lack this pathway (see also Jenkins and Lew 1973): in ferret (*Mustela putorius furo*) red cells, A23187 + Ca treatment inhibited rather than stimulated potassium transport (Flatman 1983), and in turkey red cells had no effect (Ueberschär and Bakker-Grunwald 1983).

7.3 Anion Exchange

Band 3 (see Sect. 4.2.2.4), the most abundant protein of the cell membrane, facilitates the exchange of especially monovalent anions across the red cell membrane. The rate of transport decreases with the increasing size of the ion. Divalent anions are also transported, although at a much slower rate and, at least in mammals, with a different pH sensitivity (see below). In physiological situations the transporter mainly carries chloride and bicarbonate ions across the red cell membrane. However, band 3 is also the major pathway for the transport of inorganic phosphate across the red cell membrane (e.g. Shoemaker et al. 1988). In addition, the anion exchange pathway transports phosphoenolpyruvate (Hamasaki et al. 1987), some monocarboxylic acids and amino acids across the red cell membrane.

As required for carrier-mediated transport, anion exchange is saturable, with K_D values for the transfer site of 67 and 10 mM for chloride (pH 7.2) and bicarbonate (pH 8.7) ions in human red cells at 0°C respectively (see Passow 1986). Band 3-mediated chloride exchange is 10 000-fold faster than the conductive chloride transport in man, and up to six orders of magnitude greater than the conductive chloride transport in *Amphiuma* (Lassen 1977). At physiological temperatures and pH values, the half-times of chloride equilibration are ca. 100 ms for man (37°C), and ca. 1 s for trout (15°C; see Romano and Passow 1984).

The temperature dependence of the anion exchange of mammals is greater than that of the poikilotherms so far studied: the activation enthalpy for the anion exchange in human red cells was 30–35 kcal mol^{-1} (Brahm 1977), and in rainbow trout 14–16 kcal mol^{-1} (Romano and Passow 1984). The pH-sensitivity of DIDS-sensitive anion fluxes in man depends markedly on the valency of the anion, whereas in fish the DIDS-sensitive chloride and sulphate fluxes may have similar pH sensitivities.

On the basis of sulphate fluxes for different fish it appears that there are marked differences in the anion equilibration across the red cell membrane of different species (Pasternack and Nikinmaa 1988). These conclusions are supported by the fragmentary data on chloride fluxes (see Table 7.3). However, at

Table 7.3. Approximate halftimes for chloride equilibrium exchange in some vertebrate red cells

Species	$t^{1/2}$	pH	T	Source
Human	17.2 s	7.2	0°C	Brahm 1977
	20	7.4	0	Wieth et al. 1980
	17.5	7.6	0	Wieth et al. 1980
	2.32	7.2	10	Brahm 1977
	0.89	7.2	15	Brahm 1977
Tilapia mossambica	10	7.5	0.5	Haswell et al. 1978
Salmo gairdneri	3.4	7.4	0	Romano and Passow 1984
	1.3	7.4	10	Romano and Passow 1984
	0.8	7.4	15	Romano and Passow 1984
Cyprinus carpio	10–15	6.8	1	Pasternack and Nikinmaa 1988
Lampetra fluviatilis	7.4×10^5	7.6	20	Nikinmaa and Railo 1987
Eptatretus stouti	3.8×10^6	7.5	11	Ellory et al. 1987

present it is not known if these differences reflect different numbers of anion exchange proteins in the red cells of the different species or if they are a result of a different turnover number of individual exchangers. Both a decreased number of anion exchangers (e.g. in primitive erythrocytes of mammalian embryos, see Koury et al. 1987), and a changed turnover number of individual anion exchange protein (in a family with altered band 3 protein, see Kay et al. 1988a) have been observed. Anion transport increases markedly during erythroid development. Kirk and Lee (1988) have measured bromide uptake in the erythroid cells of rabbits, and found that the uptake increases from early erythroblasts all through to circulating erythrocytes.

In terms of red cell function, it is important to remember the difference between the anion exchange and conductive movements of chloride. Movements of chloride ions to re-establish electrochemical equilibrium for the ion after a perturbation of membrane potential require the much slower conductive fluxes rather than the electroneutral anion exchange. This is illustrated by the results on valinomycin-treated red cells of several species. In valinomycin-treated (4×10^{-6} M) *Amphiuma* red cells, the membrane potential was a function of potassium concentration gradient instead of chloride concentration gradient, although the conductive potassium movements must have been at least 100 times slower than the anion exchange (see Lassen 1977). Similarly, it took 80–100 min for the lipophilic ion tetraphenylphosphonium (TPP^+) to reach a steady-state distribution (i.e. for the membrane potential to reach a constant value) after salmonid red cells were treated with 10^{-6} M valinomycin (Fig. 7.1). If anion equilibration via the anion exchange pathway had played a role in the control of membrane potential, steady-state distribution of TPP^+ should have been reached in seconds.

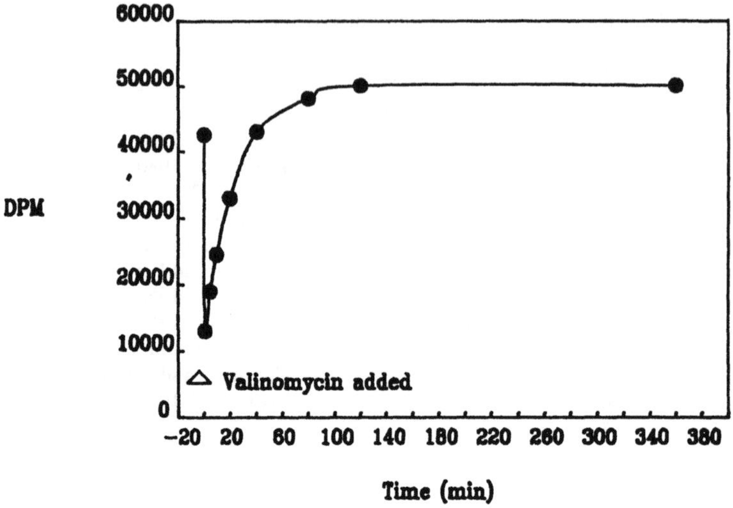

Fig. 7.1. Changes in the external concentration (in DPM) of tetraphenylphosphonium (^{14}C-TPP^+) in rainbow trout red cell suspension before and after the potassium permeability of the cells was increased with 10^{-6} M valinomycin. A decrease in external TPP concentration indicates hyperpolarization

7.4 Ion Transport Coupled to Sodium or Potassium Gradient

7.4.1 Sodium/Potassium/Chloride Cotransport

The sodium/potassium/chloride cotransport has been most intensively studied using bird red cells. Avian red cells appear unique in that the transport pathway can be activated by both cyclic AMP-dependent (β-adrenergic agonists), and cyclic AMP-independent (hypertonicity, sodium fluoride, deoxygenation) stimuli. In other red cells studied, the activity of the transporter can be affected only by cyclic AMP-independent stimuli.

Orskov (1956) observed that pigeon erythrocytes, incubated with noradrenaline, lowered the plasma potassium concentration. Davoren and Sutherland (1963) showed that avian red cells possess a catecholamine-sensitive adenylate cyclase system. Riddick et al. (1971) found that the catecholamine-induced stimulation of cation transport was mimicked by dibutyryl cAMP, linking together the adrenergically activated ion transport and function of adenylate cyclase. After these observations, β-adrenergic stimulation of both sodium and potassium transport across the red cell membrane was observed in several bird species (see Kregenow 1973; Kregenow et al. 1976; Gardner et al. 1975). Schmidt and McManus (1977a, b) also showed that sodium/potassium cotransport only occurs in a medium containing chloride. Thus, it is now well established that catecholamines activate sodium/potassium/2 chloride cotransport across bird red cell membrane (for reviews see, e.g. McManus and

Schmidt 1978; Kregenow 1978; Palfrey and Greengard 1981). The potency of catecholamines in activating the transport is isoproterenol $\gg$ adrenaline = noradrenaline (Gardner et al. 1975). This cotransport mechanism is electrically silent (McManus and Schmidt 1978), and is inhibited by bumetanide and furosemide, but not by DIDS or SITS, ouabain or amiloride (see Palfrey et al. 1980; Palfrey and Greengard 1981). Bumetanide and furosemide act on the actual transport site and not on the β-adrenergic receptor-adenylate cyclase system, since they do not affect cAMP accumulation, but inhibit the transport (see McManus and Schmidt 1978). Although, at present, the molecular nature of the transport site is not known, adrenergic stimulation is associated with phosphorylation of goblin (ankyrin; see e.g. Alper et al. 1980; Palfrey and Greengard 1981). This observation suggests that an alteration in the cytoskeleton-membrane protein interactions may be involved.

The Na-K-2Cl transport pathway in avian red cells can also be activated by cAMP-independent stimuli; hypertonicity, sodium fluoride and deoxygenation (see Palfrey and Greengard 1981). The transport sites in both cases must be the same, since maximal activation by one set of stimuli precludes further activation by the other set (e.g. Kregenow et al. 1976; Palfrey and Greengard 1981; Ueberschär and Bakker-Grunwald 1983).

The net movement of ions through the transport pathway and the effects of stimulated ion transport on the cell volume depend on the sum of chemical potential gradients of the transported ions (see e.g. McManus and Schmidt 1978). If the medium has plasma-like concentrations of the transported ions, the driving force is zero, and no net movements via the pathway occur either after hormonal stimulation or after osmotic shrinkage. When extracellular potassium concentration is elevated, net inward movement of the transported ions occurs and cells swell (see e.g. Palfrey and Greengard 1981; Haas et al. 1982). If the potassium concentration of the extracellular medium is decreased, net loss of ions occurs (e.g. Schmidt and McManus 1977b).

Characteristically, catecholamine-stimulated cotransport activity declines to basal level within 1–2 h in the presence of catecholamines in high-potassium (20 mM) incubation medium (e.g. Kregenow 1973). These catecholamine-refractory cells exhibit increased cotransport if transferred to hypertonic medium (e.g. Ueberschär and Bakker-Grunwald 1983). These observations suggest that the "shutdown" of the cotransporter is a volume-dependent response (e.g. Kregenow 1973).

The cotransport pathway (as determined by inhibition and activation studies) is also capable of carrying out obligatory K/K (or K/Rb) exchange in high potassium red cells, and Na/Na (or Na/Li) exchange in high sodium red cells (see Lauf et al. 1987). Although the K/Rb exchange does not require external sodium, it requires potassium and chloride on the outside, and potassium, sodium and chloride in the inside; similarly, the Na/Na exchange requires sodium, potassium and chloride on the outside. Lauf et al. (1987) have explained these observations by supposing that the carrier can be transported across the red cell membrane either in fully loaded or empty form, the fully loaded being the more mobile of the two. Further, there must be an ordered "first on, first off" loading and unloading of the ions to the transporter.

The furosemide/bumetanide-sensitive sodium-potassium-chloride cotransport is also observed in mammalian red cells. In ferret red cells, which appear not to have a sodium pump on the membrane (Flatman and Andrews 1983), the bumetanide-sensitive potassium transport is quite pronounced, approaching the rates observed in bird red cells (see Flatman 1983; Lauf et al. 1987). Most of the bumetanide-sensitive potassium transport is associated with the movements of sodium (Flatman 1983). In addition, a requirement for chloride has been demonstrated (Hall and Ellory 1985). However, the transport appears to have a different stoichiometry for the ions from the avian system, i.e. 2Na-K-3Cl (see Hall and Ellory 1985). The activity of the sodium-potassium-chloride cotransport across ferret red cell membrane increases with increasing internal magnesium concentration (Flatman 1988).

The furosemide-sensitive potassium fluxes in human and rat red cells are only 1/15th of those observed in duck red cells (see Ellory et al. 1983). In human red cells the inward coupled transport of sodium, potassium and chloride can be fitted to the Na-K-2Cl cotransport model (Chipperfield 1980; Dunham et al. 1980; Kracke et al. 1988). The results (based on net fluxes) of Kracke et al. (1988) strongly suggest that the sodium-potassium-2 chloride cotransport is asymmetric, only taking place in the inward direction. However, the furosemide-sensitive pathway may also carry out several other modes of cation transport in human red cells, since Na-K stoichiometries varying from 1:1 to 1:5 have been reported (see e.g. Duhm 1987). In physiological concentrations of sodium, potassium and chloride, the furosemide-sensitive ion transport (based on tracer fluxes) in human red cells appears to have a stoichiometry of 2 Na:3 K (Brugnara et al. 1986). According to Canessa et al. (1986), furosemide-sensitive transport pathways in human red cells carry out, in addition to the inward sodium/potassium/chloride cotransport, sodium/sodium and potassium/potassium exchange, and uncoupled sodium and potassium effluxes. The sodium-dependent furosemide-sensitive rubidium fluxes in human red cells are practically insensitive to cell volume, but in rat are markedly accelerated by cell shrinkage (see Duhm and Gobel 1984; Duhm 1987).

The sodium/potassium/chloride cotransport pathway appears to be absent at least in sheep (e.g. Lauf 1985b) and *Amphiuma* (e.g. Cala 1985a) red cells.

7.4.2 Potassium/Chloride Cotransport

Cell swelling activates a potassium/chloride cotransport pathway in many red cells, including fish (Lauf 1982; Cossins and Richardson 1985; Borgese et al. 1987), bird (see e.g. Kregenow 1981), and many mammalian, e.g. sheep (e.g. Dunham and Ellory 1981; Lauf and Valet 1983; Fujise and Lauf 1987; Lauf and Bauer 1987), rabbit, horse, guinea-pig, rat, hedgehog, ground squirrel, hamster, cat, dog, ferret and ox (see Ellory et al. 1985) erythrocytes. In low-potassium sheep red cells, a significant portion of the ouabain-resistant potassium flux is via this transport pathway (e.g. Lauf 1985a), whereas the pathway is quiescent in high-potassium sheep red cells in iso-osmotic conditions, but can be activated by exposing the cells to hyposmotic conditions (Fujise and Lauf 1987). The basal activity of this transport pathway in normal human red cells is very low, but is markedly increased in sickle cell anaemia (e.g. Brugnara et al. 1986; Canessa et al. 1986; Brugnara and Tosteson 1987; Berkowitz and Orringer 1987). The properties of the K/Cl cotransport pathway, especially in sheep red cells, have been reviewed by Lauf (1985a, b) and Ellory et al. (1985).

In addition to being activated by increased cell volume, the potassium/chloride cotransport can be activated by high hydrostatic pressure (Hall et al. 1982), by sulfhydryl reagents (e.g. N-ethyl maleimide and methane-methyl-thiosulphonate; Lauf and Theg 1980; Lauf et al. 1985; Lauf 1985a; Kaji and Kahn 1985) and by the ionophore A23187 in the presence of divalent ion chelator EGTA (ethylene glycol tetracetic acid; Lauf 1985a; Fujise and Lauf 1987). At least the N-ethylmaleimide-activated potassium-chloride cotransport activity decreases with cell age (see e.g. Berkowitz et al. 1987).

At least two requirements must be fulfilled before a transport system can be defined as KCl cotransport (see Fig. 7.2 for distinction between potassium/chloride cotransport, electrical coupling of conductive potassium and chloride transport, and coupled potassium/proton and chloride/bicarbonate exchanges). First, the transporter must have an absolute requirement for chloride (or bromide). Potassium fluxes via the cotransport pathway can be inhibited by anion substitution: only bromide ions can substitute chloride in the transport – nitrate, iodide and thiocyanate inhibit potassium movements through the system (e.g. Dunham and Ellory 1981; Ellory et al. 1985; Lauf 1985b). If the transport is speeded up when thiocyanate or nitrate is substituted for chloride, it is likely that the transport proceeds via electrical coupling of conductive potassium and chloride transport, since thiocyanate and nitrate have higher conductive permeability across the red cell membrane than chloride. Second, potassium chloride efflux via the cotransport pathway will cause a temporary disequilibrium for chloride ions. Consequently, chloride ions will re-enter the cell via the anion exchange pathway in exchange for bicarbonate. As a result, the extracellular compartment will be alkalinized and the intracellular compartment acidified. These pH changes should be inhibited by the blockade of the anion exchanger, if the potassium and chloride transport occurs via K/Cl cotransporter. If the pH changes are accentuated, it is probable that the potassium and chloride transport proceeds via coupled potassium/proton and chloride/bicarbonate exchanges. Both of these requirements have not been tested in some of the studies which have operationally defined the pathway leading to coupled efflux of potassium and chloride from the cells as K/Cl cotransport.

The volume-sensitive potassium/chloride cotransport can be inhibited by furosemide and bumetanide. However, much higher concentrations of the drugs are required than for the inhibition of the sodium/potassium/chloride cotransport (see Ellory et al. 1985; Lauf 1985b). Potassium/chloride cotransport of sheep red cells can be drastically reduced by the use of the anti-L_L antibody, present in alloimmune sera prepared from high-potassium sheep injected with low-potassium sheep red cells (see e.g. Lauf 1985b). In addition, transport is reduced by calcium (A23187 + Ca^{2+}; Lauf 1985a).

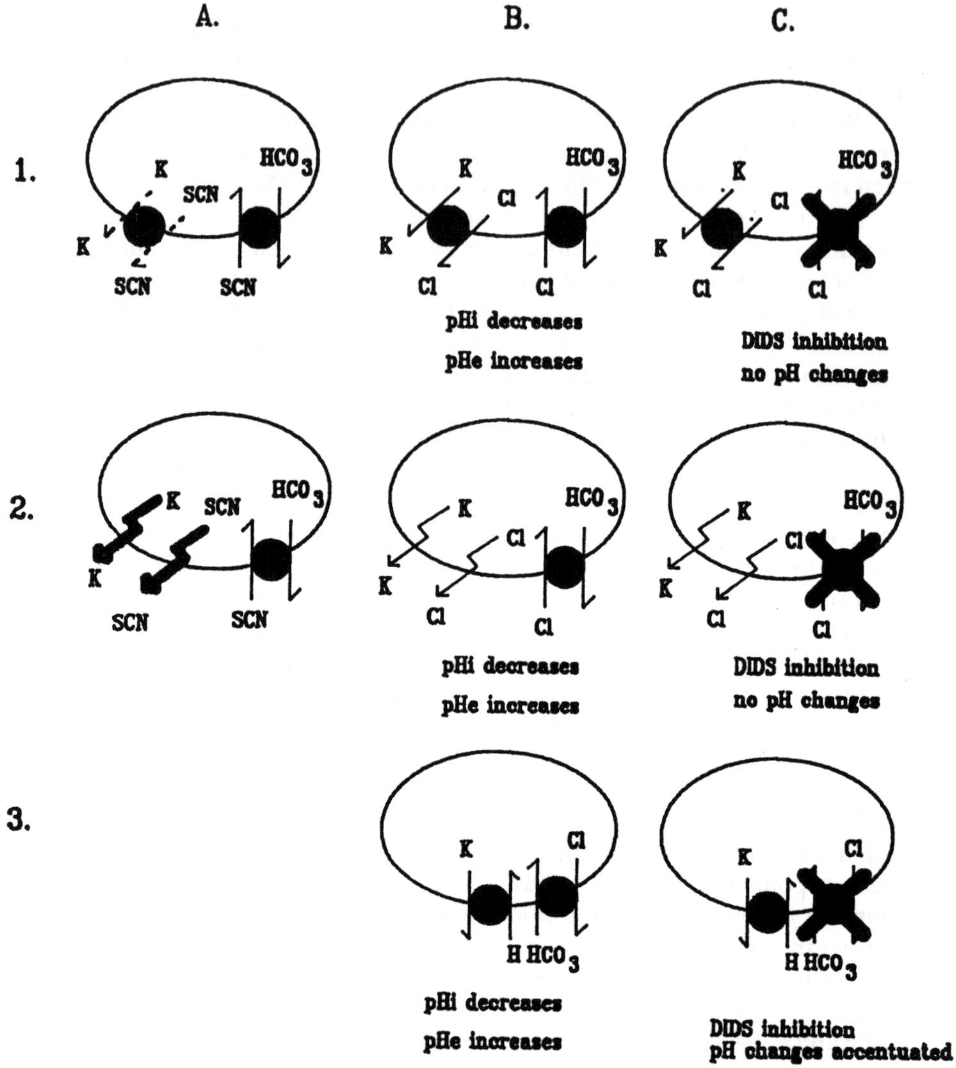

Fig. 7.2A-C. Distinction between potassium/chloride cotransport, electrically coupled conductive movements of potassium and chloride, and coupled electrosilent potassium/proton and chloride/bicarbonate exchanges. *A* Substitution of chloride by thiocyanate inhibits K/Cl cotransport but is expected to accelerate electrically coupled conductive movements. *B* When a functioning chloride/bicarbonate exchange is present in the cell membrane, and bicarbonate, chloride and potassium are present in the medium, all three mechanisms cause an increase in extracellular pH and a decrease in intracellular pH. *C* Inhibition of the anion exchange pathway by DIDS prevents the pH changes if the net KCl transport is either K/Cl cotransport or electrically coupled conductive transport of K and Cl, but accentuates the pH changes if the net KCl transport is the result of coupled K/H and Cl/HCO_3 exchange. For further details, se text

On the basis of the activation and inhibition studies, Lauf (1985b) has proposed a two-domain model for the potassium/chloride cotransporter. A volume-sensitive domain is responsive to volume perturbations, and a chemically stimulated domain, containing sulfhydryl groups, is sensitive to thiolalkylation. Both domains can be inhibited by divalent cations and the anti-L_L antibody.

7.4.3 Alkali Metal/Proton Exchange

Volume disturbances activate alkali metal/proton exchange in *Amphiuma* red cells. Osmotic shrinking activates sodium/proton exchange and osmotic swelling potassium/proton exchange (for reviews see Cala 1983b; 1985b). Cala's recent studies (Cala et al. 1988) have further indicated that the alkali metal/proton exchanges can be induced to function in "pH-regulatory mode" e.g. by intracellular acidification. Acidification activates the sodium/proton exchange. At least the sodium/proton exchanger is operating in steady-state volume and pH conditions (Tufts et al. 1987b). Cala's (1983b, 1985b) studies have indicated that the sodium/proton and potassium/proton exchange across *Amphiuma* red cell membrane are mediated by the same protein.

Osmotic shrinkage activates sodium/proton exchange in rabbit red cells (Jennings et al. 1986b) and in dog red cells (Parker 1983a). Sodium/proton exchange has also been described for human red cells (Escobales and Canessa 1986). The sodium-dependent acid extrusion pathway, described for lamprey (*Lampetra fluviatilis*) red cells (Nikinmaa 1986a; Nikinmaa et al. 1986), may also be a sodium/proton exchanger.

In frog (*Rana pipiens;* Rudolph and Greengard 1980) and in many teleost red cells (see e.g. Nikinmaa and Huestis 1984b; Baroin et al. 1984a; Cossins and Richardson 1985; Salama and Nikinmaa 1988, 1989), sodium/proton exchange can be activated by β-adrenergic drugs. In the frog, activation requires an inclusion of phosphodiesterase inhibitor in the incubation, whereas in teleost fish adrenergic alkalinization of the intracellular compartment occurs in physiological medium (e.g. Nikinmaa 1982b).

Several criteria must be fulfilled before sodium and proton transport can conclusively be characterized as sodium/proton exchange (for details see e.g. Cala 1985b):

1. Transport must be inhibited by treating the cells with amiloride analogues and by removing sodium from the incubation medium. Amiloride analogues act on the outward-facing transport site, as shown by the observation that internal amiloride did not inhibit the sodium/proton exchange of acid-loaded dog red cell ghosts (Grinstein and Smith 1987). Interestingly, the potassium/proton exchange of *Amphiuma* red cells is normally not sensitive to amiloride, but if the cells are shrunk in the presence of amiloride and thereafter swollen, still in the presence of amiloride, potassium/proton exchange is inhibited (Cala 1983a).
2. The transport must proceed even in the absence of potassium from the incubation medium.

3. Net sodium movements must be associated with extracellular acidification and intracellular alkalinization, which are inhibited by amiloride analogues.

4. The transport must proceed in the absence of chloride. Baroin et al. (1984b) suggested that the catecholamine-activated sodium transport would proceed via a sodium-chloride cotransport pathway, because the transport was inhibited by replacing nitrate for chloride. However, Cala (1983a) on Amphiuma red cells and Parker (1983a) on dog red cells showed that nitrate inhibited the apparent flux via sodium/proton exchanger. This was later verified on trout red cells by Borgese et al. (1986). Parker's (1986) studies on glutaraldehyde-fixed dog red cells have shown that the chloride requirement of the sodium/proton exchanger involves the activation of the mechanism rather than the transport function. The transporter can be fixed either in on or off position with glutaraldehyde. If the cells are incubated in hypertonic medium containing chloride, there is a marked amiloride-sensitive sodium flux. If the cells are thereafter transferred to hypertonic medium containing thiocyanate as a principal anion, the amiloride-sensitive sodium flux is inhibited. If the experiment is repeated in media containing glutaraldehyde, transferring the cells to a medium containing thiocyanate does not inhibit the amiloride-sensitive sodium fluxes. Thus, the sodium/proton exchanger is poorly activated in the absence of chloride, but once activated, chloride is not required for transport.

5. The transport as such must not affect membrane potential. The electroneutrality of the sodium-dependent acid extrusion in *Amphiuma* red cells was shown by Cala (1980). Red cells which were treated with the anion exchange inhibitor DIDS and thereafter osmotically shrunken gained sodium with no net flux of chloride or potassium. If sodium in this situation were to traverse the membrane via a conductive pathway, the membrane potential should have changed by hundreds of volts, which would have led to a dielectric breakdown (for a more detailed discussion, see e.g. Cala 1983b). In fact, practically no changes in membrane potential were observed. Changes in the membrane potential are not a necessary component in the adrenergic activation of the sodium/proton exchange in teleost red cells. Nikinmaa et al. (unpublished data) treated rainbow trout red cells with valinomycin, and allowed the membrane potential to "relax" to potassium equilibrium potential (see Fig. 7.1). The cells were then stimulated with isoproterenol. As indicated by the extracellular TPP^+ activity, membrane potential did not change as a result of the adrenergic stimulation. However, the red cells swelled, and their chloride distribution ratio increased as a result of the adrenergic stimulation.

6. The transport must be insensitive to bumetanide and furosemide. Baroin et al. (1984a) based their original suggestion that the adrenergically stimulated sodium and chloride accumulation would be due to sodium/chloride cotransport partly on the apparent inhibition of the response by furosemide. However, it has later been shown (Nikinmaa et al. 1987b) that in this case furosemide inhibits the anion exchange pathway. After adrenergic stimulation of the sodium/proton exchange and consecutive increase in intracellular pH, net influx of chloride occurs through the anion exchanger (see Sect. 8.6.1).

7. The transport must be dependent on the concentration gradients of both hydrogen ions and sodium. Funder et al. (1987) have shown that when the driving forces for sodium and protons are equal in dog red cells, flow through the sodium/proton exchanger stops. Similarly, Escobales and Canessa (1986) were able to manipulate the sodium and proton fluxes through the exchanger by altering the sodium and proton gradients across the human red cell membrane. Because of the dependence on the proton gradient, the transporter is activated by a decrease and inactivated by an increase in intracellular pH (see Escobales and Canessa 1986; Borgese et al. 1987).

It is likely that sulfhydryl groups are involved in the activation of the sodium/proton exchange. The exchanger can be "locked" in the active state by treating it with sulfhydryl reagents; in *Amphiuma* red cells by N-ethyl maleimide (Adorante and Cala 1987) and in dog red cells by N-phenyl maleimide (Parker and Glosson 1987).

Calcium appears to be involved in the activation of volume-sensitive alkali metal/proton exchange. The regulatory volume increase is greater in the presence of A23187 and calcium than in cells not treated with the ionophore. Furthermore, calcium induced a potassium loss from *Amphiuma* red cells, similar to that observed in swollen cells even in isotonic medium (Cala 1986a). However, sodium/proton exchange is not affected by calcium in isotonic medium. The effect of calcium on the alkali metal/proton exchanger may be mediated via the diacylglycerol-protein kinase C system. Cells exposed to the diacylglycerol analogue 12-O-tetradecanoyl-4β-phorbol-13-acetate (PMA) gain sodium and lose potassium in a calcium-dependent fashion, whereas the ion content of cells treated with an inactive diacylglycerol analogue 4α-phorbol-12,13-dodecanoate was not affected (Cala 1986b).

Cyclic AMP is the second messenger in the adrenergic activation of the sodium/proton exchanger of teleost red cells. Mahe et al. (1985) showed that the sodium/proton exchanger of rainbow trout red cells could be activated by forskolin, which activates adenylate cyclase without interacting with the receptor, and by extracellular addition of 8-bromo cyclic AMP, which crosses the red cell membrane. Salama, Marttila and Nikinmaa (unpublished data) have since observed that these treatments also cause red cell swelling in carp.

The potency of different β-adrenergic agonists in mediating the adrenergic effects on fish red cells depends both on the species and on the oxygen tension. Noradrenaline is a much more potent activator of the adrenergic proton extrusion than adrenaline in rainbow trout red cells; Tetens et al. (1988) obtained EC_{50} values of 1.3×10^{-8} M and 7.6×10^{-7} M for noradrenaline and adrenaline respectively. In normoxic carp red cells, noradrenaline induces adrenergic red cell swelling at a lower concentration than adrenaline, with an EC_{50} value of ca. 5×10^{-8} M for noradrenaline and ca. 5×10^{-7} M for adrenaline (Salama, Marttila and Nikinmaa, unpublished data). In hypoxic carp red cells (oxygen tension 8 mmHg), the difference between noradrenaline and adrenaline is even greater; the dose-response curve for adrenergic swelling in the presence of noradrenaline is shifted to the left, giving an EC_{50} value of ca. 5×10^{-9} M, whereas the dose-response curve for adrenaline is not affected. The measured plasma concentrations of noradrenaline in stressed (chasing, hypoxia, acid loads) fish are adequate to cause the adrenergic response of red cells (see Tetens et al. 1988). In fact, some stimulation is apparent even under "resting" conditions.

The activity of the adrenergic sodium/proton exchange is influenced by pH (e.g. Borgese et al. 1987; Heming et al. 1987; Nikinmaa et al. 1987b) and oxygen tension (Motais et al. 1987; Salama and Nikinmaa 1988). Figure 7.3 gives the pH-dependence of adrenergic sodium accumulation in four species of fish. There are marked species differences in the total accumulation of sodium and in the pH-dependence of the response. The accumulation of sodium is greatest in the two *Salmoniformes*, and much smaller in the *Cyprinid* and the *Perciformes*. Borgese et al. (1987) have studied the pH dependence of the adrenergic response of rainbow trout red cells in detail, and found that in the extracellular pH range 8–7.3, the increase in the sodium/proton exchange is mainly due to the activation by internal protons. At lower pH values, the inhibitory effect of external protons becomes predominant and the activity of the exchanger decreases.

The dependence of adrenergic response on oxygen tension is very pronounced in carp. The sodium/proton exchange is not activated by catecholamines (concentration $< 10^{-5}$ M) at atmospheric oxygen tension at pH values above 7.5 (Salama and Nikinmaa 1988). However, marked sodium accumulation occurs when the oxygen tension is decreased below 30 mmHg at pH 7.5. Motais et al. (1987) have shown that, in trout, the net sodium uptake of adrenergically stimulated red cells is ca. 4-fold greater in nitrogen atmosphere than in oxygen atmosphere at external pH of 7.6.

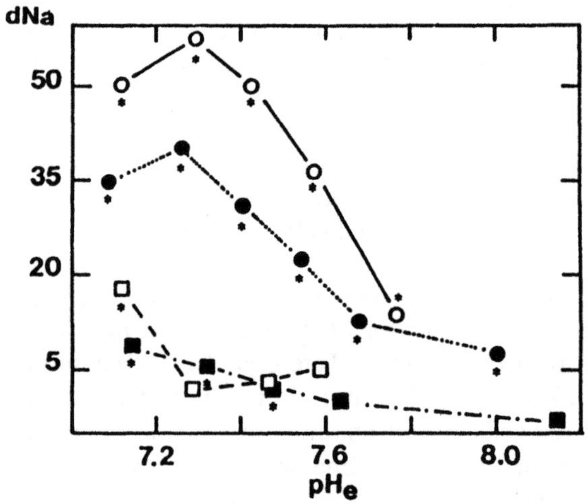

Fig. 7.3. The adrenergic increase in red cell sodium concentration (dNa in mmol l^{-1} red cell water) in whitefish (*Coregonus pallasi*; ○——○, N = 4–5), rainbow trout (*Salmo gairdneri*; ●····●, N = 8), pikeperch (*Stizostedion lucioperca*; □ ——□, N = 4) and carp (*Cyprinus carpio*; ■ —·—·■, N = 6–8) as a function of external pH (pH$_e$). The statistical significance of the adrenergic increase was calculated on the original sodium concentration data for isoproterenol-treated and control cells using paired t-test (* $P < 0.05$). Data from Salama and Nikinmaa (1989)

The dependence of the adrenergic response on pH and oxygen tension is not caused by an effect of pH and oxygen tension on the accumulation of cAMP. Cellular cyclic AMP levels were similar in nitrogen and air-treated rainbow trout red cells (Motais et al. 1987). Both adrenaline- and noradrenaline-induced accumulation of cyclic AMP were independent of pH in carp red cells (Salama, Marttila and Nikinmaa, unpublished data).

Oxygen affects the function of sodium/proton exchanger either directly or via a "transducer". One possible transducer is haemoglobin, the T-form of which appears to bind to the anion exchange protein in intact human red blood cells (Chetrite and Cassoly 1985). Both decreasing pH and oxygen tension increase the proportion of T-form (deoxyform) of haemoglobin, which, by binding to the anion exchanger, may alter the cytoskeletal-membrane protein interactions to allow the activation of sodium/proton exchange (see Motais et al. 1987; Salama and Nikinmaa 1988).

The temperature sensitivity of the transport pathway has not been investigated in detail. Studies at different temperatures (e.g. Salama 1986; Nikinmaa et al. 1987b) indicate that the sodium/proton exchanger of rainbow trout red cells can be activated by adrenergic drugs at a temperature range 2–20 °C. However, the K_d values and the maximal fluxes at the different temperatures are not known.

7.4.4 Sodium/Calcium Exchange

So far the presence of sodium/calcium exchange has only been demonstrated in carnivore red cells. The information presently available comes from the studies by Parker (1978, 1979) on dog red cells, which lack the sodium pump (see below). In dog red cells, net outward movement of sodium is coupled to the active transport of calcium via the calcium pump (see below). The calcium gradient generated will move sodium ions out of the cell (for a review see e.g. Parker 1977). The sodium/calcium exchanger is electrogenic in many cell types (see e.g. Schatzmann 1986). Although it is not known if this is true for the dog red cell system, data on the effects of anions on the sodium/calcium exchange suggest that this might be the case: sodium/calcium exchange increases when thiocyanate or nitrate is replaced for chloride in the incubation medium (Parker 1983b). The conductive permeability of these anions is greater than that of chloride (Parker 1983b). If the sodium/calcium exchange were electrogenic, then calcium influx would increase in proportion to the membrane mobility of the prevalent anion, as is experimentally observed. The sodium-dependent calcium transport is inhibited by low pH, cell shrinkage and quinidine (Parker 1983b).

7.5 Active Transport

7.5.1 Sodium Pump

Structural aspects of the sodium pump have been outlined in Section 4.2.2.4. The function of sodium pump is twofold. First, it generates the ion gradients for sodium and potassium. The gradients can then be used for transport of solutes (e.g. amino acids and other ions). The actively maintained ion gradients are also the major factor behind the membrane potential (see Sect. 7.6.2). Second, the electrogenic nature of sodium pump as such influences the membrane potential (see Heinz 1981). The sodium pump is activated by internal sodium and external potassium, and can be selectively inhibited by cardiac glycosides, e.g. ouabain (for a detailed review on the function of sodium pump see e.g. Cavieres 1977; Glynn 1985). The ouabain sensitivity of the sodium pump varies between species. Willis and Ellory (1983), studying rodent red cells, observed that 50% inhibition of the sodium pump was achieved at 0.1–1 μM concentrations in woodchucks and *Sphermophilus* ground squirrels, at ca. 3 μM concentration in flying and grey squirrel and at > 100 μM concentration in hamsters, rats, lemmings and house mice. Other inhibitors of the sodium pump include vanadate, oligomycin, thimerosal (ethylmercurithiosalicylate), arsenite, N-ethylmaleimide, butanedione, fluorescein isothiocyanate, eosin, quercetin and N-acetylimidazole.

In addition to its normal function, the sodium pump can perform sodium/sodium exchange, potassium/potassium exchange, pump reversal (sodium in, potassium out) and electrogenic sodium efflux, either as uncoupled sodium efflux or as 2Na/3Na exchange, both accompanied by ATP hydrolysis (see Glynn 1985). In addition, the pump can carry out potassium/proton exchange in the absence of cytoplasmic sodium, and sodium/proton exchange in the absence of extracellular potassium (Polvani and Blostein 1988).

Although the presence of sodium pump is almost universal in vertebrate red cells, the red cells of carnivores are an exception (see Parker 1977). The red cells of adult dogs are characterized by the absence of ouabain-sensitive sodium or potassium movements, and the absence of sodium/potassium ATPase activity (Parker 1977). Some ouabain-sensitive potassium transport is observed in puppies (Lee and Miles 1972; Miles and Lee 1972). Furthermore, dog reticulocytes are high in potassium (see Parker 1977). Thus, sodium pump activity is lost during the maturation of the red cells.

Ruminants are characterized by two distinct types of red cells, LK and HK (see Table 7.1; Ellory 1977; Ellory and Tucker 1983). In goat red cells, the different ion concentrations are primarily caused by differences in the kinetic properties of the sodium pump: although in both HK and LK goat red cells the number of specific ouabain binding sites are the same, the ouabain-sensitive sodium efflux and potassium influx in LK cells is only one-third to one-quarter of that in HK cells. In sheep, the number of ouabain specific binding sites of LK red cells is also smaller than that of HK red cells (e.g. Ellory and Tucker 1983).

Newborn ruminants have HK red cells, regardless of the adult red cell type (e.g. Ellory 1977). After birth the ouabain-sensitive potassium uptake of LK animals gradually decreases, reaching adult values after ca. 45 days. Also, reticulocytes have much higher sodium pump activity than mature red cells (e.g. Ellory 1977). It appears that the macrocytic reticulocytes, appearing after massive blood loss, can already be characterized as either LK or HK red cells on the basis of their ion contents, although they have much higher pumping activity than adult red cells (see Lauf and Valet 1983). The increased pumping rate of the reticulocytes appears to be mainly caused by the increased turnover rate, rather than increased number of pumping sites (Lauf and Valet 1983).

In hibernating mammals, which lower their body temperature to ca. 5°C, the temperature-sensitivity of the sodium pump is smaller than in nonhibernators (see Willis et al. 1980): the percentage ratio of ouabain-sensitive potassium influxes at 5° and 37°C ($J_5/J_{37} \times 100$) is characteristically less than 1% for nonhibernators, and 2–5% for hibernators. The decreased temperature sensitivity of the sodium pump of hibernators can largely explain why hibernators do not lose potassium at low body temperatures, as do nonhibernators. In nonhibernators, the passive leak of potassium across the cell membranes decreases less than the active pumping rate (see e.g. Willis 1979).

Studies on poikilothermic animals suggest that the temperature-dependence of the sodium pump of cold-acclimated specimens is lower than that of the warm-acclimated specimens (e.g. Bourne and Cossins 1981). Raynard (1987) measured the potassium influx of red cells from 6°C and 21°C acclimated trout at 10°C, and observed that the pumping rate was greater in the cold-acclimated specimens. The difference was entirely due to the increased turnover number of the pump, since the number of ouabain binding sites was similar in both groups of fish. The increased turnover number may result from decreased order of the red cell membrane (homeoviscous adaptation) of the cold-acclimated fish.

7.5.2 Calcium Pump

An overview of the structure of the calcium pump has been given in Section 4.2.2.4. The major function of the calcium pump is to maintain the intracellular free Ca^{2+} level near or below 1 μM (e.g. Yoshida et al. 1986), despite the 1–2 mM external calcium concentrations. In dog red cells the calcium pump has an additional, important role in volume regulation in the absence of active sodium pumping (see e.g. Parker 1977, 1979). The calcium pump exchanges protons for calcium (e.g. Villalobo and Roufogalis 1986). In human red cells at steady-state, the pump transports ca. 50 μmol Ca^{2+} l^{-1} cells h^{-1}. Only a minor fraction of the maximal capacity of the pump is used to do this: at saturating calcium concentrations (> 10 μM), the pump is capable of transporting 10–20 mmol Ca^{2+} l^{-1} cells h^{-1} (e.g. Schatzmann 1986).

Similar to the activity of the sodium pump, the calcium pumping activity of ruminant (cattle) red cells appears to decrease markedly with the age of the animal (Zimmermann and Schatzmann 1985). Also, as with the sodium pump, the calcium pump of a hibernating mammal, Richardson's ground squirrel (*Spermophilus richardsonii*), was much less temperature-sensitive than that of a nonhibernator, the guinea-pig (*Cavia porcellus*, Hall et al. 1987). Because of the high activity of the calcium pump of hibernators at low temperatures, they can maintain intracellular calcium concentration at a low level. Thus, the calcium-dependent potassium channel (see Sect. 7.2) is not activated under cold conditions. As a result, the red blood cells of hibernators can maintain a stable intracellular potassium concentration even at low temperatures, in contrast to nonhibernators, which lose massive amounts of potassium (see also Ellory and Hall 1987).

7.6 Membrane Potential

Membrane potentials in biological systems have been treated in detail by, e.g. Heinz (1981). In the following sections, a short account is given of the Gibbs-Donnan equilibrium, which is commonly referred to when the ion concentrations and intracellular pH of red cells are investigated, and factors influencing the membrane potential in the presence of active ion transport are later briefly discussed.

7.6.1 Gibbs-Donnan Equilibrium

In a situation in which there are impermeable intracellular ions (X^-) and the membrane is permeable to water and diffusible ions (C^+ and A^-), the equilibrium distribution of diffusible ions is such that

(a) electroneutrality on both sides of the membrane is maintained, i.e.

$$C_i^+ = A_i^- + X_i^- \text{ and } C_o^+ = A_o^-; \tag{7.1}$$

(b) at equilibrium, the electrochemical potential of all the diffusible ions is the same on both sides of the membrane. The electrochemical potential of an ion ($\tilde{\mu}_i$) is given by the equation

$$\tilde{\mu}_i = \tilde{\mu}_i^\circ(T) + RT\ln\tau_i c_i + z_i F ¥, \tag{7.2}$$

in which $\tilde{\mu}_i^\circ(T)$ is the standard potential of i at the temperature T, τ_i the activity coefficient, c_i the concentration, z_i the valency, R the universal gas constant, F the Faraday constant and ¥ the electrical potential. At equilibrium at constant temperature, and assuming that the activity coefficients on both sides of the membrane are the same, the electrochemical potential difference across the membrane for univalent ions, e.g. chloride and protons, simplifies to

$$\delta\tilde{\mu}_{Cl} = RT\ln([Cl_i]/[Cl_o]) - F\delta ¥ \tag{7.3}$$

and

$$\delta\tilde{\mu}_H = RT\ln([H_i]/[H_o]) + F\delta ¥. \tag{7.4}$$

Since at equilibrium the electrochemical potential difference is zero, there is an electrical potential difference (Donnan potential = V_m = $F\delta¥$) across the membrane, which is given from

$$V_m = -RT\ln([H_i]/[H_o]) = RT\ln([Cl_i]/[Cl_o]). \quad (7.5)$$

Also, at Donnan equilibrium, there is a concentration gradient for protons and chloride ions across the membrane. The concentration ratio (r) for protons is the inverse of the concentration ratio for chloride.

The Donnan equilibrium is not an equilibrium state for the solvent. If no other factors are involved, water will be drawn into the cell to level out the osmotic gradient, but simultaneously diffusible ions will redistribute according to the Donnan rule. The influx of water and ions will go on until the cell bursts. For more detailed treatments of the Donnan equilibrium, see e.g. Macknight and Leaf (1977) and Heinz (1981).

Notably, a true Donnan distribution of ions (and Donnan potential) is possible only if none of the diffusible ions are actively transported. As stated by Heinz (1981)

"the terms 'Donnan distribution' and 'Donnan potential' are often applied incorrectly to systems which as a whole are transient or in a steady state rather than in true equilibrium. There the ion treated as impermeant is only virtually so: it either penetrates very slowly, much slower than the other permeant ions, or it may be maintained at an constant distribution by an active pump."

7.6.2 Membrane Potential in the Presence of Active Transport

The membrane potential for cells (with active ion transport) in the steady state is a function of two processes: first, the electrogenic ion pumps create a potential across the cell membrane, and second, the ions, which are maintained at disequilibrium by the active ion pumps, have diffusion potentials which are a function of the conductive permeancies and concentrations of the ions. The membrane potential is affected by changes in the transport rates of electrogenic pumps. For example, activation of the sodium pump should lead to hyperpolarization of the membrane (the inside should become more negative). An increase in the conductive permeability of the ions held at disequilibrium will affect membrane potential. Finally, changes in the concentrations of ions held at disequilibrium will affect membrane potential (e.g. Reuss et al. 1986).

The membrane potential resulting from disequilibrium distribution of permeable ions can be calculated from the Goldman-Hodgkin-Katz equation, which relates the electrical potential (V_m), and the concentrations ([ion]) and current-carrying (conductive) permeancies (P_{ion}) to each other

$$V = RTF^{-1}\ln\{(P_K[K_o]+P_{Na}[Na_o]+P_{Cl}[Cl_i]+\ldots)/(P_K[K_i]+P_{Na}[Na_i]+P_{Cl}[Cl_o]+\ldots)\} \quad (7.6)$$

(see e.g. Hladky 1977; Heinz 1981).

It is important to note that in the Goldman-Hodgkin-Katz equation, terms referring to ions which have reached complete equilibrium distribution can be omitted. This applies to passive ions in the presence of ion pumps (e.g. Heinz

1981). Thus, in the steady state and in the absence of active transport of anions, the membrane potential of the red cells must be created mainly by the concentration gradients for sodium and potassium, maintained by the sodium pump (see e.g. Lassen 1977). However, as long as chloride ions and protons are passively distributed, the membrane potential can be estimated from their intra- and extracellular concentrations, using the same equation as for the Donnan equilibrium. In fact, in the steady state, the permeant, passively distributed ions (e.g. protons and chloride) behave largely in the same way as in a Donnan system. For this reason the steady-state system is often called "double-Donnan" system (e.g. Macknight and Leaf 1977). It is important, however, to remember that passive distribution of an ion requires that its permeation is not coupled to any other permeation process directly or indirectly (see Heinz 1981). If chloride or proton movements are coupled to actively maintained ion gradients, as in the sodium/proton exchange, potassium/chloride cotransport and sodium/potassium/chloride cotransport, they cannot be considered passively distributed on theoretical grounds (although their distribution may be so close to passive that for practical purposes the term "passive distribution" can be used).

The Goldman-Hodgkin-Katz equation can be used to determine the membrane potential of red cells also in transient states. In this case, the ions having the greatest transient concentration gradients and greatest conductive permeancies are the most important determinants of the membrane potential. For human red cells, the conductive permeability for chloride is approximately 100 times greater than for sodium or potassium (see e.g. Hladky and Rink 1977). As a result, the membrane potential is in transient states mainly determined by the diffusion potential for chloride, which is largely determined by chloride distribution (see e.g. Hladky 1977). However, in *Amphiuma* red cells, the conductive permeancies for sodium, potassium and chloride are similar (see Lassen 1977). Therefore, the permeation and concentration of all these ions contributes to the membrane potential of *Amphiuma* red cells in transient states.

In a situation in which chloride is at equilibrium distribution, and the cell is in steady state, the contribution of the electrogenic sodium pump can be incorporated into the Goldman-Hodgkin-Katz equation. In this situation the membrane voltage becomes

$$V_m = -RTF^{-1} \ln\{(rP_K[K]_i + P_{Na}[Na]_i)/(rP_K[K]_o + P_{Na}[Na]_o)\}, \qquad (7.7)$$

where r is the coupling ratio of the pump (see Mullins and Noda 1963). According to Mullins and Noda, working on frog muscle, the maximum contribution of the electrogenic sodium pump on the membrane potential is up to 10 mV at steady state.

The membrane potentials of red cells have mainly been determined from the chloride distribution ratio, supposing that chloride ions are passively distributed. Direct measurements of the membrane voltage with microelectrodes are available mainly for the giant, nucleated red cells of *Amphiuma* (e.g. Lassen 1977; Stoner and Kregenow 1980). In both studies, the measured membrane potential and the potential calculated from chloride distribution were somewhat different. Similarly, measurements of the membrane potential of primitive bird erythrocytes with fluorescent dye (Baumann and Engelke 1987) gave much higher membrane potential than that obtained from the chloride distribution ratio. These data suggest that chloride ion is not necessarily distributed according to the membrane potential in the red cells. Because of this, an independent measurement of membrane potential should be made, at least in nucleated red cells, before concluding that chloride ions are in electrochemical equilibrium. Kimmich et al. (1985) and Montrose and Kimmich (1986) have described a method for measuring membrane potential in small cells. The method is based on the unidirectional influx of the lipophilic cation, tetraphenylphosphonium (TPP^+). ^{14}C-labelled TPP^+ is added to the cell suspension, after which the amount entering the cells is measured as a function of time. The greater the membrane potential, the faster the initial influx. Heinz et al. (1975) have used the steady-state distribution of TPP^+ to obtain information on the membrane potential of Ehrlich ascites cancer cells. The cell/incubation medium concentration ratio was considerably higher than was expected, and could not be used to calculate the membrane potential directly. However, the distribution ratio varied in the predicted direction upon alterations of the membrane potential, making it possible to follow relative changes in the membrane potential. Relative changes in membrane potential of red cells can also be followed by following the changes in the extracellular TPP^+ concentration at constant haematocrit (Nikinmaa, Tiihonen and Paajaste, unpublished data, see Fig. 7.1). Although the use of fluorescent carbocyanine dyes (e.g. Hoffman and Laris 1974) for measuring membrane potential in red cells is complicated by the influence of these dyes e.g. on the sodium and potassium fluxes across the red cell membrane, reproducible results have also been obtained by this method (e.g. Hoffman and Laris 1974; Baumann and Engelke 1987).

In human red cells, the measured membrane potentials in the steady state range from −10 to −14 mV (Lassen 1977). In primitive chicken erythrocytes, the membrane potential was ca. −40 mV (Baumann and Engelke 1987). In *Amphiuma* red cells, the measured membrane potential is approximately −20 mV (Hoffman and Laris 1974; Cala 1980).

Chapter 8

Control of Volume and pH

Jacobs and Stewart (1947) presented a detailed description of the acid-base and volume behaviour of human red cells, based on the Donnan equilibrium theory. The red cell was considered to be impermeable to cations and large intracellular anions such as haemoglobin, but permeable to small anions, water and carbon dioxide. In addition, the hydration/dehydration reactions of carbon dioxide were known to be catalysed by carbonic anhydrase within the cell.

Subsequently, data accumulated showing that the red cells were not cation-impermeable, but that their characteristic disequilibrium distribution for the major cations, sodium and potassium, was maintained by the sodium pump. In view of this, Tosteson and Hoffman (1960) formulated the "pump-leak" concept for cell volume maintenance. The cell membrane was made "functionally impermeable" to external sodium by the fact that all the sodium entering the cells through the "leak" pathways was removed from the cell by the sodium pump.

The studies, largely on nucleated red cells during the 1970s and 1980s, have shown that what were originally lumped together as "leak" pathways for the transport of cations across the red cell membrane are actually a wide variety of ion transport pathways (see Chap. 7). Many of these pathways link the movements of protons, bicarbonate or chloride to actively maintained ion gradients. In this situation, protons and small anions will be passively distributed only if the secondarily active movements of protons and/or chloride are markedly slower than the slowest step of passive proton equilibration across the red cell membrane.

Figure 8.1 depicts the factors involved in the control of red cell pH and volume. How these factors interact in pH and volume regulation depends largely on their relative reaction and transport rates. Thus, for a given type of red cell, the system may be simplified to take into account only the factors likely to be important in the control of steady-state volume or pH, or in the regulation of volume or pH after disturbances. In brief, the properties of the major factors involved in the control of red cell pH and volume are the following:

1. Transport of water across the red cell membrane. Since the permeability of red cells to water, both under osmotic flow and under tracer diffusion, is one to two orders of magnitude greater than the permeability to the most permanent ions via the anion exchange pathway (e.g. Sha'afi 1977), movements of water across the cell membrane are not the limiting factor in volume and pH regulation.

2. Transport of carbon dioxide across the red cell membrane, and the hydration/dehydration and acid-base reactions involving carbon dioxide. The diffusion of CO_2 is a very rapid reaction, with a half-time of ca. 1 ms at 37 °C, (see

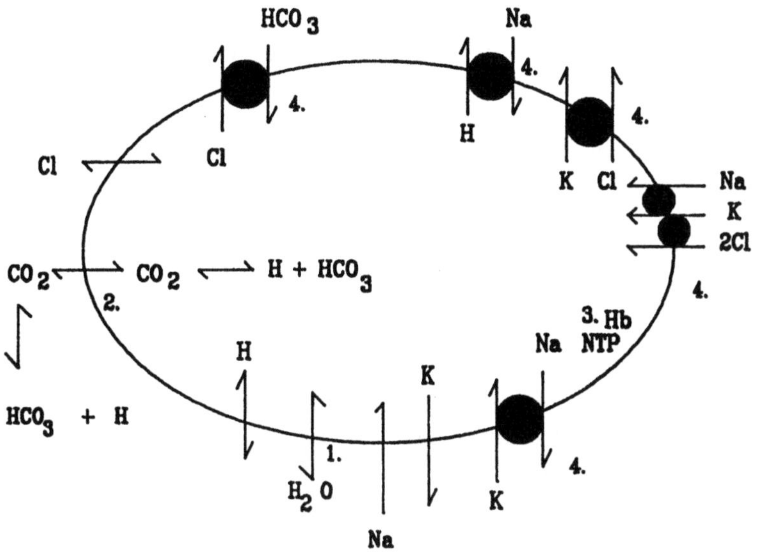

Fig. 8.1. Factors involved in volume and pH regulation of vertebrate red cells: *1* water movements across the red cell membrane, *2* movements of carbon dioxide across the red cell membrane, and hydration/dehydration reactions involving carbon dioxide, *3* the properties of cellular buffers, *4* ion transport pathways across the red cell membrane

e.g. Wagner 1977) and thereby 100 times faster than the chloride/bicarbonate exchange at similar temperature. The hydration/dehydration reactions of carbon dioxide and carbonic acid in the red cells are catalyzed by carbonic anhydrase. They are more rapid than the anion exchange, with a half-time of less than 20 ms at 37°C, and, as a result, not likely to be a limiting factor. In contrast, the uncatalyzed hydration/dehydration reactions of carbon dioxide and carbonic acid in plasma are relatively slow – 100 times slower than the anion exchange with a half-time ranging from 10 to 20 s at 37°C to several minutes at 0°C (see Sect. 9.2.1). Thus, these reactions can limit the net transfer of acid/base equivalents from plasma to the red cells.

3. The concentration and properties of intracellular impermeable anions/buffers. The major groups of these solutes are organic phosphates and haemoglobin. The concentration of tetrameric haemoglobin varies between 3 and 5 mmol l^{-1} RBC^{-1} (200–300 g l^{-1} RBC^{-1}). The concentration of organic phosphates varies markedly between species, but can be up to 20 mmol l^{-1} RBC^{-1}. Since both organic phosphates and haemoglobin have ionizable groups near the physiological pH range, they are intimately involved in the volume and pH changes associated with acid and alkaline loads.

4. Transport of ions across the red cell membrane. At least the following ion transport pathways have been experimentally shown to be involved in the control of red cell volume and pH: (1) the anion exchange, (2) the sodium pump, (3) the calcium pump (in carnivores, see e.g. Parker 1977), (4) sodium/calcium exchange (in carnivores, see e.g. Parker 1977), (5) sodium/proton exchange, (6) sodium/potassium/chloride cotransport, (7) potassium/chloride cotrans-

port and (8) potassium/proton exchange. By far the most rapid ion transport is the anion exchange, with a half-time of ca. 0.1 s at 37 °C in man, (see e.g. Lowe and Lambert 1983; Passow 1986) and ca. 1 s in rainbow trout at 15 °C (Romano and Passow 1984). Since the anion exchange is ca. 100 times more rapid than the uncatalyzed hydration/dehydration reactions of carbon dioxide in the extracellular medium, the equilibration of protons across the red cell membrane is limited by the rate of extracellular hydration/dehydration reactions, and not by the anion exchange (see e.g. Hladky and Rink 1977). Thus, the secondarily active ion transport pathways present on the red cell membrane, the sodium/proton exchange, potassium/proton exchange, potassium/chloride cotransport and sodium/potassium/chloride cotransport are able to generate a disequilibrium for protons across the red cell membrane if their transport rate is greater than the passive, conductive permeability of the membrane for protons, bicarbonate and/or chloride, and approaches the rate of uncatalyzed hydration/dehydration reactions of carbon dioxide.

In the following chapter, the control of red cell volume and pH in different vertebrates is examined, using models which take into account the secondarily active transport pathways in addition to the passive phenomena and the sodium pump.

8.1 The Basic Model: Control of Red Cell pH and Volume in the Absence of Significant Secondarily Active Transport

In this case most of the passive cation transport occurs via electrodiffusive leak. The sodium pump and the rapid anion exchange pathway are present. Thus, the secondarily active transport pathways do not play a role in the control of cellular volume and pH, whereby the red cell resembles a Donnan system (for detailed treatments see e.g. Jacobs and Stewart 1947; Siggard-Andersen 1974; Hladky and Rink 1977).

8.1.1 Maintenance of the Steady-State Volume

As evident from the discussion on Gibbs-Donnan equilibrium (Sect. 7.6.1), the presence of intracellular, impermeable solutes creates a considerable colloid-osmotic pressure difference across the red cell membrane. Unopposed, this would lead to continuous entry of permeable solutes and eventual bursting of the cells. The mechanisms by which colloid osmotic swelling at steady-state can be prevented can be defined as cell volume maintenance.

The mechanism of volume maintenance can be successfully explained by the pump-leak concept (Tosteson and Hoffman 1960; for a review see e.g. Macknight and Leaf 1977). The sodium pump generates and maintains sodium and potassium gradients across the red cell membrane of most erythrocytes. In the steady state, the sodium pump actively extrudes sodium ions at the same rate as they passively enter the cell (leak). In this situation, sodium can be treated as a "functionally impermeable" solute, and provides the osmotic force outside required to prevent swelling of the cell (see e.g. Tosteson and Hoffman 1960). In this case, if either the leak permeabilities increase or active transport rates decrease, the cell will swell. The "pump-leak concept" has been tested, using mammalian red cells (see e.g. Dunham and Hoffman 1980), Erlich ascites tumor cells (see e.g. Hoffmann 1983) and kidney cortex cells (see e.g. Macknight and Leaf 1977). In each case, the cell volume changes can be predicted on the basis of decreased sodium pumping after metabolic inhibition or on the basis of increased leak permeabilities for sodium.

The steady-state volume of red cell depends on the charge of the intracellular impermeable solutes. Both haemoglobin and organic phosphates have ionizable groups within the physiological pH-range. The secondary phosphate group of ATP and GTP has a pK value in the range 6–7 (e.g. Dawson et al. 1987). The proton charge of haemoglobins decreases markedly with decreasing pH. At a given oxygen saturation the buffer values of teleost haemoglobins are much lower than those of elasmobranch or mammalian haemoglobins (see e.g. Jensen 1988). Between pH 7.5 and 6.5 1 mol of carp (*Cyprinus carpio*) deoxyhaemoglobin tetramer takes up 4–5 mol protons, whereas 1 mol of dogfish (*Scyliorhinus canicula*) haemoglobin takes up 10–15 mol and pig haemoglobin 8–10 mol protons (cf. Jensen 1988). As long as the pH changes are not associated with changes in the relative rates of the sodium pump vs other cation transporting pathways, the volume changes are as expected from Donnan-like behaviour. The pH-induced decrease in the negative charge of haemoglobin and organic phosphates increases the water content of the red cells (e.g. Van Slyke et al. 1923). The volume increase is due to the net influx of negative diffusible anions, mainly chloride, in order to maintain the electroneutrality and the consecutive influx of osmotically bound water. However, as illustrated, e.g. by Dalmark (1975), the cellular water content at constant osmolarity increased much less with decreasing pH than expected from the changes in intracellular chloride concentration if the red cells behaved as perfect osmometers. Similar behaviour is also observed if red cell volume is changed by changing the tonicity of the environment (e.g. Savitz et al. 1964). Thus, a significant fraction of the red cell water behaves as "nonosmotic", i.e. not serving as a solvent for cell solutes. This apparent nonosmotic water partly depends on the concentration dependence of the osmotic coefficient of the haemoglobin (e.g. Freedman and Hoffman 1979). However, Solomon et al. (1986) observed that even after taking into account the change of the osmotic coefficient of haemoglobin some "nonosmotic" water, of presently unknown origin, remains.

In contrast to the increase in cell volume induced by an isothermal decrease in extracellular pH, the decrease in pH (in a closed system) induced by an increase in temperature does not alter either the chloride distribution ratio across the red cell membrane or the red cell water content (see e.g. Reeves 1976). The temperature-induced change in plasma and red cell pH is such that the net charge of the major impermeable ions does not change with temperature (see Reeves

1976; for an account on the temperature-induced changes in blood pH and the alphastat hypothesis, see e.g. Reeves and Rahn 1979).

Oxygenation-deoxygenation reactions of haemoglobin are also associated with cell volume changes. Within the physiological pH range, oxyhaemoglobin is more acidic than deoxyhaemoglobin (e.g. Kilmartin and Rossi-Bernardi 1973). In tench, the number of protons taken up during deoxygenation at physiological ATP concentration and intracellular pH is ca. 4 (Jensen and Weber 1985a). As a result of the uptake of protons upon deoxygenation, the charge on haemoglobin decreases, and chloride plus osmotically obliged water is drawn into the cell (see e.g. Hladky and Rink 1977).

8.1.2 Determinants of the Steady-State pH

In the absence of significant secondarily active proton, bicarbonate and chloride movements, all these ions are distributed according to the membrane potential at the steady-state, and the intracellular pH can be calculated from the chloride distribution ratio using the formula

$$\text{pHi} = \text{pHe} + \log[\text{Cl}]\text{i} - \log[\text{Cl}]\text{e}. \tag{8.1}$$

The distribution ratio for exchangeable anions is given from the relation (for derivation of the formula, see Hladky and Rink 1977).

$$\text{Ai}/\text{Ao} = 2(\text{mol cation} + z \cdot \text{mol impermeable polyions})/(2 \cdot \text{mol cation} + (z+1) \cdot \text{mol impermeable polyions}), \tag{8.2}$$

in which z is the charge of impermeable polyions.

From these two relations, the steady-state intracellular pH at different extracellular pH values can be estimated. An increase in the extracellular pH will increase the steady-state intracellular pH, whereby the charge of the impermeable polyions decreases (becomes more negative). This will decrease the anion ratio and cell water. Since the difference between pHe and pHi depends on the anion ratio, pHe-pHi will increase. With decreasing pHe, the charge on impermeable polyions increases (becomes less negative), the anion ratio increases, the cell swells and pHe-pHi becomes smaller.

The relation between pHe and pHi at the steady-state depends mainly on the concentrations of the impermeable intracellular polyions and the pK values of their ionizable groups. Duhm (1972, 1976) has studied the effects of variations in the red cell organic phosphate concentration on the relation between extra- and intracellular pH in human red cells. The intracellular pH decreases with increasing concentration of organic phosphates (see Fig. 8.2). The change in intracellular pH per unit change of extracellular pH (ΔpHi/ΔpHe) depends on both the organic phosphate concentration and the extracellular pH. An increase in extracellular pH and an increase in the cellular 2,3-DPG concentration decrease the change in intracellular pH per unit change in extracellular pH (see Duhm 1972, 1976). On the basis of his data, Duhm (1976) has formulated an empirical equation describing the relation between the Donnan distribution

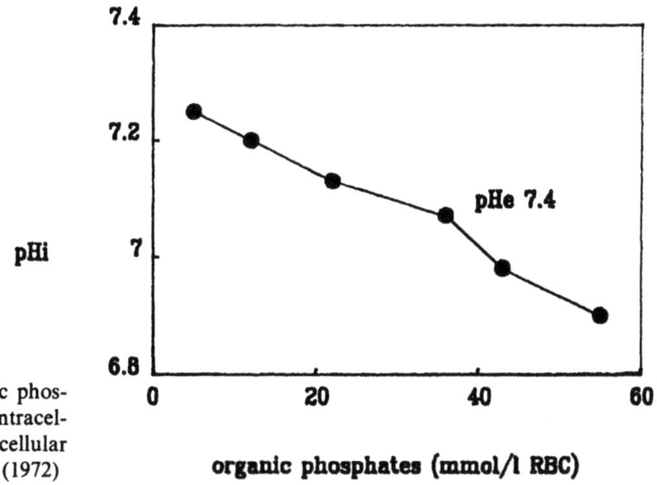

Fig. 8.2. Effect of organic phosphate concentration on intracellular pH at constant extracellular pH. Data are from Duhm (1972)

ratio of protons and the change in intracellular pH per unit change in extracellular pH:

$$\Delta pHi/\Delta pHe = 1 + \log r_H = 1 + pHi - pHe. \qquad (8.3)$$

This equation is valid when the Donnan ratio is altered between 0.3 and 1, by altering either the concentration of intracellular buffers or the extracellular pH.

Based on the above relation between the intracellular organic phosphate concentration and intracellular pH, and the observed decrease in the ATP/GTP concentration of the red cells from hypoxic and stressed fish (e.g. Wood and Johansen 1973b; Weber et al. 1976a; Weber and Lykkeboe 1978; Greaney and Powers 1978; Nikinmaa and Soivio 1982; Jensen and Weber 1982, 1985b, c; Nikinmaa 1986b), it has been suggested that the hypoxia-induced increase in intracellular pH of fish erythrocytes would be caused by a change in the concentration of intracellular impermeable polyions. However, none of the above papers presented conclusive evidence on the subject, since all the major organic phosphates were not measured. The hydrolysis of ATP and GTP as such is not associated with a decrease in the total negative charge of intracellular phosphates. For example, at pH 7.0, the charge of ATP is -3.7 and the charges of its hydrolysis products ADP and inorganic phosphate are -2.4 and -1.4 respectively (calculated from the dissociation constants of the different compounds in aqueous solution). Thus, if the change in ATP/GTP concentration were only due to hydrolysis, the Donnan ratio would not be expected to change. Tetens (1987; see Fig. 6.4), however, has shown that the total pool of adenine and guanine nucleotides within the red cells diminishes in hypoxic conditions, thus conclusively confirming earlier suggestions.

The effects of oxygenation reactions on the red cell volume have been outlined in Section 8.1.1. Since the deoxygenation decreases the negative charge on haemoglobin, the anion ratio will increase and the intracellular pH will be shifted to an alkaline direction.

8.1.3 pH and Volume Changes in Response to Carbon Dioxide Addition and Removal, and Acid and Base Loads

Physiologically, the most important anion and pH disequilibria are caused by the release of carbon dioxide from the tissues to blood, and the liberation of carbon dioxide from blood to the environment in the lung/gill capillaries (see Sect. 9.4). The CO_2 produced in the tissues diffuses down its partial pressure gradient from tissues to plasma and from plasma to red cells within a few milliseconds. Within the red cells, some of the CO_2 reacts with haemoglobin to form carbamino compounds, but most of it is rapidly hydrated to carbonic acid in a reaction catalyzed by carbonic anhydrase. Carbonic acid then dissociates to bicarbonate and protons. Most of the protons are taken up by the intracellular buffers, whereas bicarbonate is exchanged for chloride via the anion exchange pathway.

Anions are rapidly equilibrated across the red cell membrane, whereas proton equilibration takes considerably longer. Thus, after the anions have reached similar distribution ratios, the distribution ratio for protons is different from the anion distribution ratio (see e.g. Roughton 1964; Forster and Crandall 1975; Hladky and Rink 1977). The CO_2 entering the cell is rapidly hydrated to carbonic acid, which dissociates to bicarbonate and protons, whereby intracellular pH decreases. In contrast, the extracellular pH is initially not affected, because its rate of change is limited by the uncatalyzed hydration/dehydration rates of carbon dioxide and carbonic acid (see below for the equilibration of acid/base loads). Because of this, it is possible that protons do not reach equilibrium distribution in circulation.

The proton equilibration after an extracellular acid load is depicted in Fig. 8.3. Proton equilibration is limited by the speed of the uncatalyzed hydration/dehydration reactions of carbon dioxide and carbonic acid. This had been observed already by Jacobs and Stewart (1942), who showed that addition of external carbonic anhydrase into the incubation markedly speeded up the volume changes of human red cells associated with ammonium chloride treat-

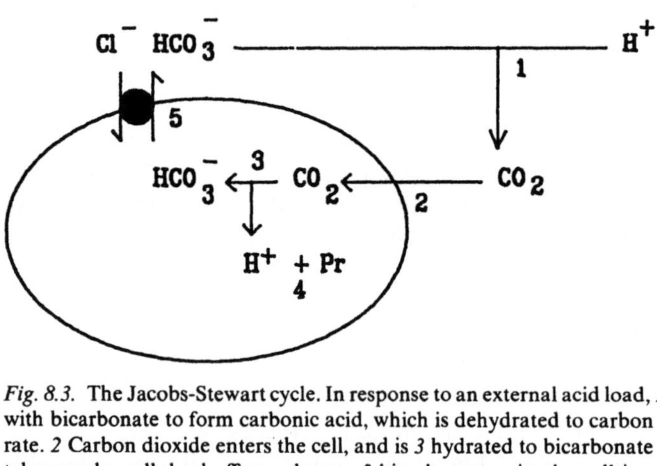

Fig. 8.3. The Jacobs-Stewart cycle. In response to an external acid load, *1* external protons combine with bicarbonate to form carbonic acid, which is dehydrated to carbon dioxide at the uncatalyzed rate. *2* Carbon dioxide enters the cell, and is *3* hydrated to bicarbonate and protons. *4* Protons are taken up by cellular buffers, whereas *5* bicarbonate exits the cell in exchange for chloride. The sequence of events is reversed in response to alkaline load

ment. When acid is added to a suspension of red cells, protons react with bicarbonate to form carbonic acid, which is converted to carbon dioxide at the uncatalyzed reaction rate. The carbon dioxide formed enters the cell and is hydrated to form carbonic acid, which dissociates to form bicarbonate and protons. The protons are taken up by intracellular buffers, and bicarbonate ions exchanged for chloride. The net result of this Jacobs-Stewart cycle is the transport of a proton into the cell, where it becomes buffered by intracellular buffers, mainly haemoglobin. When an alkaline load is added to the incubation medium, the sequence of events is reversed. Carbon dioxide must first be hydrated – in the uncatalyzed reaction – to carbonic acid, which then dissociates to protons and bicarbonate. Protons react with hydroxyl ions forming water. Bicarbonate enters the cell in exchange for chloride and takes up protons from intracellular buffers, after which it can re-enter the cycle by diffusing out of the cell as carbon dioxide.

Recovery of extracellular pH after an addition of a bolus of acid or base to a suspension of red cells in unbuffered medium has often been used to demonstrate the presence of rapid anion exchange in the red cells (e.g. Cossins and Richardson 1985; Kregenow et al. 1985; Nikinmaa and Railo 1987). However, in a medium containing even trace amounts of CO_2 and bicarbonate, as in nominally bicarbonate-free, air-equilibrated medium, the rate of recovery may be limited by the uncatalyzed extracellular reactions and not by the anion exchange rate. The anion exchange rate is limiting only if its half-time is greater than the half-time of the uncatalyzed reaction.

The equilibration of acid loads across the membrane is faster than the equilibration of alkaline loads (see e.g. Jacobs and Stewart 1942; Hladky and Rink 1977). An addition of acid sufficient to decrease external pH by 1–2 units will increase the amount of carbonic acid in proportion with the increase of proton concentration. Since the dehydration of carbonic acid is an equilibrium reaction, the reaction rate will also increase proportionally. In contrast, the alkalinization of the medium by base addition will decrease the carbonic acid concentration, and very little bicarbonate is formed in the dissociation of H_2CO_3. Although the bicarbonate formed can rapidly be converted to carbon dioxide in the remaining steps of the Jacobs-Stewart cycle, the amount of carbon dioxide added to the external pool is so small that it hardly affects the initial hydration rate. If, in this situation, the external bicarbonate concentration is increased, the initial rate of dehydration will be increased, as observed in the study by Kregenow et al. (1985).

The effect of an acid or alkaline load on the intracellular pH (after reaching the steady state) depends on the buffering power of the blood. Duhm (1976) has given an equation for the nonbicarbonate buffering power in red cell suspensions

$$\beta H^+(B) = \beta H^+(E) \cdot (\Delta pHi/\Delta pHe) \cdot \phi(E) + \beta H^+(P) \cdot \phi(P), \tag{8.4}$$

in which βH^+ = the buffer value (mol H^+ l^{-1} pH unit^{-1}), B = blood, E = erythrocytes, P = plasma, ϕ = the volume fraction. In the equation, $\beta H^+(B)$ is given in terms of extracellular pH and $\beta H^+(E)$ in terms of intracellular pH. An important point emerges from the equation: although an increase in the concentration of intracellular impermeable organic phosphates increases the erythrocytic buffering capacity, the apparent buffering power (in terms of moles of protons/unit change of extracellular pH per litre of blood) may actually decrease, owing to the effect of organic phosphates on the relation pHi and pHe (Duhm 1976, see also Sect. 8.1.2).

Any increase in the anion ratio, as will result from carbon dioxide or acid load (which leads to a decrease in the negative charge of the intracellular impermeable ions), will increase the red cell volume, whereas any decrease in the anion ratio, as will result from alkaline load or carbon dioxide excretion in the respiratory epithelia, will decrease the red cell volume.

8.1.4 Volume Changes in Response to Osmotic Disturbances

When placed in a hypertonic medium, the red cells will shrink. Conversely, in hypotonic medium, they will swell. In the absence of secondarily active, volume-responsive transport pathways, the red cell will not be able to regulate its volume after osmotic disturbances within physiological time scale.

8.2 Mammalian Red Cells

Although the basic model for the control of red cell volume and pH is useful as a starting point, it is doubtful if any red cells actually behave as expected from the model. Lew and Bookchin (1986) pointed out that even a weak cotransport markedly affects the volume regulatory behaviour of the red cells. It is now known that all the studied mammalian red cells show either sodium/proton exchange, sodium/potassium/chloride cotransport or potassium/chloride co-transport activity (see e.g. Ellory et al. 1982, 1985; Jennings et al. 1986b; Escobales and Canessa 1986).

8.2.1 Human Red Cells

The presence of sodium/proton exchange in human red cells has been demonstrated, using cells in which proton equilibration via the Jacobs-Stewart cycle is inhibited by the use of DIDS and methazolamide, an inhibitor of carbonic anhydrase activity (Escobales and Canessa 1986). The presence of sodium/potassium/chloride cotransport and potassium/chloride cotransport across human red cell membrane has also been demonstrated (see Sects. 7.4.1 and 7.4.2). However, the role of these transport pathways in the control of cell volume and pH in normal, mature erythrocytes appears to be small. The degree of activation of potassium/chloride cotransport out of the cell after cell swelling is so modest that mature human red cells are unable to restore original volume after an osmotic disturbance (Berkowitz and Orringer 1987). The volume-activated potassium/chloride cotransport is more active in young erythrocytes, and may play a role in the decrease of volume of young erythrocytes during erythrocyte maturation (Brugnara and Tosteson 1987).

In human red cells with abnormal haemoglobins C (CC cells) and S (SS cells), the volume sensitive potassium/chloride cotransport is much more active after osmotic swelling than in normal erythrocytes, being ca. 20 mmol l^{-1} cells h^{-1} for SS and CC red cells, and 2–10 mmol l^{-1} cells h^{-1} for normal erythrocytes (Brugnara et al. 1986; Canessa et al. 1986; Berkowitz and Orringer 1987). It is probable that the volume-sensitive potassium/chloride cotransport is responsible for the reduced potassium and water content of these abnormal erythrocytes (see Brugnara et al. 1986).

8.2.2 Ruminant Red Cells

Much of the information on the volume-sensitive potassium/chloride cotransport stems from work on low-potassium ruminant red cells, especially those of sheep (see Sect. 7.4.2). In low-potassium ruminant red cells, the requirement for a volume-activated potassium-chloride cotransport may stem from the need to protect the cells against an osmotic shock, because of the low activity of the sodium pump (see Ellory and Tucker 1983). Very small increases in the relative cell volume (e.g. from 1.00 to 1.05) are adequate to increase the potassium efflux from LK sheep red cells fivefold (Dunham and Ellory 1981). The volume-sensitive potassium/chloride cotransport also plays a major role in the reduction of cell volume during transition from the reticulocyte to the mature red cell in sheep (Lauf and Bauer 1987).

8.2.3 Carnivore Red Cells

In carnivore red cells, which lack the sodium pump, there is still an electrochemical gradient of sodium into the cell (Parker 1977). In the absence of sodium pump, this gradient is generated by extrusion of sodium ions via the sodium/calcium exchange coupled to calcium pump (see Sects. 7.4.4 and 7.5.2). This secondarily active sodium/calcium exchange thereby prevents colloid osmotic swelling (for a review see e.g. Parker 1977).

Volume regulation in dog red cells involves sodium/proton exchange in addition to the sodium/calcium exchange. In shrunken cells sodium fluxes via the sodium/proton exchange pathway increase markedly, from ca. 20 mmol Na^+ kg^{-1} dry weight of red cells h^{-1} in cells with a water content of 2 kg kg^{-1} dry weight of the red cells to > 300 mmol Na^+ kg^{-1} dry weight of red cells h^{-1} in cells shrunken to a water content of 1.4 kg kg^{-1} dry weight of red cells. Thus, the cell volume is regulated by regulating the activity of the sodium/proton exchange. In swollen cells, the net sodium efflux via the sodium/calcium exchange is greater than the net sodium influx via the sodium/proton exchange, whereby sodium and osmotically obliged water leave the cell, shrinking the cells back towards the original volume. In contrast, in shrunken cells the net sodium influx via the sodium/proton exchange pathway far exceeds the efflux via the sodium/calcium exchange, whereby the cells gain sodium and osmotically obliged water, and swell back towards the steady-state volume (for details see Parker 1977, 1983a, 1986).

Because of the coupling between the sodium and proton movements, the net influx of sodium is associated with a net efflux of protons. Thus, the system may be used in pH regulation of the red cells. Indeed, Parker (1986) has shown that, in addition to cell shrinking, acidification of the cytoplasma activates the exchanger. Furthermore, the proton effects can overwhelm the influences of cell volume (Parker 1986). The maximal fluxes observed (i.e. over 300 mmol Na^+ kg^{-1} dry red cells h^{-1}; e.g. Parker 1983a) are similar to those observed in catecholamine-stimulated red cells of rainbow trout (e.g. Borgese et al. 1987). In the latter species, the sodium/proton exchanger plays an important role in the control of intracellular pH (see below). However, significant effects of sodium/proton exchange on the pH gradient across the red cell membrane of dog are not very likely, because the equilibration of protons across the red cell membrane via the Jacobs-Stewart cycle is much more rapid at the body temperature of a mammal (37°C) than of a teleost (15°–20°C).

8.2.4 Rodent Red Cells

Rat red cells possess a volume-activated, furosemide-sensitive sodium and potassium transport pathway. The maximal transport via this pathway was activated tenfold by cell shrinkage (Duhm and Göbel 1984). In normal rat red cells, the net sodium and potassium movements were in an inward direction, thus increasing the red cell volume towards normal.

8.2.5 Lagomorpha Red Cells

In rabbit red cells, cell shrinkage activates sodium/proton exchange (Jennings et al. 1986b; Escobales and Rivera 1987). The exchange can also be activated by acid-loading and is under allosteric control by intracellular protons (Escobales and Rivera 1987). These properties would make the transport pathway suited for control of intracellular pH. However, the large reduction of intracellular pH (to values below 6.5) that is required for full activation of the sodium/proton exchanger (see Escobales and Rivera 1987) is unlikely to occur in physiological situations, precluding the role for this transport system in pH regulation.

8.3 Avian Red Cells

In avian red cells, the sodium/potassium/2 chloride cotransport (see Sect. 7.4.1) is activated after osmotic shrinking (e.g. Kregenow 1971; Schmidt and McManus 1977a) and upon stimulation with catecholamines (e.g. Kregenow 1973;

Gardner et al. 1975; Schmidt and McManus 1977b). This transport pathway allows the net transfer of salt and water to occur in either direction across the red cell membrane, according to the direction and magnitude of the sum of the chemical potential gradients of the involved ions (Haas et al. 1982). In hypertonic medium with a plasma-like composition, the driving force is zero (Schmidt and McManus 1977a), whereby no net movements of the ions and water occur although the transport pathway is activated. RVI (regulatory volume increase) only occurs in the presence of elevated potassium concentration (e.g. Kregenow 1978, 1981). Thus, the importance of the function of sodium/potassium/chloride cotransport in physiological situations is, as yet, uncertain. One possibility is that the cotransport would be used in extrarenal potassium regulation. Since the system responds linearly to extracellular potassium in the physiological range, during transient hypokalemia the cotransport might help mobilize potassium from the red cells to the extracellular medium. Conversely, during hyperkalemia excess extracellular potassium could be taken up by the red cells (see e.g. Haas and McManus 1985).

In osmotically swollen cells, in the absence of catecholamines, a potassium/chloride cotransport pathway is activated (e.g. Kregenow 1981). The net loss of potassium and chloride draws osmotically obliged water out of the cell, and cells shrink towards the original volume. A similar net loss of potassium and chloride is observed if cells are allowed to swell in catecholamine-containing solutions at high extracellular potassium concentration, whereafter the catecholamine and excess potassium are removed (Kregenow 1973). Again, the physiological function of this transport pathway is unknown. If catecholamines are present in the incubation, hypotonicity-induced loss of potassium and chloride is prevented (Haas and McManus 1985), because catecholamine activation of the sodium/potassium/2 chloride cotransport appears to prevent, or override, the swelling-induced potassium/chloride cotransport.

Both regulatory volume increase (activation of sodium/potassium/2 chloride cotransport at high extracellular potassium concentration) and regulatory volume decrease (activation of potassium/chloride cotransport in the absence of catecholamines) are associated with changes in extracellular pH in minimally buffered solutions; RVI is associated with acidification and RVD (regulatory volume decrease) with alkalinization of the medium (Kregenow 1981). In strongly buffered salines the effect of adrenergic stimulation on the pH gradient across bird red cell membrane is small (Nikinmaa and Huestis 1984b). The reason for the pH changes, associated with the cotransport pathways, is depicted in Fig. 7.2 for K/Cl cotransport. Owing to the cotransport, there is a net flux of chloride into or out of the cell. As a result, the anion transporter is momentarily at disequilibrium and will carry out net bicarbonate transport until the distribution ratios for chloride and bicarbonate are the same. Net bicarbonate fluxes in one direction are equivalent to net proton fluxes in the opposite direction. Thus, the extracellular acidification or alkalinization depends on the function of the anion exchange pathway, and any such effect should be abolished by treating the cells with anion exchange inhibitors, SITS or DIDS, as is experimentally observed (Kregenow 1981).

In the primitive bird red cells, the ion transport appears to be different from that in the red cells of adult birds. The membrane potential calculated from chloride distribution ratio across the red cell membrane and the measured membrane potential are different (Baumann and Engelke 1987). Apparently, the conductance of the membrane for protons is high, whereby they are distributed according to the membrane potential. Primitive bird red cells have an ineffective anion exchange, since the treatment of cells with the electrosilent chloride/hydroxyl-ion exchanger tributyltin abolishes the difference between the membrane potential calculated from chloride distribution and measured membrane potential. Thus, chloride distribution may be actively maintained, e.g. by the cotransport (sodium/potassium/2 chloride or potassium/chloride) mechanisms coupled to actively maintained ion gradients. Since the anion exchanger is nonfunctional, there is no linkage between chloride and bicarbonate ions, whereby chloride and proton distribution ratios can be different.

8.4 Reptilian Red Cells

At present, the volume and pH regulation of reptilian red cells has not been studied in detail. However, the data of Maginniss and Hitzig (1987) show that the chloride and proton gradients are markedly different in the turtle, *Chrysemys picta bellii.* These data suggest that the anion exchanger is ineffective, as in primitive bird red cells.

8.5 Amphibian Red Cells

Cala's studies (e.g. Cala 1983b, 1985b) have shown that *Amphiuma* red cells are capable of volume regulation after osmotic disturbances by activating the alkali metal/proton exchanger. Furthermore, his recent studies (Cala et al. 1988) have indicated that the alkali metal/proton exchanges can be induced to function in "pH-regulatory mode" by intracellular acidification. The exchange is operating to some extent even in steady-state volume and pH conditions (Tufts et al. 1987b).

The alkali metal/proton exchanger is activated in the sodium/proton exchange mode by intracellular acidification and cellular shrinking (e.g. Cala 1980, 1983b, 1985a, b; Siebens and Kregenow 1985; Kregenow et al. 1985; Cala et al. 1988).

It is likely that the activation of sodium/proton exchange in response to cellular acidification is a direct effect of protons interacting with the transporter at the transport and modifier sites, as in sodium/proton exchangers from other cell types (for a review see Aronson 1985). As a response to osmotic shrinking, the sodium/proton exchange is activated with a finite lagtime (5–25 min) between osmotic shrinkage and maximal rate of sodium uptake (Siebens and Kregenow 1985). The signal by which osmotic shrinking activates the sodium/proton exchange is not fully clarified. Cala (1980) suggested that it would be due to intracellular acidification associated with osmotic shrinking. Grinstein et al. (1986) observed that in thymic lymphocytes which also show RVI, the sodium/proton exchanger was activated at a higher intracellular pH in shrunken than in normal cells, suggesting that the intracellular modifier site for protons which affects the sodium/proton exchanger was influenced by osmotic shrinkage. Grinstein et al. (1986) hypothesize that the readjustment of the setpoint results from the following sequence of events: (1) shrinking activates a phosphodiesterase which breaks down phosphoinositides, (2) the resultant release of diacylglycerol stimulates protein kinase C, and (3) the kinase phosphorylates the sodium/proton exchanger or a neighbouring protein, which affects the operation of the antiport, shifting the pHi setpoint for the activation of the exchanger. Cala (1986a) has observed that the diacylglycerol-protein kinase C pathway is capable of activating the alkali metal/proton exchanger also in Amphiuma red cells. Thus, it is possible that the activatory step is similar to that in lymphocytes.

The operation of sodium/proton exchange leads to a net efflux of protons, decreasing the pH of a weakly buffered incubation medium by 0.1–0.2 units (e.g. Cala 1980; Kregenow et al. 1985), and increasing intracellular pH. As a result, the intracellular bicarbonate concentration increases, creating a net flux of bicarbonate out of the cell and a net flux of chloride into the cell. The extruded protons and bicarbonate combine in the extracellular medium to form carbon dioxide and water. Carbon dioxide then diffuses into the cell and is hydrated to protons and bicarbonate. Protons and bicarbonate continue to be extruded via the sodium/proton exchanger and chloride/bicarbonate exchanger as long as the sodium/proton exchanger is active. The RVI is a result of continuous accumulation of sodium and chloride into the cell (see Fig. 8.4). The detailed ionic events occurring after the activation of sodium/proton exchange in Amphiuma red cells are essentially the same as those occurring in adrenergically stimulated red cells of teleost fish. These are discussed in detail in Section 8.6.1.

When the exchanger is activated by cell swelling, it functions in the potassium/proton exchange mode (e.g. Cala 1986b), causing an increase in the extracellular pH and a net efflux of potassium and chloride (see Fig. 8.4).

Catecholamines, which are involved in the activation of pH- and volume-regulating ion exchange pathways in teleost fish (see below) and in birds (see Sect. 8.3), do not affect the red cell function of amphibians in physiological conditions (see Tufts et al. 1987a,b); neither stress nor β-adrenergic drugs affected the red cell water content or pH gradient in the toad, *Bufo marinus*, or in the aquatic salamander, *Amphiuma tridactylum*. β-adrenergic drugs activate the sodium/proton exchange only if cellular phosphodiesterase is inhibited (e.g. Rudolph and Greengard 1980). Notably, however, all the data gathered thus far are on adult lung/skin-breathing animals.

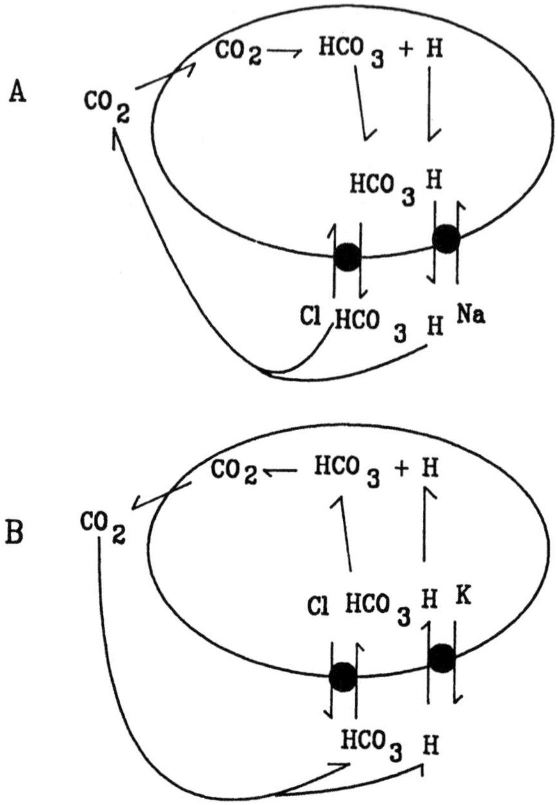

Fig. 8.4 A,B. Volume regulation in *Amphiuma* red cells. *A* In response to osmotic shrinking or acidification, the sodium/proton exchange is activated. The proton extrusion via the sodium/proton exchanger is followed by net bicarbonate efflux via the anion exchange pathway. As a result, sodium and chloride enter the cell, and the cell swells (RVI). Protons and bicarbonate combine to form carbon dioxide and water in the extracellular medium. Carbon dioxide enters the cell and is hydrated to bicarbonate and protons, which are, again, extruded in exchange for chloride and sodium. *B* In response to cell swelling, the potassium/proton exchange is activated. Following the proton influx via the potassium/proton exchange, bicarbonate enters the cells in exchange for chloride. The net potassium and chloride efflux cause the cell shrinkage (RVD). The consequent steps are a reversal of the RVI

8.6 Fish Red Cells

8.6.1 Teleost Red Cells

Fugelli (1967) showed that the red cells of flounder (*Platichtys flesus*) could shrink back to original volume after osmotic swelling, i.e. they exhibited RVD. Associated with this change was a net efflux of ninhydrin-positive substances. Fugelli and Zachariassen (1976) and Fugelli and Rohrs (1980) later showed that the RVD was associated with a decrease in cellular taurine, γ-aminobutyric acid, potassium and chloride. Cala (1977) showed that the red cells of winter flounder

(*Pseudopleuronectes americanus*) shrank after osmotic swelling and swelled after osmotic shrinking, thus showing both RVD and RVI in synthetic medium. RVD was characterized mainly by a loss of potassium from the cell, followed by osmotically obliged water. RVI was characterized by an uptake of sodium, potassium and chloride. It was further shown by Cala (1977) that the uptake of potassium during RVI appeared to be due to the activation of the sodium pump as a response to the increased sodium permeability. Thus, both organic and inorganic effectors are involved in the regulation of cell volume after osmotic disturbances.

The volume-sensitive pathways for the transport of taurine, the most important of the organic osmotic effectors in teleost red cells, have been studied in detail by Fugelli and Thoroed (1986) and Fincham et al. (1987). Two volume-sensitive pathways appear to be present: the first, a saturable, high-affinity, sodium-dependent β-amino-acid transport system (see Sect. 6.2.1) which is responsible for the very high (25–45 mmol l^{-1} cell water; Fincham et al. 1987) taurine concentrations in eel (*Anquilla japonica*) and starry flounder (*Platichthys stellatus*) red cells. The activity of this transport system decreases with decreasing osmolality of the medium. The second pathway for the transport of taurine is sodium-independent, and is activated in hypoosmotic medium. Consequently, there is a net efflux of taurine from the cells in reduced osmolality (Fugelli and Thoroed 1986; Fincham et al. 1987). Quantitatively, the decrease in cellular organic osmolyte levels accounts for ca. 45% of the total intracellular osmolality reduction.

Another pathway important in RVD appears to be the volume-sensitive potassium/chloride cotransport, demonstrated for the teleost, toadfish (*Opsanus tau*) by Lauf (1982). When the red cells of the toadfish were placed in hypotonic environment, they lost potassium and chloride, and shrank back to the preperturbed volume within 1.5 h. RVD did not occur if the cells were placed in a high potassium medium, which abolished the driving force for potassium efflux, or if nitrate, iodide or thiosulphate was substituted for chloride. Also, RVD was inhibited by furosemide and DIDS. These experiments could not, however, exclude the possibility that coupled potassium/proton and chloride/bicarbonate exchanges would be responsible for the potassium and chloride loss. Similarly, Bourne and Cossins (1984) showed that cell swelling increased net potassium efflux (both unidirectional potassium influx and efflux increased) from rainbow trout red cells. The volume-sensitive potassium fluxes in this case were also inhibited by furosemide and SITS.

The pathways involved in the RVI in the absence of catecholamines are less well characterized. The accumulation of sodium and chloride in the red cells of winter flounder (Cala 1977) suggests that sodium permeability may be volume-sensitive. Bourne and Cossins (1984) showed that furosemide-sensitive sodium influx was sensitive to cell volume. The transport increased in shrunken cells.

Catecholamines cause red cell swelling in various fish erythrocytes, as first observed by Nikinmaa (1982b) for trout red cells. Subsequently, adrenergic volume changes have been observed, e.g. in striped bass (*Morone saxatilis*; Nikinmaa and Huestis 1984b), carp (*Cyprinus carpio*; Salama and Nikinmaa 1988; Fuchs and Albers 1988), whitefish (*Coreqonus pallasi*; Salama and Nikinmaa 1989) and pikeperch (*Stizostedion lucioperca*; Salama and Nikinmaa 1989) red cells.

As discussed in Section 7.4.3, catecholamines activate sodium/proton exchange across the red cell membrane of many teleost fish. The activation of the exchanger leads to intracellular alkalinization and extracellular acidification, as first observed by Nikinmaa (1982b) and later confirmed by Nikinmaa (1983b); Nikinmaa and Huestis (1984b); Baroin et al. (1984a); Cossins and Richardson (1985); Borgese et al. (1986); and Heming et al. (1987). See also the review by Nikinmaa (1986b).

The mechanism of adrenergic volume and pH changes is depicted in Fig. 8.5. The sequence of ionic events occurring after the activation of sodium/proton exchange is essentially the same in *Amphiuma* red cells:

1. The sodium/proton exchange is activated (Nikinmaa and Huestis 1984b; Baroin et al. 1984a; Cossins and Richardson 1984, 1985).

2. Activation of the sodium/proton exchange causes a net efflux of protons. The removal of protons from the cells shifts the reaction $CO_2 + H_2O \rightleftharpoons H^+ + HCO_3^-$ to the right, whereby more bicarbonate is formed. Bicarbonate accumulates within the cell because the catalyzed reaction between carbon dioxide, bicarbonate and protons within the red cell is more rapid than the anion exchange. Thus, intracellular pH and bicarbonate concentration increase and extracellular pH decreases.

3. The net proton efflux depends on the relative rates of proton extrusion via the sodium/proton exchanger, and passive influx of protons. Net proton efflux occurs as long as the passive fluxes of protons (or acid equivalents) back into the red cell are smaller than the efflux of protons via the sodium/proton exchanger. Because of this, intra- and extracellular pH changes should be diminished by speeding up passive proton (or hydroxyl ion) movements. Indeed, if salmonid

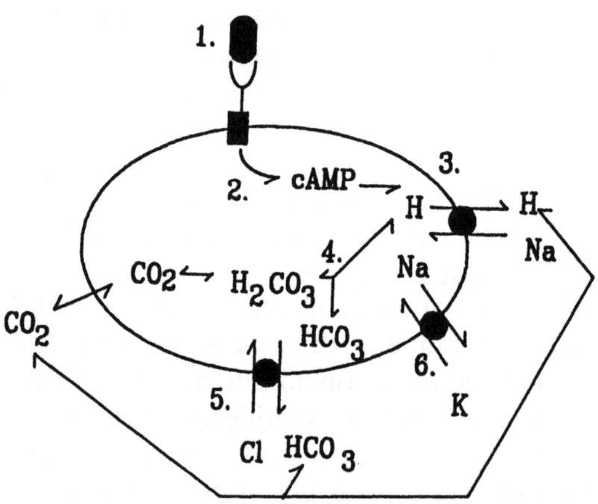

Fig. 8.5. The events occurring during adrenergic stimulation of teleost red cells. *1* Binding of β-adrenergic agonists to receptor activates adenylate cyclase, whereby *2* cellular cAMP levels increase. *3* The sodium/proton exchange is activated and increases intracellular pH. As a result, *4* more bicarbonate is formed from carbon dioxide, and the bicarbonate ratio across the red cell membrane increases. *5* Thereafter, chloride enters the cell in exchange for bicarbonate. *6* The increased intracellular sodium concentration increases the activity of the sodium pump. For further details, see text

red cells were treated with a protonophore, 2,4-DNP, which speeds up passive proton movements, adrenergic stimulation had no effect on the pH-gradient across the red cell membrane. Also, when rainbow trout red cells were treated with tributyltin chloride, which permits chloride/hydroxyl ion exchange, adrenergic stimulation did not cause a decrease in pH gradient across the red cell membrane. In both cases the sodium/proton exchanger was activated by adrenergic stimulation (Nikinmaa, Tiihonen and Paajaste, unpublished data). Similarly, Borgese et al. (1986) have shown that when trout red cells are treated with tripropyltin chloride, adrenergic stimulation does not cause an increase in cellular volume.

The passive proton flux consists of three components: the conductive movements of acid equivalents, the Jacobs-Stewart cycle (Jacobs and Stewart 1942) and the chloride/hydroxyl ion exchange via the anion exchange pathway. In physiological situations the most important of these dissipative pathways is the Jacobs-Stewart cycle. This also applies to in vitro incubations of red cells in which they are equilibrated with air; sufficient carbon dioxide and bicarbonate is present to "fuel" the Jacobs-Stewart cycle (see Lew and Bookchin 1986). However, in nitrogen-equilibrated blood the chloride/hydroxyl ion exchange may be the dominant proton-equilibrating pathway across the red cell membrane (see e.g. Borgese et al. 1986).

Given that the Jacobs-Stewart cycle is the predominant passive pathway for protons across the red cell membrane, the net proton efflux should be diminished by speeding up the slowest step of the Jacobs-Stewart cycle, the uncatalysed formation of carbon dioxide from bicarbonate and protons. This statement has also been experimentally verified: when we added 100 mg l^{-1} carbonic anhydrase (specific activity 2 500 W-A units per mg protein) to a suspension of rainbow trout red cells, adrenergic stimulation did not affect the pH gradient across the red cell membrane (Nikinmaa, Tiihonen and Paajaste, unpublished data).

4. The degree of external acidification for a given efflux of protons depends on the external buffers present. If red cells are air-equilibrated in media buffered with Tris or HEPES, the changes in pH are essentially a function of the buffer capacity of these substances. If bicarbonate/carbon dioxide buffer is used, the adrenergic activation of sodium/proton exchange initially causes a marked drop in extracellular pH. This is due to the fact that, initially, no carbon dioxide can be formed from carbonic acid, owing to the slow uncatalyzed dehydration reaction. Because carbonic acid is a relatively strong acid with a pK value of 3–3.5 (e.g. Truchot 1987), carbonic acid/bicarbonate buffer cannot buffer the protons excreted. The buffering only proceeds with the uncatalyzed hydration rate of carbonic acid, and, therefore, an increase is expected in the extracellular pH after the initial drop. This suggestion has been experimentally verified by Motais et al. (1989).

5. The chloride influx observed is secondary to the accumulation of bicarbonate within the cell. Thus, any treatment that prevents the formation of initial disequilibrium for bicarbonate ions should prevent the chloride influx. Bicarbonate accumulation is prevented if intracellular formation of bicarbonate and protons is slower than the anion exchange. No chloride accumulation occurred in adrenergically stimulated salmonid red cells, if they were treated

with the carbonic anhydrase inhibitor, acetazolamide, before the stimulation (Nikinmaa, Tiihonen and Paajaste, unpublished data).

6. The increase in intracellular pH depends, in addition to the activity of the sodium/proton exchange, on the buffering properties of intracellular buffers, mainly haemoglobin, and on the availability of carbon dioxide for intracellular hydration. A significant complicating factor in this regard is the marked Haldane effect of teleost haemoglobins (see e.g. Jensen and Weber 1985a; Jensen 1988). Net proton extrusion will tend to increase intracellular pH, whereby the oxygen affinity of the haemoglobin increases. This, in turn, will tend to increase the proportion of oxyhaemoglobin within the cell. Up to four protons are liberated by one molecule of tench haemoglobin (Jensen and Weber 1985a) upon oxygenation, whereby intracellular pH tends to decrease. The role of carbon dioxide, available for intracellular hydration, is also important. This is shown by our acetazolamide experiments (Nikinmaa, Tiihonen and Paajaste, unpublished data). When the cells were treated with the carbonic anhydrase inhibitor, acetazolamide, and thereafter adrenergically stimulated, the pH changes were accentuated. This finding shows that the rapid hydration of intracellular carbon dioxide to bicarbonate and protons can significantly limit the increase in intracellular pH. As a consequence, if experiments are carried out with a constant input of CO_2, most of the protons excreted will be replenished by the rapid hydration reaction, and the effect of adrenergic stimulation on the intracellular pH will be small. If, on the other hand, experiments are carried out in conditions in which there is a net loss of carbon dioxide to the environment, protons and bicarbonate cannot be readily formed from carbon dioxide, whereby the increase in intracellular pH will be greater. The in vivo data by Milligan and Wood (1986) support this suggestion. Physical disturbance caused about 0.5 unit decrease in both ventral and dorsal aortic plasma pH. However, the red cell pH of dorsal aortic blood (i.e. blood that had just passed the gills, in which a net loss of carbon dioxide occurs) was not affected, whereas the red cell pH of ventral aortic blood (i.e. blood that had flowed from the tissues in which net carbon dioxide influx occurs) decreased by 0.25 pH units.

7. The net flux of protons stops when the apparent proton fluxes via the sodium/proton exchanger and the Jacobs-Stewart cycle become equal. However, both protons and bicarbonate continue to cycle across the red cell membrane, causing intracellular sodium and chloride concentrations to increase at a rate determined by the flux via sodium/proton exchanger.

8. The upper steady-state volume reached depends on two factors. First, the sodium/proton exchanger appears to be self-inactivated by a process dependent on external sodium concentration (Garcia-Romeu et al. 1988). Second, an increase in volume activates the potassium/chloride cotransport pathway (see Cossins and Richardson 1985), with net potassium and chloride effluxes similar to RVD.

9. Since the ionic exchanges involved are electrically neutral, changes in the membrane potential should not be required for the adrenergic response. We (Nikinmaa, Tiihonen and Paajaste, unpublished data) have confirmed this prediction using "voltage-clamped" rainbow trout red cells. The cells were treated with valinomycin, and potassium ions were allowed to equilibrate. After

potassium ions had reached equilibrium, the cells were adrenergically stimulated. The membrane potential did not change after adrenergic stimulation, as shown by the unchanged extracellular TPP^+ concentration. However, the cells swelled and their chloride ratio increased.

10. Secondary to the adrenergic activation of sodium/proton exchange, and the increase in intracellular sodium, the activity of the sodium pump increases (see e.g. Bourne and Cossins 1982). Since the sodium pump consumes up to 25% of the total energy consumption of erythrocytes (see Sect. 6.1), the 250% increase in the activity of sodium pump, associated with adrenergic stimulation, leads to a marked increase in the ATP consumption, whereby cellular NTP levels drop (Nikinmaa 1983b; Milligan and Wood 1987; Ferguson and Boutilier 1988). During the initial stages of adrenergic stimulation, the drop represents almost completely the hydrolysis of ATP to ADP, as shown by the data of Tetens (1987; see Fig. 6.4). The sum of ATP and ADP remains constant for 80 min during adrenergic stimulation.

The effects of catecholamines on the red cell pH of teleost fish have been demonstrated in exercised striped bass (Nikinmaa et al. 1984) and rainbow trout (Primmett et al. 1986; Milligan and Wood 1986), in acid-infused trout (Boutilier et al. 1986; Tang et al. 1988), in hypoxic trout (Tetens and Christensen 1987), carp (Nikinmaa et al. 1987a) and hypercapnic trout (Vermette and Perry 1988). However, there are marked species differences in the magnitude of the adrenergic response. Jensen (1987) observed that neither exercise nor adrenaline injection elicit a red cell response in normoxic tench, *Tinca tinca*; Milligan and Wood (1987) showed that exercise does not affect the red cell function in the flatfish, *Platichtys stellatus*; and Hyde et al. (1987) observed that aerial exposure does not induce a preferential maintenance of red cell pH in the American eel, *Anguilla rostrata*. Furthermore, Nikinmaa et al. (1987b) showed that the adrenergic sodium accumulation in hypoxic carp was critically dependent on the arterial oxygen tension, observed only if the dorsal aortic oxygen tension dropped below 10 mmHg. This result shows that the dependence of the activity of the sodium/proton exchanger on the oxygen tension is an important modulator of the in vivo responses.

8.6.2 Elasmobranch Red Cells

Only fragmentary information is available on the volume regulatory behaviour of elasmobranch red cells. Boyd et al. (1977) have shown that when the skate, *Raja erinacea*, is acclimated to dilute seawater, the concentrations of β-amino-acids, β-alanine and taurine decrease. Goldstein and Boyd (1978) showed that the decrease of β-alanine in hypotonic medium is due to an increased efflux, whereas influx remained unaffected. As in the teleosts, the efflux of β-amino acids was sodium-independent, whereas sodium was required for the influx component. Recent data by Leite and Goldstein (1987) and McConnell and Goldstein (1988) suggest that the activation of β-amino acid efflux in hypotonic media may be controlled by the diacylglycerol-protein kinase C pathway. Taurine efflux from skate red cells was increased by both the calcium ionophore

A23187 and the phorbol ester, phorbol 12-myristate 13-acetate. Furthermore, hyposmotic shock was associated with changes in the diacylglycerol content of the cells.

Catecholamines appear not to influence the behaviour of elasmobranch red cells. Tufts and Randall (1988) observed that isoproterenol had no effect on either the red cell water content or on the pH gradient across the red cell membrane in the elasmobranchs, *Squalus suckleyi* and *Raja binoculata*.

8.6.3 Agnathan Red Cells

The most significant difference between agnathan red cells and the red cells of other vertebrate groups so far studied is the lack of the anion exchange pathway in agnathan red cells. Ellory et al. (1987) and Nikinmaa and Railo (1987) have shown that the equilibration of chloride across the red cell membrane of the hagfish, *Eptatrerus stouti*, and the river lamprey, *Lampetra fluviatilis*, respectively, requires several hours. Nikinmaa and Railo (1987) further showed that the red cell membrane is extremely impermeable to extracellularly added protons and hydroxyl ions (i.e. extracellular acid or alkaline loads are not buffered by intracellular buffers; see Fig. 8.6). Tufts and Boutilier (1989) have shown, by incubating *Petromyzon marinus* red cells at different carbon dioxide tensions, that the cells are also impermeable to bicarbonate.

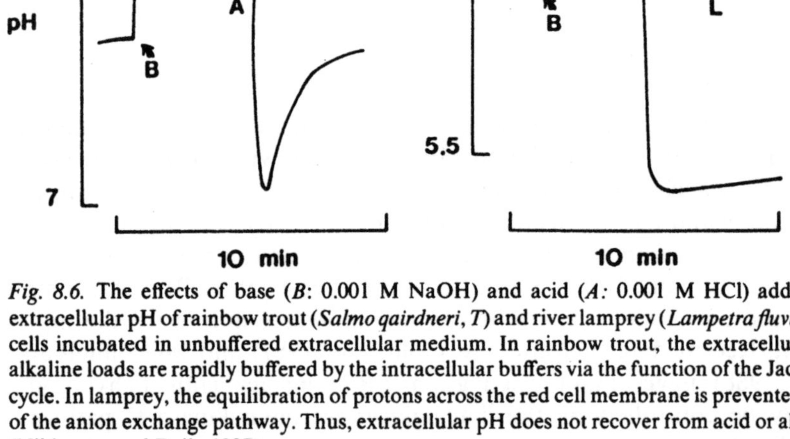

Fig. 8.6. The effects of base (*B*: 0.001 M NaOH) and acid (*A*: 0.001 M HCl) addition on the extracellular pH of rainbow trout (*Salmo qairdneri, T*) and river lamprey (*Lampetra fluviatilis, L*) red cells incubated in unbuffered extracellular medium. In rainbow trout, the extracellular acid and alkaline loads are rapidly buffered by the intracellular buffers via the function of the Jacobs-Stewart cycle. In lamprey, the equilibration of protons across the red cell membrane is prevented by the lack of the anion exchange pathway. Thus, extracellular pH does not recover from acid or alkaline loads (Nikinmaa and Railo 1987)

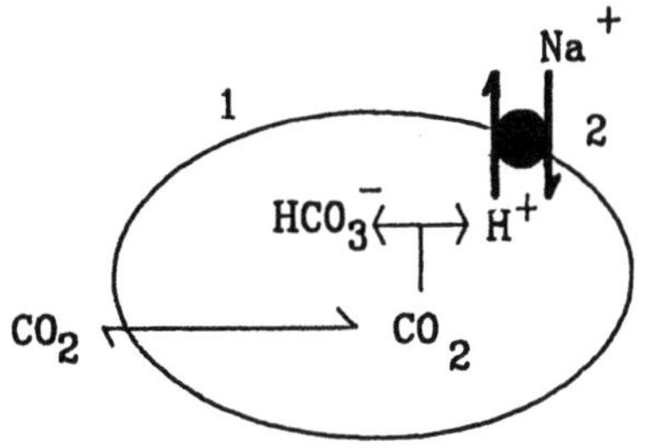

Fig. 8.7. pH regulation in lamprey (*Lampetra fluviatilis*) red cells. In the absence of the anion exchange pathway (*1*), the sodium-dependent acid extrusion (*2*) is the most important proton transporting pathway

The lack of anion exchange pathway has important implications, both in the control of red cell volume and red cell pH. The volume regulatory behaviour of agnathan red cells has been little studied. Tufts, Nikinmaa and Boutilier (unpublished data) have observed that the red cells of *Myxine glutinosa* do not show regulatory volume decrease after osmotic swelling. In contrast, *Lampetra fluviatilis* red cells shrink towards the original volume after swelling in hypotonic medium. In addition, the changes in red cell volume and ion concentrations, associated with oxygenation-deoxygenation cycles of haemoglobin in other vertebrates, are not expected to take place in lamprey red cells, since rapid movements of small anions across the red cell membrane do not take place.

The mechanisms of pH regulation in cyclostome red cells have so far been studied in detail only using *Lampetra* red cells. Nikinmaa (1986a) showed, using the protonophore, 2,4-dinitrophenol and the metabolic inhibitor potassium cyanide, that the proton gradient across *Lampetra* red cell membrane was actively maintained: intracellular pH decreased when the red cells were treated with either of the two compounds. Nikinmaa (1986a) and Nikinmaa et al. (1986) further showed that the pH-regulation required sodium: in the absence of sodium, the steady-state intracellular pH decreased, and cellular pH did not recover after acidification induced by the ammonium chloride prepulse technique described by Thomas (1984). Furthermore, the recovery from acidification was slowed down by amiloride, the commonly used inhibitor of the sodium/proton exchange. Thus, it appears that the intracellular pH in *Lampetra* red cells is maintained by sodium/proton exchange (see Fig. 8.7). However, the data to date cannot exclude the possibility that the sodium-dependent acid extrusion would be bicarbonate-dependent.

In the absence of rapid proton equilibration across the red cell membrane, and in the presence of sodium-dependent acid extrusion, the red cell pH in lamprey is often higher than the extracellular pH (Nikinmaa and Weber 1984; Mattsoff and Nikinmaa 1988). Since extracellular acid loads cannot be buffered by intracellular buffers in blood, a given acid load causes a much more marked drop in the plasma pH of lamprey than in the plasma pH of teleost fish (Mattsoff and Nikinmaa 1988).

Somewhat different results were obtained by Tufts and Boutilier (1989), working on *Petromyzon marinus*. 2,4-dinitrophenol did not affect the steady-state proton distribution in *Petromyzon* red cells. However, the responses of *Petromyzon* red cells to acid loads have not been studied as yet. Thus, it has not been shown conclusively whether these cells can actively respond to pH changes or not.

Chapter 9

Carbon Dioxide Transport

Several reviews are available on different aspects of blood CO_2 transport in various vertebrates (e.g. Roughton 1964; Bauer 1974; Wagner 1977; Albers 1985; Perry 1986; Klocke 1987, 1988; Bidani and Crandall 1988). Carbon dioxide excretion is a passive phenomenon in which metabolically produced CO_2 is transferred down an electrochemical gradient from the site of production to the environment. Initially, carbon dioxide diffuses from metabolizing tissue to the blood. It is then transported in the blood stream to the gas exchange organs, and thereafter diffuses to the environment. The amount of carbon dioxide transported from tissues to the gas exchangers per unit of time is a function of blood flow velocity (minute volume) and blood carbon dioxide content (capacitance coefficient × carbon dioxide tension).

Carbon dioxide can be transported as molecular CO_2, as bicarbonate (and carbonate) or as carbamino compounds. At physiological pH values most of the carbon dioxide in the blood is in the form of bicarbonate (see Tables 9.1 and 9.2). Since blood is a two-compartment system consisting of plasma and red cells, there is a characteristic distribution of the different forms of carbon dioxide between plasma and red cells. In addition, the following properties of red cells markedly affect the blood carbon dioxide transport: intracellular carbonic anhydrase facilitates the hydration/dehydration reactions of CO_2; intracellular buffers, most importantly haemoglobin, take up or release protons formed/consumed in the interconversion between carbon dioxide and bicarbonate; and the uncharged, free amino groups of intracellular proteins, mainly haemoglobin, form carbamino compounds with carbon dioxide. The role of red cells in blood carbon dioxide transport is treated in more detail in the following sections.

9.1 Distribution of Total Carbon Dioxide Content Between Red Cells and Plasma

Chloride and bicarbonate are the major exchangeable anions across the red cell membrane. The concentration of these ions in the intracellular compartment of most vertebrates is lower than in the extracellular compartment. Since bicarbonate is the major species of carbon dioxide in blood at physiological pH values, the total carbon dioxide content of red cells is lower than that of plasma. The difference between red cells and plasma is smaller than expected solely from the

Table 9.1. In vivo carbon dioxide content in the red cells and plasma of man (data from Roughton 1964 and Klocke 1987)

	Venous	Arterial
Plasma		
pCO_2	46 mmHg	40 mmHg
Dissolved CO_2	1.47 mmol l^{-1}	1.29 mmol l^{-1}
HCO_3^-	27.1 mmol l^{-1}	25.1 mmol l^{-1}
Carbamino CO_2	0.5 mmol l^{-1}	0.5 mmol l^{-1}
Total CO_2	29.1 mmol l^{-1}	26.9 mmol l^{-1}
Red cells		
pCO_2	46 mmHg	40 mmHg
Dissolved CO_2	1.24 mmol l^{-1}	1.07 mmol l^{-1}
HCO_3^-	13.5 mmol l^{-1}	12.9 mmol l^{-1}
Carbamino CO_2	3.81 mmol l^{-1}	2.43 mmol l^{-1}
Total CO_2	18.6 mmol l^{-1}	16.4 mmol l^{-1}
Whole blood		
Total CO_2	24.1 mmol l^{-1}	21.9 mmol l^{-1}
Haematocrit	0.451	0.448

Table 9.2. The in vivo carbon dioxide content in the plasma of rainbow trout at 10°C, pHe ca. 7.9. Data are from Heming (1984)

	Arterial	Venous
Plasma		
pCO_2	2.3 mmHg	3.0 mmHg
Dissolved CO_2	0.12 mmol l^{-1}	0.16 mmol l^{-1}
HCO_3^-	7.93 mmol l^{-1}	9.01 mmol l^{-1}

distribution of exchangeable anions, because carbamino formation, discussed in Section 9.3, is greater in red cells than in plasma.

The red cells of *Agnatha*, e.g. those of *Lampetra fluviatilis*, are a notable exception. In these animals, bicarbonate (and chloride) does not permeate red cell membrane (Nikinmaa and Railo 1987), and the red cell pH is often higher than the extracellular pH (Nikinmaa 1986a; Nikinmaa et al. 1986). As a result, the intracellular concentration of bicarbonate is often higher than the extracellular concentration.

9.2 Carbon Dioxide – Bicarbonate Equilibria

9.2.1 Uncatalyzed Interconversion Between Carbon Dioxide and Bicarbonate

Interconversion of carbon dioxide and bicarbonate can be achieved by the following hydration/dehydration reaction schemes (see e.g. Klocke 1987):

$$CO_2 + H_2O \leftrightharpoons H_2CO_3 \leftrightharpoons H^+ + HCO_3^- \quad (9.1)$$
$$CO_2 + H_2O \leftrightharpoons H^+ + HCO_3^-, \quad (9.2)$$

and direct combination of hydroxyl ions with carbon dioxide

$$CO_2 + OH^- \leftrightharpoons HCO_3^-. \quad (9.3)$$

The relative importance of the hydration/dehydration pathways and the hydroxyl pathway depend on the pH, the higher the pH the greater the role of hydroxyl pathway. At pH 7.2 (37 °C) its contribution is ca. 6%, at pH 7.4 ca. 9% and at pH 7.6 ca. 14% (Klocke 1987). The uncatalyzed hydration/dehydration rate is slow; the half-time for HCO_3^- dehydration is ca. 6 min at 1 °C, about 30 s at 25 °C, and 10–15 s at 37 °C (see Fig. 9.1). Since the residence time of blood in the capillaries of lungs and gills is 0.3–3 s (e.g. Wagner 1977; Hughes et al. 1981), the uncatalyzed reaction is clearly too slow to allow for efficient conversion of bicarbonate to carbon dioxide in the respiratory epithelium.

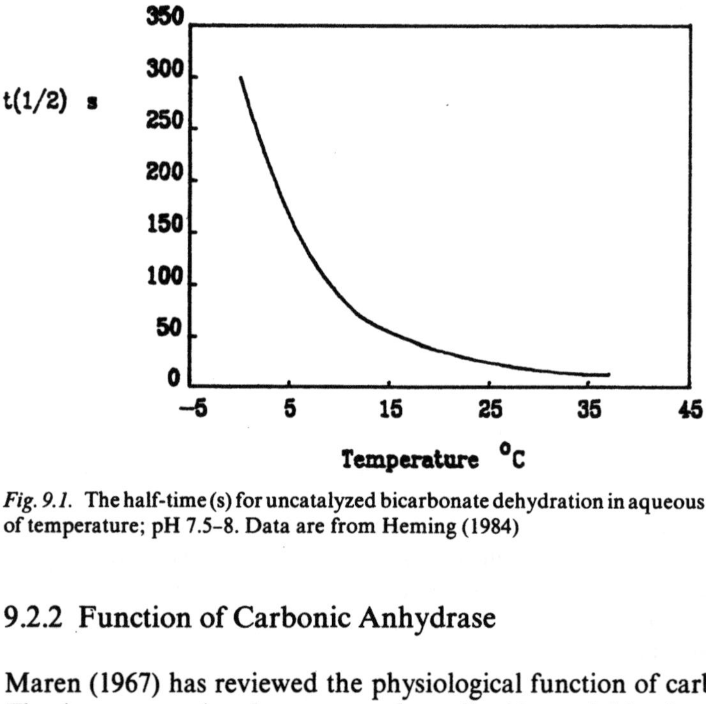

Fig. 9.1. The half-time (s) for uncatalyzed bicarbonate dehydration in aqueous solution as a function of temperature; pH 7.5–8. Data are from Heming (1984)

9.2.2 Function of Carbonic Anhydrase

Maren (1967) has reviewed the physiological function of carbonic anhydrase. The interconversion between carbon dioxide and bicarbonate is markedly facilitated by the presence of intracellular carbonic anhydrase. In human red cells, carbonic anhydrase is capable of increasing the speed of bicarbonate dehydration by 13 000 times (see e.g. Klocke 1987). Carbonic anhydrase activity has been observed in all red cells studied, including those of the agnathans, *Myxine* (Carlsson et al. 1980) and *Lampetra fluviatilis* (Nikinmaa et al. 1986), and elasmobranchs (e.g. Maynard and Coleman 1971; Bergenhem et al. 1986; Swenson and Maren 1987). High activity of carbonic anhydrase is also found in salivary glands, gastric mucosa, gills, the eye and kidney. However, the enzyme is present in most tissues, including both muscle and liver (e.g. Maren 1967; Klocke 1987; Swenson and Maren 1987; Gros and Dodgson 1988).

Three isozymes, carbonic anhydrase I, II and III, have been described in man. All the isozymes have a molecular weight of ca. 30 000, and contain one zinc atom as the active group. The three-dimensional structure has been determined for human isozymes I and II, which are present in the red cells. In both molecules, the active zinc is situated in a 1.2 nm deep pocket in the enzyme, and is liganded to the protein through three histidine groups (Liljas et al. 1972; Kannan et al. 1975). The fourth coordination position of zinc is occupied by water or hydroxyl ion (e.g. Silverman and Vincent 1983).

Isozyme II has the highest activity: the rate constant for the conversion of CO_2 to bicarbonate is $7 \times 10^5\ s^{-1}$ for isozyme II, $4 \times 10^4\ s^{-1}$ for isozyme I and $4 \times 10^3\ s^{-1}$ for isozyme III (Khalifah 1971; Sanyal et al. 1982). The difference between isozymes I and II may be correlated with the presence of two additional histidines close to the active site in isozyme I, which decrease the size of the active site of isozyme I as compared to isozyme II. In addition, His 200 interacts with the ligands of the zinc (Silverman and Vincent 1983). Many mammalian red cells contain carbonic anhydrase I and II, but the red cells of ruminants, dogs and dolphins only contain the high activity isozyme II (e.g. Carter 1972).

Of the nonmammalian species studied, the turtle, *Malaclemys terrapin centrata* (Hall and Schraer 1979), has both a low-affinity and a high-affinity carbonic anhydrase isozyme in the red cells. In contrast, the agnathan, *Myxine glutinosa*, and the elasmobranch, *Squalus acanthias* only have a low-affinity isozyme within the red cell (Maren et al. 1980), and the teleosts, *Archosargus probatocephalus, Oncorhynchus gorbushka* and *Salmo gairdneri* (Sanyal et al. 1982; Kim et al. 1983; Hall and Schraer 1983), the amphibian, *Rana catesbeiana* (Bundy and Cheng 1976), and the birds, *Gallus domesticus* (Bernstein and Schraer 1972) and *Meleagris gallopavo* (Lemke and Graf 1974), appear to have only the high-activity enzyme, with a somewhat smaller turnover rate than that of the high activity isozyme of man (see Table 9.3 for rate constants of the carbonic anhydrases of selected vertebrates). Similar to the mammalian high

Table 9.3. The approximate turnover numbers (s^{-1}) of red cell carbonic anhydrases from selected vertebrates (pH 7.5–7.6)

Species	k_{cat} (s^{-1})	Source
Man isozyme I	29 000 (1°C) 52 000 (25°C)	Sanyal and Maren 1981
Man isozyme II	236 000 (1°C) 860 000 (25°C)	
Gallus domesticus (Aves)	100 000 (4°C) 500 000 (25°C)	Bernstein and Schraer 1972
Archos. probatocephalus (Teleosta)	77 000 (1°C) 140 000 (25°C)	Sanyal et al. 1982
Salmo gairdneri	700 000 (0°C)	Maren et al. 1980
Squalus acanthias (Elasmobranchii)	25 000 (0°C)	
Myxine glutinosa	13 000 (0°C)	

Table 9.4. The inhibition constants of red cell carbonic anhydrases for sulphonamides and anions

Species	Sulfanilamide (μM)	Acetazolamide (μM)	Ethoxzolamide (μM)	Cl^- (mM)	I^- (mM)	Source
Human CAI	50	0.2	0.002	6	0.3	Sanyal 1984
CAII	2	0.01	0.002	200	26	
Chicken	22	0.03	0.002	50	–	Maren and Sanyal 1983
A. probatoc.	5	0.03	0.002	200	9	Sanyal et al. 1982
S. acanthias	90	0.2	0.015	70	9	Maren et al. 1980
M. glutinosa	–	0.01	–	300	–	Carlsson et al. 1980

activity isozyme, the nonmammalian carbonic anhydrases are only weakly inhibited by anions, and the inhibition pattern of the enzymes designated as high-activity enzymes resembles that of human isozyme II (see Table 9.4). The molecular weight of nonmammalian enzyme is similar to that of mammals, and, as in mammals, there is one zinc atom per molecule. The high molecular weight (36 000–39 000) reported for sharks (Maynard and Coleman 1971) appears to result at least partly from disulphide links between the carbonic anhydrase and glutathione (see Bergenhem et al. 1986). An important difference between mammalian and nonmammalian carbonic anhydrase appears to be the requirement of a sulfhydryl-protecting agent for full activity in nonmammalian but not in mammalian enzymes (Kim et al. 1983).

The catalytic mechanism of carbonic anhydrase has been reviewed in detail by Silverman and Vincent (1983); Coleman (1984); and Lindskog et al. (1984). It appears that carbonic acid is not a substrate or product in the reaction – the catalyzed rate of reaction (of at least the high-activity enzyme) is too rapid to be explained by a diffusion-limited bimolecular reaction between the enzyme and the very low concentration of carbonic acid present in physiological fluids (see Silverman and Vincent 1983). In contrast, there are ample amounts of bicarbonate present to account for the high turnover number of the enzyme. However, if bicarbonate is a substrate or a product for carbonic anhydrase, as depicted in Eq. (9.4), the ionization state of the enzyme must change during catalysis:

$$EH^+ + HCO_3^- \rightleftharpoons E + CO_2 + H_2O. \qquad (9.4)$$

For continued dehydration, the enzyme must be able to return to the original protonated state and, for continued hydration, to the unprotonated state. If free protons and hydroxyl ions were to donate protons to or accept protons from the enzyme, their rate of diffusion should be orders of magnitude greater than that observed experimentally. Thus, protonation and deprotonation cannot be achieved by diffusion of free protons and hydroxyl ions. However, the red cell contains high concentration of buffers, which are able to donate or accept the protons. The role of buffers in the catalysis has been shown, e.g. by Jonsson et al. (1976). Jonsson et al. (1976) observed that when the total buffer concentration in the reaction medium was decreased below 5 mM, the initial velocity of the catalyzed hydration of carbon dioxide decreased markedly. At high external buffer concentrations, the rate-limiting step in the catalyzed hydration/dehydration reaction appears to be proton transfer between the active site and an internal group in the enzyme molecule, from which proton transfer between the enzyme and the solvent can then be carried out. The steps that convert carbon dioxide to bicarbonate appear not to be rate-limiting for isozyme II (see Silverman and Vincent 1983).

The catalyzed hydration reaction of carbon dioxide would then involve three separate steps (see Fig. 9.2):

1. A nucleophilic attack of the zinc-bound hydroxide (in the active site of the enzyme) on carbon dioxide to form metal-bound bicarbonate (see Silverman and Vincent 1983; Lindskog et al. 1984). Water then replaces the bicarbonate molecule at the active site, and bicarbonate is liberated in solution.

2. Formation of metal-bound hydroxyl ion from metal-bound water. This reaction step involves an intramolecular transfer of proton between the metal site and a titratable histidine residue via a number of hydrogen-bonded water molecules. In isozyme II, this step limits the maximal rate of catalysis.

3. A rapid proton transfer between the titratable histidine residue of the active site and the buffer molecules of the solvent (see Lindskog et al. 1984).

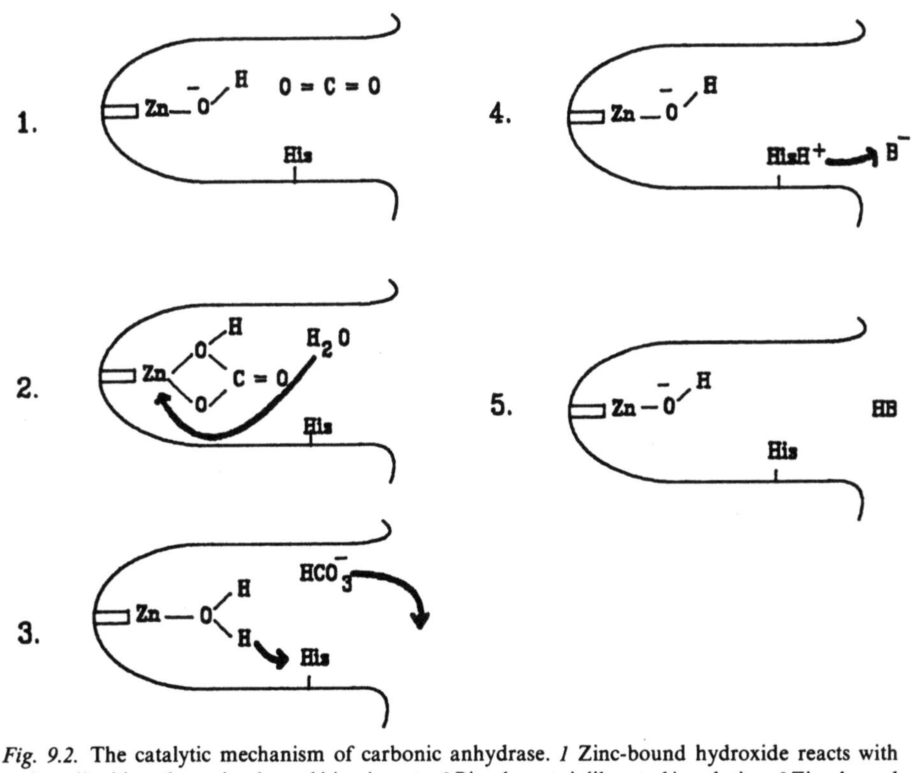

Fig. 9.2. The catalytic mechanism of carbonic anhydrase. *1* Zinc-bound hydroxide reacts with carbon dioxide to form zinc-bound bicarbonate. *2* Bicarbonate is liberated in solution. *3* Zinc-bound water is converted to zinc-bound hydroxyl ion by intramolecular transfer of proton from the metal site to a titratable histidine residue. *4* The proton is removed from the histidine by a buffer molecule. *5* Thereafter the active site is ready for another catalytic cycle. For details, see text

The activity of red cell carbonic anhydrase may be affected by calcitonin and parathyroid hormone. Arlot-Bonnemains et al. (1985) have observed that physiological concentrations of calcitonin increase the activity of carbonic anhydrase isozyme II of human red cells in vitro. In contrast, physiological concentrations of parathyroid hormone decrease its activity.

9.2.3 Role of Buffers

Because the CO_2/bicarbonate system behaves as a weak acid, any addition of carbon dioxide to blood generates significant quantities of protons. In the absence of buffers, the protons produced markedly lower the pH of the solution. This shifts the equilibrium between carbon dioxide and bicarbonate towards carbon dioxide, and little bicarbonate will be formed per unit increase of carbon dioxide tension. Thus, the increase in total blood CO_2 content as a function of carbon dioxide tension is small. If the buffer concentration is increased, pH changes per unit change in carbon dioxide tension are reduced, and formation of bicarbonate increased.

Because the red cell is permeable to carbon dioxide, the internal buffers, notably haemoglobin, can buffer the protons formed in the conversion of carbon dioxide to bicarbonate within the cell (in fact, as stated in Sect. 9.2.2, the buffers are necessary for the rapidly catalyzed reaction). Because of the intracellular buffering, the slope of the CO_2 dissociation curve (the change in total CO_2 content/unit change in CO_2 tension) is greater in whole blood than in separated plasma (see Table 9.5).

The slope of the carbon dioxide dissociation curve of true plasma (see Table 9.5) is greater than that of whole blood in bicarbonate-permeable red cells, because the bicarbonate concentration of red cells is generally lower than that of plasma (see Sect. 9.1). The situation in bicarbonate-impermeable red cells of agnathans is different: the slope of the carbon dioxide dissociation curve of true

Table 9.5. The carbon dioxide capacitance coefficients ($\delta C_{CO_2\,tot}/\delta P_{CO_2}$, in mmol l^{-1} $mmHg^{-1}$) of cerebrospinal fluid, separated plasma, true plasma and whole blood of man between 30 and 50 mmHg carbon dioxide tension[a]

	CO_2 capacitance coefficient
Cerebrospinal fluid	0.03
Separated plasma	0.12
True plasma	0.24
Whole blood	0.22

[a] Data are from Dejours (1975). Separated plasma is plasma that has been equilibrated with different CO_2 tensions in the absence of red cells. The value for true plasma is obtained by equilibrating whole blood with different carbon dioxide tensions. Thereafter, plasma and red cells are separated and the total carbon dioxide content of plasma measured.

plasma is smaller than that of whole blood (Tufts and Boutilier 1989). This is most likely due to the fact that the bicarbonate formed in the intracellular compartment cannot exit the cell because of the lack of the anion exchange pathway and that the intracellular pH decreases much less with increasing carbon dioxide tension than the extracellular pH. As a consequence, the intracellular bicarbonate concentration increases much more with increasing carbon dioxide tension than the extracellular one.

The buffering properties of haemoglobin vary with its oxygenation state. Since the pK value of deoxyhaemoglobin is higher than that of oxyhaemoglobin, protons are taken up (at a given intracellular pH) upon deoxygenation. Because the interconversion between carbon dioxide and bicarbonate is an equilibrium reaction [see Eqs. (9.1–9.3)], the uptake of protons favours the formation of bicarbonate, whereby the total carbon dioxide content in deoxygenated blood is greater than in oxygenated blood (see Fig. 9.3). This difference was first observed in 1914 by Christiansen et al.

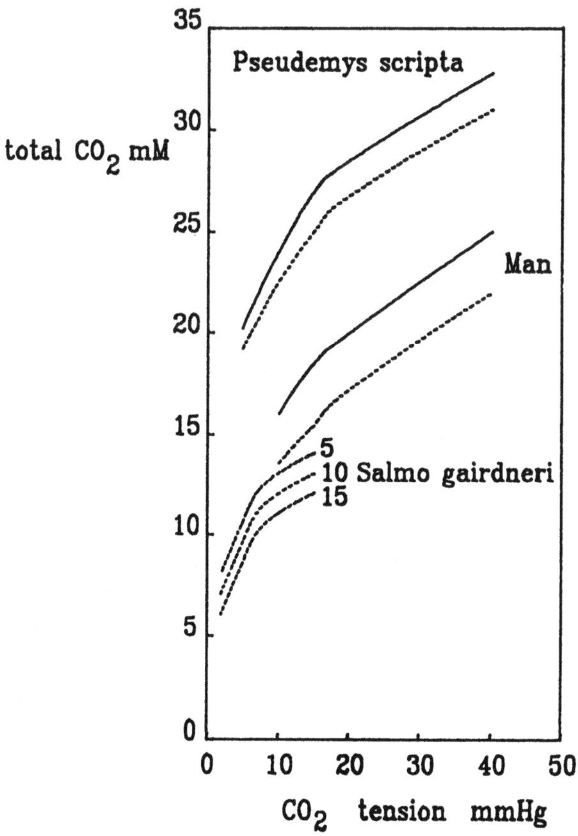

Fig. 9.3. Carbon dioxide dissociation curves of man at 37°C (Christiansen et al. 1914), an air-breathing poikilotherm (the turtle, *Pseudemys scripta*; Weinstein et al. 1986) at 25°C, and rainbow trout (Heming 1984) at 5°, 10° and 15°C. *Dotted lines* = oxygenated blood; *solid lines* = deoxygenated blood

9.3 Formation of Carbamino Compounds

Uncharged amino groups of proteins can reversibly bind either protons or carbon dioxide. With carbon dioxide, carbamino compounds of the form R-NHCOOH are formed. The major types of free amino groups in proteins the α-amino groups at the end of polypeptide chain, and the ε-amino groups in the side chains of lysine. At normal plasma pH (7.4) and carbon dioxide tension (40 mmHg) in man, ca. 60% of the carbamino formation of plasma proteins occurs at the α-amino group, and 40% at the ε-amino groups (Gros et al. 1976). In contrast, the carbamino formation of haemoglobin, e.g. reviewed by Kilmartin and Rossi-Bernardi (1973), occurs almost completely at the α-amino groups (Gros et al. 1981). Table 9.6 gives the approximate amount of carbon dioxide bound to the α and β chains of deoxy- and oxyhaemoglobin in the absence and presence of 2,3- diphosphoglycerate in man. In the absence of organic phosphates, carbon dioxide binding to the β chain is greater than the binding to the α chain. The binding of carbon dioxide to both chains of deoxyhaemoglobin is greater than the binding to oxyhaemoglobin chains. The different binding of carbon dioxide to deoxy- and oxyhaemoglobin is the basis for the specific effect of CO_2 on the oxygen affinity of haemoglobin. According to the general linkage concept of Wyman (1964), the change in haemoglobin oxygen affinity as a function of log pCO_2 at constant oxygen saturation equals the change in bound CO_2 as a function of changed oxygen saturation at constant pCO_2, i.e.

$$(\delta \log pO_2/\delta \log pCO_2)Y = (\delta C_{CO2}/\delta Y)pCO_2, \tag{9.5}$$

in which Y is the oxygen saturation and C_{CO2} is the haemoglobin bound carbon dioxide. Because of this linkage, a large CO_2 specific effect on haemoglobin oxygen affinity shows that the oxylabile carbamino formation in haemoglobin is great.

Carbon dioxide binding to the amino terminal of the β chain is markedly reduced in the presence of 2,3-diphosphoglycerate, because the amino terminal of the β chain is part of the organic phosphate binding site. Thus, carbon dioxide and organic phosphates compete for the same binding site.

Both α and β chains are available for carbon dioxide binding only in mammals, birds and reptiles. In fish and amphibians the α-amino group of the α chain is acetylated, and therefore not available for carbon dioxide binding. As

Table 9.6. Carbon dioxide bound to α and β chain of human haemoglobin (mol CO_2/mol Hb_4) at 40 mmHg CO_2 tension (values estimated from Perrella et al. 1975)

	α Chain	β Chain
DeoxyHb	0.39	0.83
DeoxyHb + 2,3-DPG	0.39	0.43
OxyHb	0.26	0.22
OxyHb + 2,3-DPG	0.26	0.22

a result, the formation of carbamate in fish haemoglobins is reduced, as compared to man (see e.g. Heming et al. 1986). Furthermore, the oxylabile carbamate formation (i.e. the difference between deoxyhaemoglobin carbamate and oxyhaemoglobin carbamate) disappears in tench, *Tinca tinca*, and carp, *Cyprinus carpio*, if organic phosphates are present (see Weber and Lykkeboe 1978; Jensen and Weber 1982). Both these species are often exposed to hypoxic-hypercapnic conditions (see e.g. Jensen and Weber 1985c).

The formation of carbamate increases with increasing pH for two reasons: first, at higher pH a greater proportion of the α-amino groups involved in carbamino formation are present in uncharged form; and second, the binding of organic phosphates to the pocket between the β chains of haemoglobin decreases with increasing pH, whereby the inhibition of carbamino formation is decreased (see e.g. Klocke 1987).

9.4 Mechanisms of Carbon Dioxide Excretion

Tables 9.1 and 9.2 give the proportions of dissolved CO_2, bicarbonate and carbamino compounds in arterial and venous blood in man and in rainbow trout. The proportion of total carbon dioxide excreted during passage through the respiratory epithelia is similar in both species.

Plasma bicarbonate forms the major source of the carbon dioxide excreted. This requires that the intracellular carbonic anhydrase is readily available for dehydration of the extracellular bicarbonate, i.e. that the bicarbonate movements across the red cell membrane are rapid. This is true for both rainbow trout and man. Perry et al. (1982), using spontaneously ventilating, blood-perfused rainbow trout, showed that the inhibition of anion exchange pathway with SITS reduced total CO_2 excretion by 70%. This indicates that plasma bicarbonate makes up more than two thirds of the total carbon dioxide excreted. The corresponding percentage in man is 52–53% (see Roughton 1964; Klocke 1987). Intracellular bicarbonate accounts for 26%, carbamate for 13%, and dissolved CO_2 for 8% of the excreted carbon dioxide in man. The role of dissolved carbon dioxide and carbamino compounds in the carbon dioxide excretion of rainbow trout is probably much smaller, because the difference between venous and arterial carbon dioxide tensions is much smaller, and carbamino formation is reduced as compared to man.

The basic principle of carbon dioxide transport and excretion, shared by most vertebrates, is shown in Fig. 9.4 (see e.g. Klocke 1987; Randall and Daxboeck 1984; Perry 1986; Swenson and Maren 1987). Carbon dioxide initially diffuses down its partial pressure gradient from the tissue to the capillaries, and rapidly enters the red cell. Within the red cell, carbon dioxide can be hydrated rapidly to bicarbonate. Haemoglobin is deoxygenated during passage through the capillaries, whereby protons are taken up and the amount of bicarbonate formed per unit change in carbon dioxide tension increases. The bicarbonate formed within the red cell is exchanged for extracellular chloride via the anion

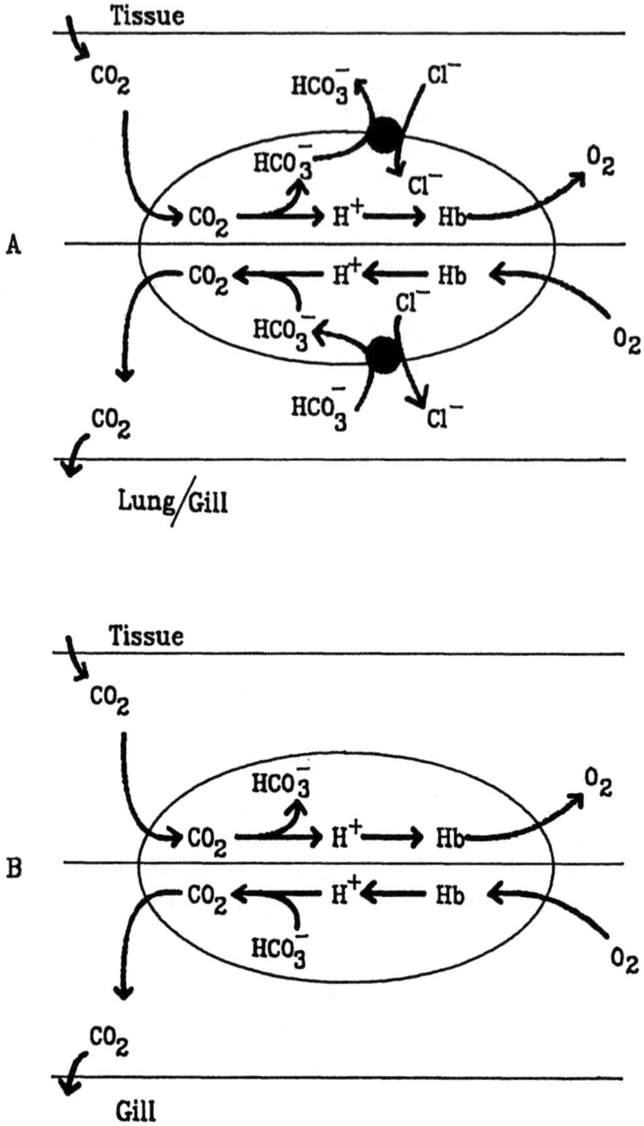

Fig. 9.4 A,B. Mechanisms of carbon dioxide excretion in vertebrates. *A* General vertebrate pattern; *B* Agnatha

exchange pathway. A small proportion of carbon dioxide binds directly to extracellular proteins, and a somewhat larger proportion to haemoglobin, forming carbamino compounds. Very little uncatalyzed hydration of carbon dioxide in the plasma occurs, since the uncatalyzed reaction rate is slow.

The reaction sequence is reversed in the lungs (or gills). Bicarbonate enters the cell via the anion exchange pathway in exchange for chloride. The transformation of bicarbonate into carbon dioxide within the red cell is facilitated by the protons liberated during the oxygenation of haemoglobin. Carbon dioxide is also liberated from haemoglobin carbamate when haemoglobin is oxygenated. In mammals, the luminar surface of lung capillary endothelial cells contains significant carbonic anhydrase activity (Crandall and O'Brasky 1978; Effros et al. 1978; Klocke 1978; Lönnerholm 1982), adequate to catalyze the extracellular hydration-dehydration reactions by a factor of 120–150 (Bidani et al. 1983; Henry et al. 1986). As a result, the half-time of the reaction in the extracellular compartment would be approximately the same as the half-time for chloride/bicarbonate exchange. Thus, it is possible that the lung carbonic anhydrase plays a role in the carbon dioxide excretion in mammals. However, the data to date suggest that the contribution is small (see Klocke 1987; Bidani and Crandall 1988). Teleost fish probably do not have carbonic anhydrase on the luminar surface of the gill capillary endothelium (Henry et al. 1988).

From the data available on the rates of the different reactions involved in the excretion of carbon dioxide (e.g. Wagner 1977; Klocke 1987), it appears that the rate-limiting step in carbon dioxide excretion is the rate of chloride/bicarbonate exchange. Thus, the carbon dioxide excretion in teleosts like pikeperch, *Stizostedion lucioperca*, and carp, *Cyprinus carpio*, may be ineffective, because their anion exchange is slow (see Sect. 7.3). At the extreme, the red cells of agnathans lack the anion exchange pathway (Ellory et al. 1987; Nikinmaa and Railo 1987), and have low intracellular carbonic anhydrase activity (e.g. Sanyal 1984). In this group, the bicarbonate formed from carbon dioxide in the red cells of tissue capillaries cannot be transported to the plasma, whereby the amount of total carbon dioxide produced for a unit increase in carbon dioxide tension is reduced. Furthermore, only the red cell bicarbonate is available for dehydration in the gills, whereby the total amount of carbon dioxide excreted must be drastically reduced (see Fig. 9.4B).

Chapter 10

Oxygen Transport

The amount of oxygen transported by unit volume of blood depends on the partial pressure of oxygen in the blood, on the number of red cells per unit volume, on the amount of functional haemoglobin within the red cells and on the oxygen affinity of haemoglobin. Haemoglobin-oxygen affinity plays a dual role in the transport of oxygen to the tissue. Haemoglobin must be able to bind oxygen effectively in the capillaries of the gas exchange organs, and unload oxygen at high partial pressures in order to maintain a large diffusion gradient for oxygen between the blood in tissue capillaries and the oxygen-consuming structures.

The blood oxygen affinity is a function of:

1. the intrinsic oxygen affinity of haemoglobin;
2. the concentration of haemoglobin within the cell;
3. the sensitivity of haemoglobin-oxygen affinity to heterotropic ligands;
4. the concentration of heterotropic ligands within the erythrocytes;
5. the temperature.

The following sections initially describe in some detail the basic principles of haemoglobin-oxygen equilibria, including methods for determining the oxygen equilibrium curves and the molecular mechanisms of haemoglobin function. Following that, adaptations of haemoglobin function to different environments are presented, including the mechanisms by which haemoglobin function can be adjusted to the varying respiratory needs of individual animals.

10.1 Haemoglobin-Oxygen Equilibria – Basic Principles

10.1.1 Oxygen Equilibrium Curve

A representative oxygen equilibrium curve is given in Fig. 10.1. The curve can be described using the empirical equations

$$Y = K_A \cdot P_{O2}^n/(1 + K_A \cdot P_{O2}^n) \text{ or} \quad (10.1)$$

$$Y/(1-Y) = K_A \cdot P_{O2}^n, \quad (10.2)$$

in which Y = the fractional oxygen saturation of haemoglobin, K_A = the equilibrium association constant for the overall haemoglobin-oxygen reaction, P_{O2} = the oxygen tension and n = Hill's coefficient, which describes the interaction in oxygen binding between globin chains in the molecule, i.e. the

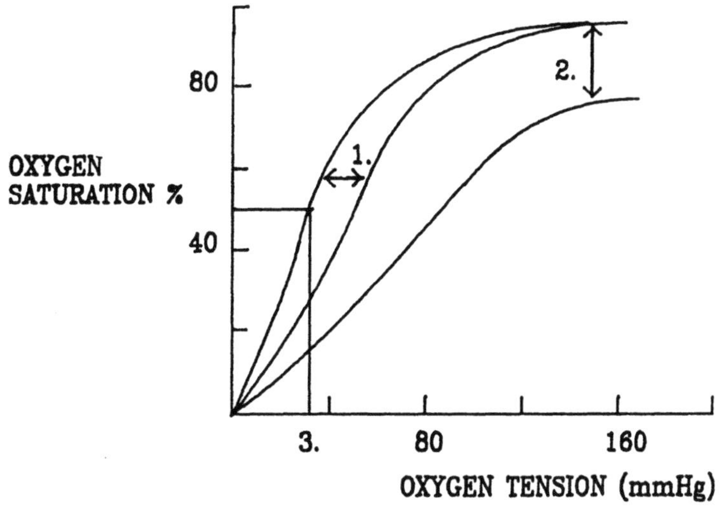

Fig. 10.1. Oxygen equilibrium curve (oxygen dissociation curve). Y-axis gives the oxygen saturation of haemoglobin (blood). X-axis gives the oxygen tension. With decreasing pH the oxygen equilibrium curve is shifted to the right, i.e. the oxygen affinity decreases (*1*). In most teleost haemoglobins a decrease in pH causes a reduction in the maximal oxygen saturation at atmospheric oxygen tension (Root effect, *2*). In many instances, the oxygen affinity of haemoglobin (blood) is given numerically as the P_{50} value, which is the oxygen tension at which haemoglobin (blood) is 50% saturated with oxygen (*3*)

degree of cooperativity. Instead of the equilibrium association constant (K_A), the equilibrium dissociation constant (K_D) can be used in the equation by substituting $1/K_D$ for K_A. Logarithmic transformation of Eq. (10.2) yields the Hill equation

$$\log Y/(1 - Y) = n \cdot \log(P_{O2}) + \log(K_A). \quad (10.3)$$

At 50% oxygen saturation (fractional saturation 0.5), $\log Y/(1 - Y)$ is zero, and the oxygen tension is the P_{50} value, whereby the value for K_A can be calculated, giving

$$K_A = 1/(P_{50})^n. \quad (10.4)$$

Figure 10.2 gives a characteristic Hill plot for a sigmoid oxygen equilibrium curve. For fractional oxygen saturation between 0.1 and 0.9, the Hill plot is nearly linear, with a slope of n. Thus, the Hill plot gives the cooperativity of haemoglobin-oxygen binding directly.

Values of n greater than one indicate that oxygen binding has positive cooperativity (i.e. the binding of an oxygen molecule facilitates the binding of consecutive oxygen molecules). At $n = 1$ no cooperative effects are present, and n values less than one indicate that the binding sites have different affinities for oxygen.

At low and high oxygen saturations, the slope of the Hill plot approaches one, even for haemoglobins with cooperative oxygen binding. This indicates that

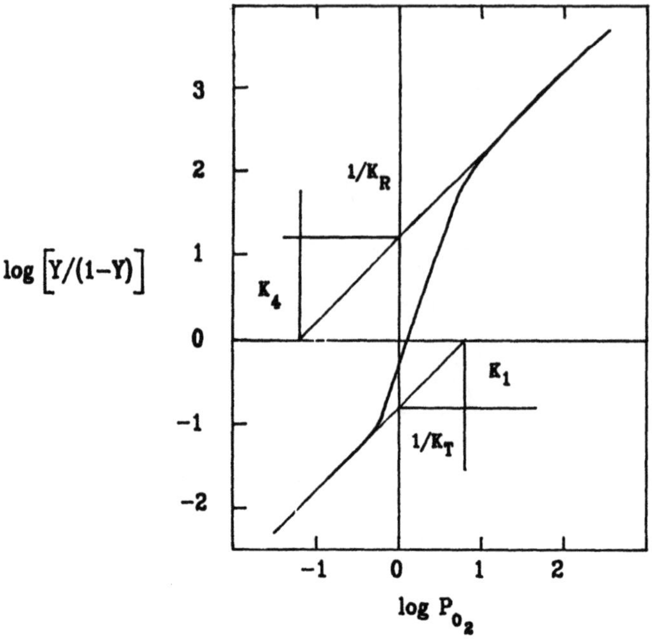

Fig. 10.2. Hill plot. Y-axis gives the logarithm of oxyhaemoglobin-deoxyhaemoglobin ratio [log Y/(1−Y)], and X – axis the logarithm of oxygen tension. The oxygen association constants for the T-form (K_T) and for the R-form (K_R) can be calculated from the intercepts of the low saturation and high saturation asymptotes of the plot with the horizontal line at 50% saturation (log oxyhaemoglobin/deoxyhaemoglobin = 0). For further details, see text

there is no subunit cooperativity during the binding of the first and the last oxygen molecule.

The value of the minimum free energy of cooperativity of haemoglobin can be determined from the distance between the asymptotes of the Hill plot, i.e. the difference in the binding affinities of the first and last ligand molecule using the equation

$$\Delta F_I = 2.3\ DRT\sqrt{2}, \tag{10.5}$$

in which ΔF_I is the minimum energy of interaction (in J mol^{-1} haem), D is the difference in binding affinities, R is the universal gas constant and T the temperature in K (Wyman 1964).

The most general expression of the haemoglobin-oxygen equilibria is the Adair equation (Adair 1925). The equation is valid as long as the haemoglobin molecule reacts in the tetrameric form, whether the four haem groups are identical or differ in their activity towards oxygen, or whether heterotropic ligands are present or absent. The Adair equation and its use have been described in detail, e.g. by Imai (1973), and used by, e.g. Imai (1973, 1981a); Imai and Yonetani (1975); Mayo and Chien (1980); Ikeda-Saito et al. (1983); Jensen and Weber (1987) and Weber et al. (1987a). The following description is a summary of the more detailed treatments.

The reaction between tetrameric haemoglobin and oxygen is expressed by the equilibrium reactions

$$Hb + 4O_2 \leftrightarrows HbO_2 + 3O_2 \leftrightarrows Hb(O_2)_2 + 2O_2 \leftrightarrows Hb(O_2)_3 + O_2 \rightleftarrows Hb(O_2)_4. \quad (10.6)$$

If the overall association constants for binding the first to fourth oxygen are taken as a1-a4, the fractional oxygen saturation of haemoglobin (Y) is given by the equation

$$Y = (a_1P_{O2} + 2a_2P^2_{O2} + 3a_3P^3_{O2} + 4a_4P^4_{O2}) / [4(1 + a_1P_{O2} + a_2P^2_{O2} + a_3P^3_{O2} + a_4P^4_{O2})]. \quad (10.7)$$

The overall association constants a_1-a_4 are functions of the intrinsic equilibrium constants of the consecutive oxygenation reactions K_1-K_4, such that $a_1 = 4K_1$; $a_2 = 6K_1K_2$; $a_3 = 4K_1K_2K_3$; and $a_4 = K_1K_2K_3K_4$. The values for K_1 to K_4 (the Adair constants) can be estimated from highly accurate data on the haemoglobin-oxygen equilibrium curve using the Scatchard plot, and are a measure of the oxygen affinity of haemoglobin at each step of the oxygenation. The median oxygen tension ($P_m = K_1K_2K_3K_4$) is the overall oxygen affinity of haemoglobin. This value is generally close to the P_{50} value.

In the Scatchard plot (as described by Imai 1981a; Fig. 10.3) the y-axis gives $\log[(Y/1\text{-}Y) \times P_{O2}]$, and the x-axis gives Y. The values for K_1 and K_4 can be read from the intercepts of the plot with the vertical axes at Y = 0 and Y = 1. (The

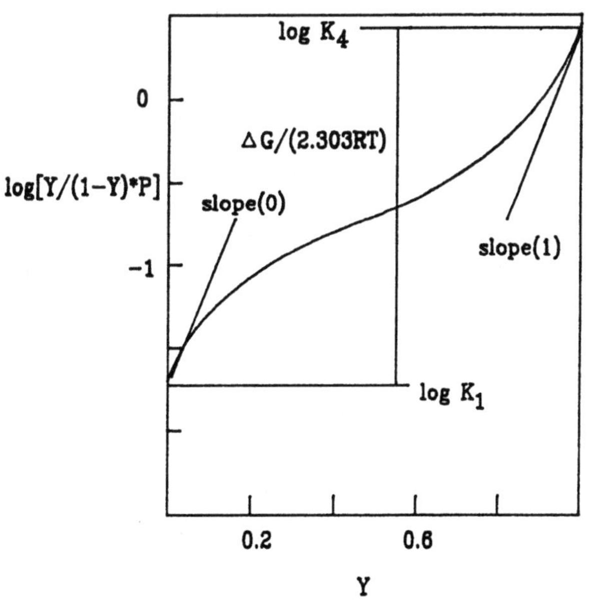

Fig. 10.3. The Scatchard plot. Y-axis gives the logarithm of the oxyhaemoglobin/deoxyhaemoglobin ratio multiplied by the oxygen tension (log [Y/(1-Y) × p]), and X- axis the oxygen saturation. The oxygen association constants of the consecutive oxygenation steps and the overall free energy change of oxygenation (ΔG) can be calculated from the graph. For details, see text

values for K_1 and K_4 can also be read from the Hill plot. They are the intercepts of the low-saturation and high-saturation asymptotes of the plot with the vertical line at log $P_{O2} = 0$). The values for K_2 and K_3 can be calculated from the slopes of the plot at the left and right ends respectively:

$$\text{slope}(0) = 3(K_2 - K_1)/2.303K_1 \quad (10.8)$$
$$\text{slope}(1) = 3(K_4 - K_3)/2.303K_3. \quad (10.9)$$

10.1.2 Homotropic and Heterotropic Interactions

10.1.2.1 Homotropic Interactions

The cooperative oxygen binding of haemoglobin is an example of homotropic interaction: binding of a ligand molecule influences the binding of consecutive molecules of the same ligand. The theory most commonly used to explain the cooperative ligand binding in tetrameric haemoglobins is the allosteric model of Monod et al. (1965). The allosteric protein (e.g. haemoglobin) can exist in two conformations designated as T (tense, e.g. deoxyhaemoglobin) and R (relaxed, e.g. oxyhaemoglobin). The ligand binding affinity of the T conformation is lower than that of the R conformation. The binding of ligand to a subunit of the T conformation induces a concerted change in the conformation of the rest of the molecule to the R conformation, such that the affinity of the unliganded subunits increases abruptly. In its simplest form, the model assumes that the equilibrium dissociation constants for ligand binding to the T-form (K_T) and R-form (K_R) are invariant.

In the absence of ligands, the equilibrium constant (L) for the allosteric protein is the concentration ratio of unliganded T-form and unliganded R-form, i.e. L = [T]/[R].

The value for L for human deoxygenated haemoglobin in normal physiological conditions is ca. 10 000. Thus, unliganded protein is almost completely in the T-form (see e.g. Bunn and Forget 1986). If the ratio of dissociation equilibrium constants of T- and R-form is denoted as c

$$c = K_R/K_T, \quad (10.10)$$

and the ratio of the amount of ligand (F) and the dissociation equilibrium constant K_R is denoted as α,

$$\alpha = F/K_R, \quad (10.11)$$

the fractional ligand saturation (Y) of the protein with n ligand binding subunits can be described as

$$Y = [Lc\alpha(1+c\alpha)^{n-1} + \alpha(1+\alpha)^{n-1}]/[L(1+c\alpha)^n + (1+\alpha)^n]. \quad (10.12)$$

If c is small, the equation simplifies to

$$Y = \alpha(1+\alpha)^{n-1}/[L+(1+\alpha)^n]. \quad (10.13)$$

Cooperativity increases with increasing values of L (i.e. the greater the proportion of unliganded T-form is) and decreasing values of c (i.e. the higher the oxygen affinity of R-form in relation to that of T-form is).

The asymptotes of the Hill plot (Fig. 10.2) with the slope of 1 represent the oxygenation of the T-state haemoglobin (low saturation asymptote) and the oxygenation of the R-state haemoglobin (high saturation asymptote). When extrapolated to 50% saturation, i.e. $\log[Y/(1 - Y)] = 0$, they give the values for K_T and K_R respectively. The value of α at 50% oxygen saturation can be denoted as $\alpha_{1/2}$, and is the ratio between the overall oxygen affinity and the oxygen affinity of R-state haemoglobin

$$\alpha_{1/2} = P_{50}/K_R. \quad (10.14)$$

Using this value, the Hill coefficient at 50% saturation can be calculated from the equation (e.g. Bunn and Forget 1986)

$$n = 1 + 3[(1 - c\alpha_{1/2})(\alpha_{1/2} - 1)]/[(1 + c\alpha_{1/2})(\alpha_{1/2} + 1). \quad (10.15)$$

10.1.2.2 Heterotropic Interactions

In heterotropic interactions the binding of a ligand to a site different from the oxygen binding site influences the oxygen affinity.

Heterotropic effects can be treated using the linked function equations (for a detailed description on the linked functions see Wyman, 1964 and, for a more recent treatment, Ackers 1979). The fundamental linkage equation for the interaction between two ligands (X, Y) binding to a macromolecule is

$$(\delta X/\delta \ln y)x = (\delta Y/\delta \ln x)y, \quad (10.16)$$

where the left side equals the change in the amount of ligand X bound per mole of the macromolecule (δX) divided by the change in the logarithm of the concentration of ligand Y in the system ($\delta \ln y$) at constant concentration of ligand X (x). The right side equals the change in the amount of ligand Y bound per mole of the macromolecule (δY) divided by the change in the logarithm of the concentration of ligand x ($\delta \ln x$) at constant concentration of ligand Y (y). In addition, the linkage equations

$$(\delta X/\delta Y)x = -(\delta \ln y/\delta \ln x)y \text{ and} \quad (10.17)$$
$$(\delta \ln x/\delta Y)_X = (\delta \ln y/\delta X)_Y \quad (10.18)$$

are important in describing the interaction between the binding of two ligands to a macromolecule. The linkage equations can also be derived for several ligands binding to the same macromolecule (for the derivation of the linkage equations for two and several ligands, see Wyman 1964, 1972).

The basis of heterotropic interactions in tetrameric haemoglobins can be explained using the allosteric model. In the simplest case, the allosteric effector, responsible for heterotropic interactions, either binds to the T-form (effector I), whereby the allosteric constant L increases and oxygen affinity decreases, or to the R-form (effector A), whereby the value for L decreases and oxygen affinity increases.

As described by Monod et al. (1965), if the allosteric effector I binds solely to the T-form and the allosteric effector A to the R-form, the new apparent allosteric constant in the presence of both ligands can be defined as

$$L' = (\Sigma T_I/\Sigma R_A). \quad (10.19)$$

The value of L′ can be related to the original allosteric constant, if the microscopic dissociation constants for ligand I with the T-form (K_I), and for ligand A with the R-form (K_A), are known. Then, $\beta = I/K_I$ and $\tau = A/K_A$, and

$$L' = L(1 + \beta)^n/(1 + \tau)^n \quad (10.20)$$

By substituting L′ to the equation describing the fractional oxygen saturation, the relation

$$Y = [\alpha(1 + \alpha)^{n-1}]/[L(1 + \beta)^n/(1 + \tau)^n + (1 + \alpha)^n] \quad (10.21)$$

is obtained.

From this equation it is obvious that the ligand binding to the T-form decreases the oxygen affinity and increases the cooperativity, whereas the ligand binding to the R-form decreases cooperativity and increases oxygen affinity.

The allosteric model in its simplest form supposes that all the globin chains are identical in their affinity for oxygen; that there is only one T-state and one R-state; and that allosteric effectors only influence the allosteric constant L. However, Pennelly et al. (1978) have shown that the haemoglobins of the teleost, *Myripristis berndti,* exhibit a marked (tenfold) difference in the binding kinetics of carbon monoxide to the α and β chains of haemoglobin. The α and β chains of human haemoglobin also exhibit dissimilar ligand binding properties (see e.g. Di Cera et al. 1987). Furthermore, binding of ligands to the subunits either in the R-state or in the T-state induces considerable changes in the tertiary structure of the globin chains, even if the quarternary structure of human haemoglobin is held rigid (Perutz 1972). Friedman (1985) observed that the binding of oxygen changes the haem pocket geometry from the unligated state to the ligated state in both the R and T conformation of haemoglobin, indicating changes in the tertiary structure of the chains. The values for K_T and K_R are affected to different degrees by allosteric effectors. Thus, the value for c ($=K_T/K_R$) is also affected (see e.g. Imai and Yonetani 1975; Mayo and Chien 1980; Weber et al. 1987a; Kister et al. 1987). In addition, e.g. Morris et al. (1981) have shown that the haemoglobin of the teleost, *Thunnus thynnus,* has multiple T-state conformations. Thus, expanded allosteric models are required to explain the observed homo- and heterotropic interactions. The model by Kister et al. (1987; see Fig. 10.4) takes into account both the observation that the haemoglobin is capable of existing in different T- and R-states, and that the equilibrium constants of these states for oxygen are different. However, the model requires that all the oxygen binding sites are equal. Di Cera et al. (1987) have presented an allosteric model in which haemoglobin exists in two alternative structures, the low-affinity T-form, and the high-affinity R-form. In the low-affinity form, only the α chains can bind oxygen in significant amounts; the oxygen affinity of β chains is negligible, because of steric hindrance at the haem (see Sect. 10.3.1). The T-state shows cooperative behaviour arising from the interaction between the α subunits

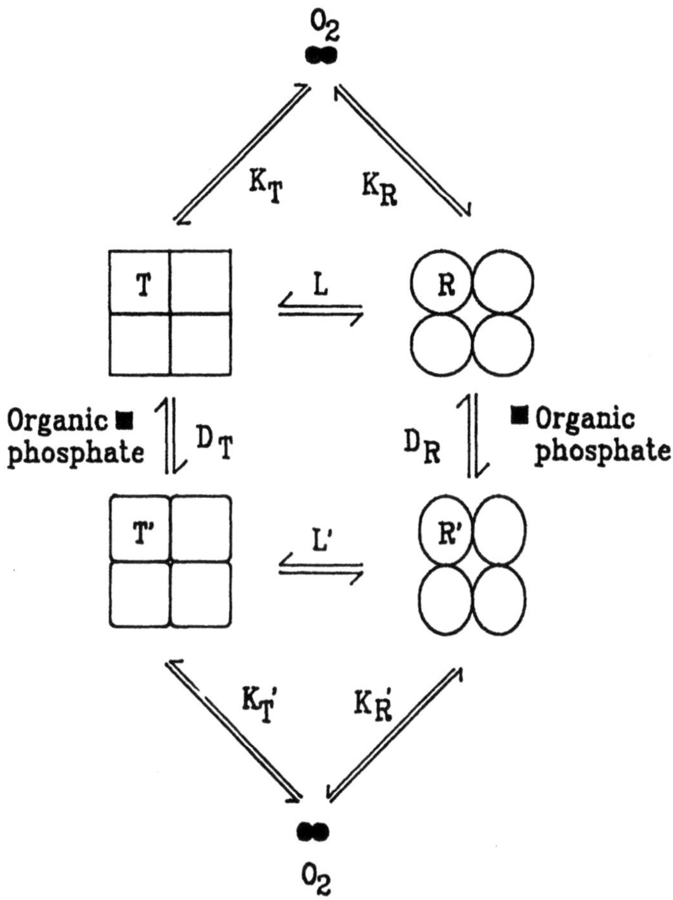

Fig. 10.4. A schematic representation of the modified allosteric model of Kister et al. (1987). Both the T- and the R-form of haemoglobin can exist in two conformations which have different oxygen association constants. Organic phosphates influence the T-T′ and R-R′ equilibria, thereby influencing both the T-form/R-form ratio and the overall oxygen association constant of both the T- and the R-form. For details, see text

(cooperativity at the first structural level). After both α subunits have bound oxygen, the T–R transition must occur, contributing cooperativity at the second structural level. The R-state, on the other hand, is noncooperative, with the α and β chains having the same affinity. Figure 10.5 gives a schematic representation of oxygen ligation of human haemoglobin in terms of the model by Di Cera et al. (1987).

The maximal value of the Hill coefficient for a tetrameric molecule is four. This value is reached if the oxygenation of one haemoglobin subunit infinitely increases the oxygen affinity of the three other subunits, whereby partially oxygenated intermediates will not exist. The Hill coefficient of one is observed for tetrameric haemoglobins which do not change conformation upon ligand binding. The n-values of haemoglobins of various birds, reptiles and amphibians

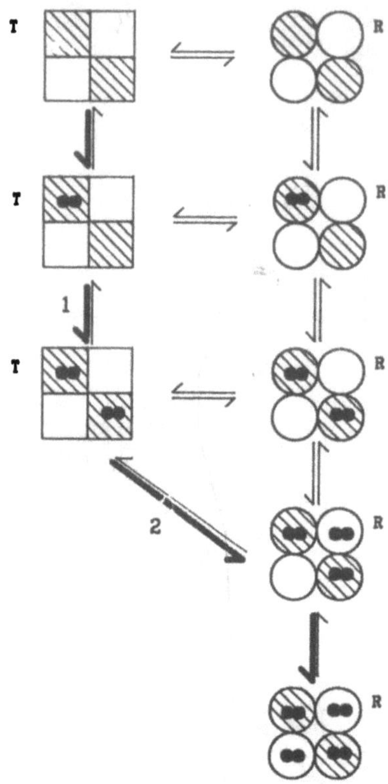

Fig. 10.5. A schematic representation of the oxygenation model of Di Cera et al. (1987). In the T-form, only the α chains (*shaded* subunits) are capable of binding oxygen. The binding of oxygen to one of the α chains influences the binding of oxygen to the second α chain, thus causing cooperativity at the first structural level (*1*). After oxygen has bound to both α chains, the quarternary structure of haemoglobin molecules changes from the T- to the R-state, whereby binding of oxygen to the ß chains is possible. This causes cooperativity at the second structural level (*2*). The normal oxygenation pathway is indicated by *bold arrows*

exceed four (see e.g. Riggs 1988). This behaviour is only possible if tetramers associate under conditions in which the oxygen equilibria are measured. This has been shown to be the case for both amphibian and bird haemoglobins (see Sect. 3.5.1 and Riggs 1988). Also, aggregation/deaggregation reactions of sickle cell haemoglobin result in n-values of 5–6 (Gill et al. 1978). The Hill coefficient depends on the concentration of certain allosteric effectors. As pointed out by Albers (1985) and Jensen (1988), the apparent cooperativity of oxygen binding for whole blood, determined at constant extracellular pH, is normally lower than that for haemoglobin solutions, because the red cell pH decreases with increasing oxygen saturation of haemoglobin (see Sect. 10.2.1).

Cooperativity and heterotropic interactions are also observed in cyclostome haemoglobins, which do not form stable tetramers. In this case the homo- and heterotropic interactions are thought to arise from oxygen-dependent association/dissociation phenomena (see e.g. Briehl 1963; Riggs 1972; Dohi et al. 1973).

It is likely that the deoxyform of lamprey haemoglobin aggregates to di- and tetramers. Aggregation is facilitated by low pH values (e.g. Dohi et al. 1973). The aggregated forms have a lower oxygen affinity than the monomer, whereby dissociation with an increase in oxygen affinity. If the haemoglobin is in monomeric form throughout the oxygenation, as in low concentration and at high pH value, no cooperativity is observed. However, if either the pH is lowered or concentration increased, the binding of oxygen causes dissociation of haemoglobin, which is seen as an increase in the Hill coefficient. If the pH is lowered and concentration increased further, the cooperativity again decreases (Dohi et al. 1973). This may indicate that no dissociation of the aggregates occurs during oxygenation, and that there is no haem-haem interaction in the aggregated form.

10.1.2.3 Bohr and Haldane Effects

The influence of protons on the haemoglobin-oxygen affinity is called the Bohr effect. Detailed reviews on the Bohr effect have been written, e.g. by Kilmartin and Rossi-Bernardi (1973) and Riggs (1988). The reciprocal of this effect, i.e. the influence of oxygen saturation on the proton binding by haemoglobin, also reviewed in detail by Kilmartin and Rossi-Bernardi (1973), is called the Haldane effect. In terms of the linkage equations, the values of the Bohr effect (i.e. Bohr coefficient) and the Haldane effect (i.e. Haldane coefficient) can be expressed as

$$(\delta \log Po_2/\delta pH)y = (\delta H^+/\delta Y)pH, \tag{10.22}$$

in which Y = haemoglobin-oxygen saturation and H^+ is the number of protons bound per haem group. Generally, the Bohr constant, which describes the Bohr effect, is calculated as

$$B.E. = \Delta \log P_{50}/\Delta pH, \tag{10.23}$$

and the Haldane coefficient can be obtained from the vertical distance between the titration curves for oxygenated and deoxygenated blood or haemoglobin (e.g. Kilmartin and Rossi-Bernardi 1973; Jensen and Weber 1985a).

If the cooperativity of oxygen binding changes with pH, the values for the Bohr constant will vary according to saturation. Furthermore, the Bohr effect at a given saturation varies with pH. Also, the magnitude of Bohr effect increases if organic phosphates are present.

In whole blood, values for Bohr coefficient have often been calculated using the formula

$$B.E. = \Delta \log P_{50}/\Delta pH_{plasma}. \tag{10.24}$$

This Bohr coefficient underestimates the pH dependence of haemoglobin-oxygen binding, since the $\delta pH_{red\ cell}/\delta pH_{plasma} < 1$. More correctly, the red cell pH should be used in the calculations of the Bohr effect (see also Albers 1985). Another complication with regards to the numerical value of the Bohr constant in teleost fish is the presence of the Root effect (defined as the decrease in the maximal oxygen saturation of haemoglobin at atmospheric P_{02}). In this case, two values for the Bohr constant can be obtained (see e.g. Albers 1985), one if P_{50} value is defined as half-saturation of haemoglobin at the pH of the measurement, and the other if P_{50} value is defined as half of the maximum saturation obtained with air equilibration at high pH (see Fig. 10.6). This can also be illustrated using the data of Nikinmaa and Soivio (1979). If the Bohr constant is calculated in terms of P_{50} value given as half of the maximum saturation obtained at high pH, the value at plasma pH range 7.85–7.40 is –1.0, whereas if it is calculated in terms of P_{50} value given as half-saturation at the pH of the measurement, the value is reduced to –0.4.

The fixed-acid Bohr effect gives the change in haemoglobin-oxygen affinity with decreasing pH when the pH is changed using a strong acid or base (e.g. HCl or NaOH). The carbon dioxide Bohr effect gives the change in haemoglobin-oxygen affinity when pH is changed by a change in the carbon dioxide tension of incubation. The carbon dioxide Bohr coefficient thus consists of two factors: the effect of protons and the effect of CO_2 on the oxygen affinity (see e.g. Lapennas 1983; Hlastala 1985; for oxylabile carbamino formation see also Sect. 9.3). The carbon dioxide Bohr effect decreases with increasing haemoglobin-oxygen saturation in the species with specific carbon dioxide binding to haemoglobin (see e.g. Garby et al. 1972; Hlastala 1985). This is most likely due to the chain differences in both oxygen and carbon dioxide binding. At low saturations oxygen binds solely to the haem groups of α-globin chains, in which carbon dioxide is also capable of binding. Thus, the interaction between oxygen and

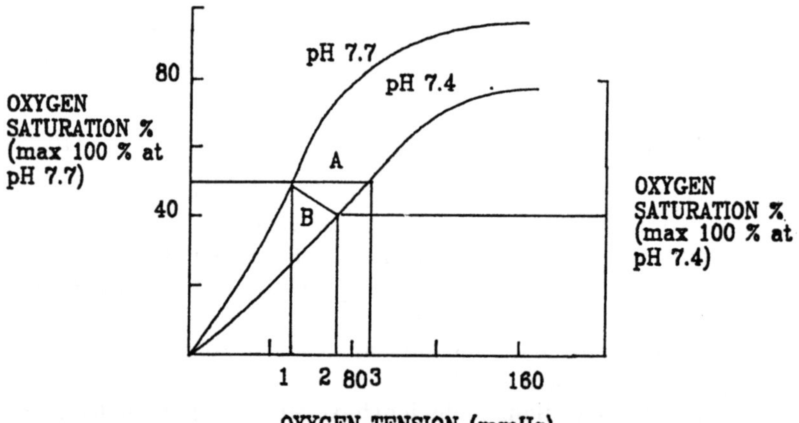

Fig. 10.6. Estimation of the numerical value of Bohr effect in Root effect haemoglobin. *A* 100% oxygen saturation is defined as the saturation asymptotically approached atmospheric oxygen tension at high pH. Thus, at pH 7.4, maximal oxygen saturation is 75–80%. P_{50} value, used in the calculation of the Bohr factor, is the oxygen tension required for 50% saturation of blood with oxygen in both cases. The Bohr factor = [(log(3)-log(1))/(7.4–7.7)], giving an approximate value of –0.85. *B* 100% oxygen saturation is defined as the saturation that is asymptotically approached at atmospheric oxygen tension at the pH of oxygen equilibrium curve determination. Thus, the maximal saturation at both pH 7.7 and 7.4 is 100%, and the P_{50} value at pH 7.4 is the oxygen tension required for 38% oxygen saturation of haemoglobin, if 100% saturation is defined as in A. In this case, the Bohr factor = [(log(2)-log(1))/(7.4–7.7)], giving an approximate value of –0.52

carbon dioxide binding is great at low saturations. At high saturations, oxygen binding occurs at the β chains of haemoglobin, where little or no carbon dioxide binding occurs in physiological conditions, because of the competition between the binding of organic phosphates and carbon dioxide. Thus, little interaction between carbon dioxide and oxygen binding occurs at high saturations, and the carbon dioxide and fixed-acid Bohr effects approach each other (see e.g. Hlastala 1985).

There is no correlation between the whole blood Bohr effect and the size of the animal; for example, the Bohr factors for hummingbirds (Johansen et al. 1987) are similar to the Bohr factor of the elephant (see Bartels 1972). Both low and high Bohr factors are associated with low cooperativity of haemoglobin (cf. carp; Weber and Lykkeboe 1978 and *Myxine glutinosa*; Bauer et al. 1975a) and with high (cf. *Arapaima gigas*; Johansen et al. 1978 and the tench *Tinca tinca*; Jensen and Weber 1982) or low oxygen affinity (e.g. *Rana temporaria*; Wells and Weber 1985, and the chicken; Lapennas and Reeves 1983).

The Bohr factor of some haemoglobin components of the teleosts, rainbow trout (e.g. Weber et al. 1976b), the eels (*Anguilla*; e.g. Gillen and Riggs 1973; Weber et al. 1976a) and *Catostomus clarkii* (e.g. Powers 1977) is positive in the absence of organic phosphates at the pH range 7–8. Thus, the oxygen affinity of haemoglobin increases with decreasing pH. In the presence of organic phosphates this positive Bohr effect disappears. In all these teleost species, haemoglobin components with a large, negative Bohr effect are also found. The oxygen affinity of the sole haemoglobin of the amphibian, *Amphiuma means*, increases with decreasing pH in the absence of organic phosphates (Bonaventura et al. 1977). Addition of ATP or IHP changes the pH dependence of oxygen binding, changing the value of the Bohr coefficient from 0.24 (in 0.3 M NaCl; pH 7–8) to −0.15 (ATP present).

The biological significance of the Bohr effect has been discussed, e.g. by Bartels (1972) and Lapennas (1983). The pH of blood in the rapidly metabolizing tissues tends to decrease, owing to the production of carbon dioxide and metabolic acids. The decrease in pH will shift the oxygen dissociation curve to the right, owing to the Bohr effect, whereby more oxygen will be given up from haemoglobin at a given partial pressure of oxygen. The importance of the Bohr effect is illustrated in Fig. 10.7. With the help of the Bohr effect, oxygen delivery can be accomplished while maintaining a large oxygen partial pressure gradient between capillary blood and the metabolizing tissue. In the absence of the Bohr effect, the delivery of the same amount of oxygen would result in a reduced partial pressure gradient for oxygen, whereby its diffusion to the sites of consumption would be slowed down.

The influence of the Bohr effect on the unloading pressure for oxygen will depend critically on the magnitude of the Bohr effect and on the arterio-capillary pH difference (which can be approximated by the arterio-venous difference). Theoretically, the greater the magnitude of the Bohr effect and the arterio-venous pH difference, the greater the unloading partial pressure for oxygen which can be achieved for a given amount of oxygen delivered. However, owing to the linkage between the Bohr and Haldane effects, it is not possible to have a large Bohr effect with a large arterio-venous pH difference. If the Bohr coefficient

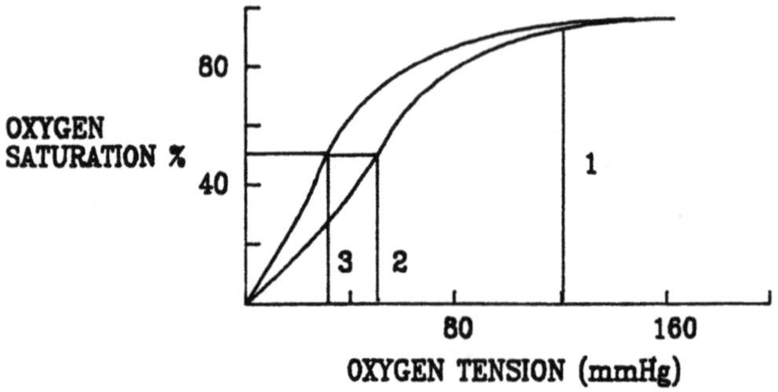

Fig. 10.7. The influence of the Bohr effect on the oxygen tension of tissue capillary blood. The shift of the oxygen equilibrium curve to the right does not affect the oxygen loading in the gas exchange organ (oxygen tension 120 mmHg; *1*). However, a right-shifted oxygen equilibrium curve unloads the same amount of oxygen at much higher (*2*; ca. 50 mmHg) oxygen tension than a left-shifted oxygen equilibrium curve (*3*; ca. 33 mmHg)

of haemoglobin is large, haemoglobin also effectively takes up protons upon deoxygenation. In tench, for example, the red cell pH increases ca. 0.3 units constant extracellular pH when oxygen saturation decreases from 90 to 50% (Jensen 1986). Similarly, in carp, the pH of deoxygenated blood is 0.20–0.25 units higher than that of oxygenated blood (Albers et al. 1983) at constant extracellular pH. This strong Haldane effect, i.e. deoxygenation-linked proton buffering, is enough to abolish any arterio-venous pH differences, as has been experimentally observed in carp from which undisturbed blood samples have been obtained both from ventral (prebranchial blood) and dorsal aorta (postbranchial blood; Nikinmaa, Salama, Mattsoff and Ryhänen, unpublished data). In this case, the Bohr-Haldane effect cannot be utilized to augment oxygen transport, but may be important in carbon dioxide transport. The strong Haldane effect may be particularly important in carbon dioxide transport of tench and carp, which appear to have a relatively sluggish anion exchange across the red cell membrane (Jensen 1988; Pasternack and Nikinmaa 1988).

Lapennas (1983) has calculated that in order to obtain maximal benefit from the Bohr effect in terms of oxygen transport to the tissues, the animal has to have haemoglobin with a Bohr coefficient of approximately half of the respiratory coefficient, and with opposite sign. Thus, to optimize oxygen transport, the Bohr factors should be between -0.35 and 0.5 for respiratory coefficients between 0.7 and 1.0. This appears to be the case for most vertebrates, with the notable exception of a few teleosts and diving animals.

The Bohr effect may either be linear, i.e. each oxygenation step is associated with a liberation of equal number of protons, or nonlinear, in which case a different number of protons are liberated at different levels of saturation. The available data suggest that both in man and in the teleosts, tench and bluefin tuna (*Thunnus thynnus*; Ikeda-Saito et al. 1983), the contribution of the different oxygenation steps to the overall proton release varies markedly. In the tench, the third oxygenation step contributes half of the protons liberated (Weber et al. 1987a) and in man, steps 2 and 3 contribute 1.6 protons out of the total 2.3 protons liberated during oxygenation of one haemoglobin molecule (Chu et al. 1984). At least in man, the overall proton release appears linear with oxygenation, because the cooperativity of oxygen binding results in the predominant species being either fully deoxygenated or fully oxygenated (see Riggs 1988).

10.1.2.4 Root Effect

Many teleost haemoglobins fail to become fully saturated even at very high oxygen tensions – in some cases the haemoglobin cannot be saturated with oxygen, even at 100 atm pure oxygen (Scholander and Van Dam 1954). This phenomenon was first observed by Root (1931), and has been named after him.

From the physiological point of view, an adequate definition of the Root effect is a decrease in the saturation of haemoglobin (blood) at atmospheric oxygen tension induced by a decrease in pH.

The Root effect has been reviewed by Brittain (1987) and Riggs (1988). In many fish species, all haemoglobin components show the Root effect (e.g. the only haemoglobin of goldfish, *Carassius auratus*, Torracca et al. 1977; the haemoglobins of the bluefin tuna *Thunnus thynnus*, Morris et al. 1981; the haemoglobins of carp *Cyprinus carpio*, Noble et al. 1970), whereas in others, only some components are sensitive to the decrease in pH (e.g. the haemoglobins of rainbow trout *Salmo gairdneri*, Brunori 1975; the eel, *Anguilla anguilla*, Bridges et al. 1983, and the menhaden *Brevatooria tyrannus*, Saffran and Gibson 1978). The Root effect is generally considered to be an extreme Bohr effect (or extension of the Bohr effect; Brittain 1987). The low-affinity T-form of haemoglobin is stabilized to such an extent that the haemoglobin fails to undergo allosteric transition from the T-form to the high-affinity R-form even at very high oxygen tension. This results in the lack of cooperativity of Root effect haemoglobins (i.e. Hill value of one). In most Root effect haemoglobins, the cooperativity of oxygen binding is even less than one at low pH values (see e.g. Brunori 1975; Pennelly et al. 1978; Ikeda-Saito et al. 1983; Noble et al. 1986), which requires that the ligand binding affinity of the different chains in the T-state is different. In fact, it appears that only half of the haem groups can bind oxygen at atmospheric oxygen tensions (Noble et al. 1986) at low pH values.

The oxygen binding curve of haemoglobin trout IV appears to approach asymptotically a saturation value lower than 100% (see e.g. Brunori 1975). Similar observations were made by Bridges et al. (1983) on eel *Anguilla anguilla* blood. These observations would indicate that some haem groups do not bind oxygen at all at low pH values.

Ligand binding to Root effect haemoglobins at low pH values has, however, been observed, using CO as the ligand (e.g. Pennelly et al. 1978; Morris and Gibson 1982; Noble et al. 1986). The n-values for CO binding often have values less than one, indicating marked chain heterogeneity. The CO affinity of the low affinity haem groups is commonly very weak: the P_{50} values of these groups for CO are ca. 100 mmHg in *Antimora rostrata, Coryphaenoides armatus* and *Coryphaenoides brevibarbis* (Noble et al. 1986). Noble et al. (1986) point out that if the partition coeffcient for carbon monoxide and oxygen is 100, P_{50} values for oxygen for these haem groups would be ca. 15 atm, indicating almost compete lack of oxygen affinity.

In view of the above observations, the Root effect can be explained in qualitative terms by modifying the model of Di Cera et al. (1987). Initially, oxygen is bound to the globin chains (probably α chains) which have some oxygen affinity even at the T-state. After the binding of two molecules of oxygen to these chains, oxygen is required to bind to the haem groups which have negligible affinity for oxygen in the T-state. Since the molecule is prevented from allosteric transition from the T-state to the R-state, the binding of oxygen can only be accomplished by using very high oxygen pressures. Thus, the asymptotes of the Hill plot would indicate the T-state oxygen affinities for the two types of chain. For many Root effect haemoglobins studied, the T-state oxygen affinity of the low-affinity chains is so low that it cannot be measured using conventional methods.

The Root effect of haemoglobin is shifted to higher pH values in the presence of organic phosphates (see e.g. Pennelly et al. 1978), and is readily observed in the whole blood of, e.g. the eel (*Anguilla australis schmidtii*; Forster 1985), rainbow trout (Nikinmaa and Soivio 1979) and tench (Jensen and Weber 1982; Jensen et al. 1983) at pH values observed in the major arteries and/or veins. As shown in Fig. 10.8, however, the onset of the Root effect occurs at different red cell pH values in different species. This difference may be related to the physiological role of the Root effect.

The possible physiological roles of the Root effect have been reviewed, e.g. by Ingermann and Terwilliger (1982b), Ingermann (1982) and Brittain (1987). The Root effect is thought to be important in oxygen secretion to the swimbladder and to the eye. The pattern of oxygen secretion to the swimbladder is given in Fig. 10.9 (see e.g. Fänge 1966). Initially, gas gland cells produce large amounts

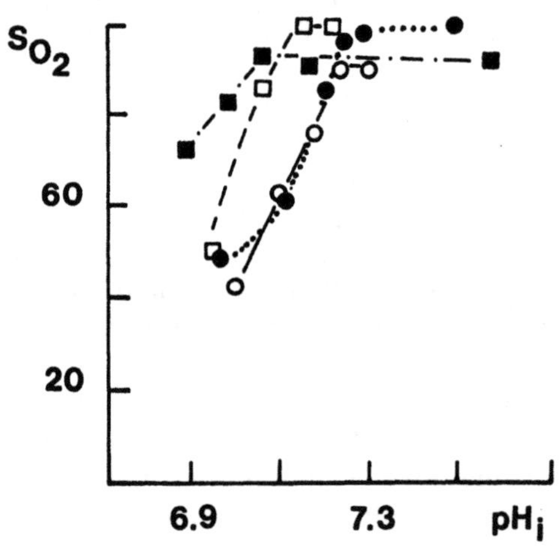

Fig. 10.8. Oxygen saturation at atmospheric oxygen tension vs red cell pH for rainbow trout (*Salmo gairdneri* ●·····●), whitefish (*Coregonus pallasi* ○ —— ○), pikeperch (*Stizostedion lucioperca* □ – □) and carp (*Cyprinus carpio* ■ –·– ■). The onset of the Root effect occurs at higher red cell pH for the salmonids than the other two species (Salama and Nikinmaa, unpublished data)

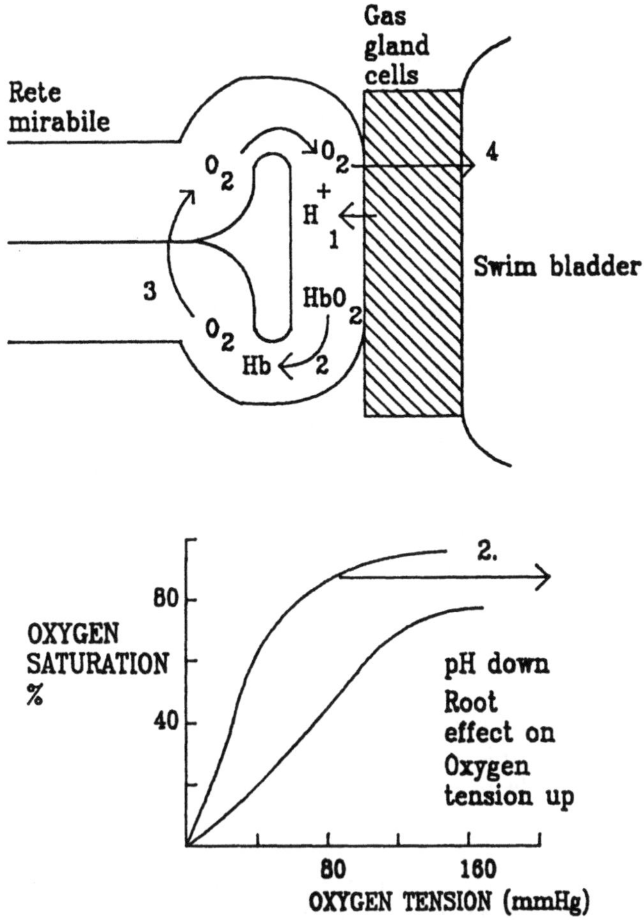

Fig. 10.9. Oxygen secretion into the swimbladder. *1* The cells of the gas gland produce metabolic acid and carbon dioxide which decrease blood pH. *2* As a result, the Root effect sets in, and, because blood is a closed system, the oxygen liberated from the haemoglobin markedly increases the blood oxygen tension. *3* The oxygen liberated diffuses down its partial pressure gradient from the efferent blood to the afferent blood in the rete mirabile of the swimbladder, increasing the oxygen tension of afferent blood. *4* Oxygen diffuses from the afferent blood into the swimbladder

of lactic acid, which acidifies the blood passing the swimbladder. As a result, oxygen is given up from haemoglobin, and, since blood is a closed system, markedly increases the oxygen tension of the blood. In the rete mirabile of the swimbladder, the oxygen given up in the afferent side diffuses to the afferent side, whereby the oxygen tension of the afferent blood increases, and oxygen can be secreted from blood to the swimbladder. Ingermann and Terwilliger (1982b) have examined many fish species without a swimbladder, yet with a pronounced Root effect. The presence of the Root effect correlated with the presence of choroid rete mirabile in all of these species. Furthermore, the data of Farmer et al. (1979) indicate that, among Amazonian fish, the presence of the Root effect could be correlated with the presence of rete mirabile in the eye. High concentrations of oxygen are required to guarantee effective diffusion of oxygen in the poorly vascularized fish eye. These concentrations can be achieved by the help of the Root effect in the rete mirabile of the eye in a manner analogous to the oxygen secretion to the swimbladder. Since trout retinas produce lactic acid (Ingermann 1982), the acidification of blood, and consecutive oxygen secretion, can be achieved.

The adrenergic sodium/proton exchange minimizes changes in the red cell pH in the blood of rainbow trout, and that of several other teleost species, thereby minimizing the Root effect (see Sect. 8.6.1). However, the activity of the sodium/proton exchanger is critically dependent on the oxygen tension: the greater the oxygen tension, the smaller the adrenergic increase in intracellular pH. In fact, stress-induced red cell swelling does not occur at 760 mmHg oxygen tension (Railo et al. 1985). Since the oxygen tension of the rete may reach 800 mmHg (Fairbanks et al. 1969), it is possible that the sodium/proton exchanger does not function in the rete mirabile of the eye, whereby oxygen tension of the eye can be maintained at a high level even when high concentrations of catecholamines are present in plasma.

10.1.2.5 Organic Phosphate Effects

Benesch and Benesch (1967) and Chanutin and Curnish (1967) showed that the major organic phosphate of mammalian red cells, 2,3-diphosphoglycerate, causes a decrease in the oxygen affinity of human haemoglobin. The major organic phosphates (see Sect. 6.7.1 for the organic phosphate distribution in vertebrate red cells) of the red cells of most vertebrates have since been shown to influence haemoglobin-oxygen affinity (for reviews, see e.g. Isaacks and Harkness 1980, 1983; Weber 1982). Exceptions are the cyclostomes (Johansen et al. 1973), at least one elasmobranch (the common sting-ray *Dasyatis sabina*; Mumm et al. 1978), crocodiles (Bauer and Jelkmann 1977) and most ruminants (Isaacks and Harkness 1983). In addition, the oxygen affinity of some haemoglobin components of many fish, e.g. rainbow trout(Brunori 1975; Weber et al. 1976b) and *Catostomus clarkii* (e.g. Powers 1977), and cats are insensitive to organic phosphates.

The mechanism of interaction between organic phosphates and haemoglobin has been reviewed, e.g. by Kilmartin and Rossi-Bernardi (1973), and Benesch and Benesch (1974). Organic phosphates bind preferentially to the T-form of haemoglobin. According to Riggs (1971), the intrinsic haemoglobin association constant for 2,3-DPG in man is 4×10^5 M^{-1} for the T-form, and 1×10^4 M^{-1} for the R-form. The effect of organic phosphates on the haemoglobin-oxygen affinity can be explained using the allosteric model. Because organic phosphates stabilize the T-form of haemoglobin, the value of L (= [T]/[R]) increases, whereby haemoglobin-oxygen affinity decreases. However, several recent studies have shown that, in addition to affecting the allosteric constant, the organic phosphates affect the value of K_T (i.e. the oxygen equilibrium constant of the T-form) at low concentrations, and both the K_T and the K_R (i.e. the oxygen equilibrium constant of the R-form) at high concentrations (see e.g. Weber et al. 1987a). Thus, any explanation of the allosteric effects of organic phosphates on haemoglobin-oxygen affinity must include these findings. Such analysis was carried out by Kister et al. (1987). In their model, there is one organophosphate binding site in haemoglobin molecule. The binding site has a characteristic organic phosphate affinity both in the T- (D_T) and in the R-form (D_R). Binding of organic phosphates changes the structure of both conformations, influencing both their relative stabilities (in favour of the T-form) and their oxygen equilibrium constants.

The effect of organic phosphates on the oxygen affinity of mammalian haemoglobins generally increases in the following order: ATP $\leq$ 2,3-DPG $<$ IPP (see e.g. Isaacks et al. 1984). In birds, the relative effects appear to be 2,3-DPG $\leq$ ATP $<$ IPP (Isaacks and Harkness 1980). In teleost fish, the effect of organic phosphates on haemoglobin-oxygen affinity decreases in the order (IHP) $>$ GTP $>$ ATP $>$ 2,3-DPG (e.g. Gillen and Riggs 1971; Weber and Lykkeboe 1978). However, the effects of ATP and GTP are equal in some fish haemoglobins, as in the organic phosphate sensitive major haemoglobin of trout (trout HB IV; see e.g. Gronenborn et al. 1984).

Two factors influence the effect of organic phosphates on the oxygen equilibrium curve (i.e. the degree by which they stabilize the T-form over the R-form): (1) the negative charge on the organic phosphate – the greater the charge, the greater the effect, and (2) the structure of the organic phosphate binding site (see Sect. 10.3.4).

The effect of organic phosphates on the haemoglobin-oxygen affinity is also influenced by complex formation with other intracellular components. ATP is readily complexed with magnesium, whereby its effect on the oxygen equilibrium curves is markedly reduced (e.g. Bunn et al. 1971). Similar complex formation between GTP and magnesium is also observed (e.g. Weber 1978). However, whereas the effect of ATP on the oxygen equilibrium curve of carp can be completely abolished by the intracellular magnesium, a considerable effect of GTP remains (see Fig. 10.10).

Direct binding of organic phosphates to haemoglobin is not the only way by which they exert an influence on the haemoglobin-oxygen affinity. In addition, an increase in intracellular organic phosphate concentration decreases intracellular pH (see Sect. 8.1.2). In human red cells the nonspecific effect plays a role in the control of haemoglobin-oxygen affinity at 2,3-DPG concentrations above 6 mmol l^{-1} red blood cells (Duhm 1972). Since the normal 2,3-DPG concentration of human red cells is ca. 5 mmol l^{-1} red blood cells, the indirect effect is small. In contrast, Wood and Johansen (1973b) have estimated that the nonspecific effect of organic phosphates on the oxygen affinity of the haemoglobins of the eel, *Anguilla anguilla*, is more important than the specific effect of NTP binding to haemoglobin.

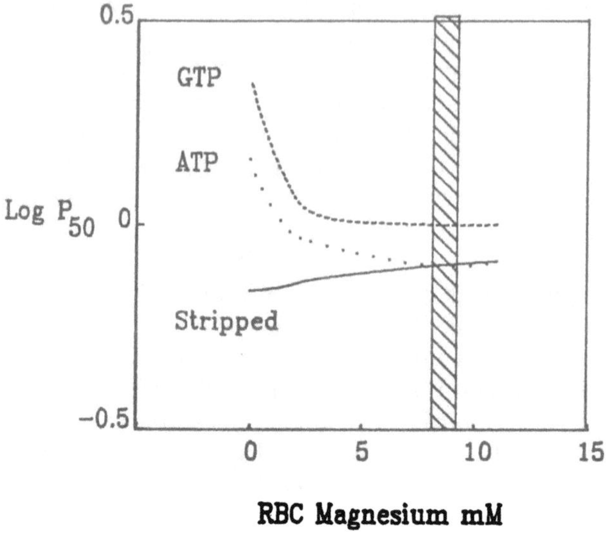

Fig. 10.10. The effect of magnesium-NTP complex formation on the oxygen affinity of carp (*Cyprinus carpio*) haemoglobin (0.1 mM). Both ATP and GTP (0.16 mM) decrease the oxygen affinity of carp haemoglobin in the absence of magnesium. However, at physiological concentrations of magnesium (*shaded area*), the effect of ATP on the oxygen affinity is abolished, whereas a significant effect of GTP persists. Data are from Weber (1978) and Houston (1985)

10.1.2.6 Effects of Other Anions

In addition to the negatively charged organic phosphates, most other anions affect the haemoglobin-oxygen affinity. For example, both lactate and chloride bind preferentially to human deoxyhaemoglobin, thereby decreasing the haemoglobin-oxygen affinity (e.g. Guesnon et al. 1979). The general anion effect is overshadowed by the organic phosphate effect in most haemoglobins: in the absence of organic phosphates a clear decrease is observed in the oxygen affinity of haemoglobin with increasing ionic strength (increasing anion concentration), but in the presence of organic phosphates, ionic strength has hardly any effect (see e.g. Benesch et al. 1969). This indicates that the oxylabile anion binding occurs at the same site(s) as organic phosphate binding. However, in some species, specific effects of certain anions are important in the regulation of haemoglobin-oxygen affinity. Examples of such effects are presented below.

Chloride
Fronticelli et al. (1984) have shown that bovine haemoglobins are sensitive to chloride ions even in the presence of 2,3-DPG: an increase of chloride concentration from 20 to 250 mM increased the P_{50} value from 28 to 39 mmHg in the presence of 10 mM 2,3-DPG. In similar experiments with human haemoglobin, only minimal changes were observed (P_{50} values were 24 and 26 mmHg respectively). Thus, Fronticelli et al. (1984) suggested that chloride ion may be an important regulator of haemoglobin-oxygen affinity in bovine red cells.

Bicarbonate and Carbon Dioxide
The haemoglobin-oxygen affinity of crocodilian blood is insensitive to organic phosphates, but sensitive to CO_2 (Bauer and Jelkmann 1977). The CO_2 effect is mainly due to the binding of two molecules of bicarbonate per deoxyhaemoglobin tetramer (Bauer et al. 1981). The specific effect of bicarbonate on the oxygen affinity of haemoglobin explains why the apparent CO_2 Bohr coefficient of crocodilian blood is so much larger than the fixed acid Bohr coefficient (-0.95 vs -0.2). In addition to bicarbonate, other small ions also lower the oxygen affinity of crocodilian haemoglobin, but much less than bicarbonate. The oxylabile binding site for the different anions appears to be the same (Perutz et al. 1981).

As discussed in Section 9.3, carbon dioxide binds preferentially to the α-amino groups of deoxyhaemoglobin, thereby decreasing the haemoglobin-oxygen affinity. The specific effect of carbon dioxide (and bicarbonate) on the oxygen affinity can be seen from the difference between the CO_2 and fixed acid Bohr coefficients. The effect of carbon dioxide depends on the availability of the α-amino groups for CO_2 binding. The α-amino groups of the α chains of many haemoglobins are acetylated, and therefore not able to form carbamino compounds. Furthermore, the α-amino groups of the β chains form a part of the organic phosphate binding site (see Sect. 10.3.4), whereby the carbon dioxide binding to these chains is reduced in the presence of organic phosphates (see Sect. 9.3). Thus, the specific effect of carbon dioxide on the oxygen affinity of most haemoglobins is absent or very small in physiological conditions. However, in some birds, carbon dioxide decreases the haemoglobin-oxygen affinity both in the absence and presence of the organic phosphate cofactors (e.g. Lutz 1980; Weber et al. 1988a). A pronounced specific effect of CO_2 is also seen in the haemolysate of the hagfish, *Myxine glutinosa,* which is insensitive to either organic phosphates or chloride ions (see Bauer et al. 1975a). It is likely that the binding of carbon dioxide to hagfish haemoglobins favours their aggregation.

10.1.2.7 *Effect of Urea*

Urea is present in high concentrations in the plasma and red cells of elasmobranchs. The effects of urea on the haemoglobin-oxygen affinity have been studied in two shark species (*Squalus acanthias*; Weber et al. 1983b; Weber 1983, *Cephaloscyllium isabella*; Tetens and Wells 1984). In both species, urea significantly increases the oxygen affinity both in the absence and the presence of organic phosphates. Furthermore, the effects of ATP on the oxygen affinity are reduced. Aschauer et al. (1985) discuss the possible reasons for the urea effect. Two major possibilities emerge: urea may bind to the C-termini of haemoglobin chains, thereby reducing the Bohr effect (and thus increasing the oxygen affinity), or may favour dissociation of haemoglobin tetramers to dimers, whereby the ATP effect would be decreased.

10.1.3 Effect of Haemoglobin Concentration

An increase in the haemoglobin concentration decreases its oxygen affinity (see e.g. Forster 1972). This effect must be considered in comparing oxygen equilibrium data obtained in dilute solutions with data obtained on whole blood. However, it is unlikely that the concentration of haemoglobin plays a significant role in the physiological regulation of haemoglobin-oxygen affinity of most animals. Cyclostomes may be an exception. In cyclostome red cells aggregation-deaggregation reactions of haemoglobin play a role in the regulation of oxygen affinity. An increase in the concentration of haemoglobin favours aggregation. This decreases the oxygen affinity of blood, and increases the Bohr effect. A decrease in the concentration of haemoglobin favours deaggregation, increasing the oxygen affinity and decreasing the Bohr effect. Nikinmaa and Weber (1984) have observed that the haemoglobin concentration of lamprey (*Lampetra fluviatilis*) red cells decreases in hypoxic conditions, contributing to the observed increase in haemoglobin-oxygen affinity and decrease in the Bohr effect.

10.1.4 Temperature Effects

Increasing temperature generally decreases the oxygen affinity of haemoglobin, because the overall oxygen binding is an exothermic reaction. The overall heat of the reaction (including the heat of oxygenation, the heat of solution of oxygen into water and the heat of liberation of any allosteric effectors) at a given saturation (ΔHy) can be calculated using the equation (see e.g. Riggs 1970)

$$\Delta Hy = -RT(\delta \ln Y/\delta \ln T). \qquad (10.25)$$

More detailed information on the effects of temperature can be obtained from the Adair constants, since they allow the calculation of heat of oxygenation for each individual oxygenation step. In addition, the entropy change associated with oxygenation can be calculated. This can be done by applying the equation

$$\log K_i = [-\Delta H_i/(2.303 \times R)] \times 1/T + [\Delta S_i/(2.303 \times R)] \qquad (10.26)$$

to the experimental data.

In the equation, ΔH_i is the enthalpy change associated with oxygenation of the i:th oxygenation step; ΔS_i is the entropy change associated with the i:th oxygenation step; R is the universal gas constant; T the absolute temperature and K_i is the i:th association equilibrium constant for oxygen with haemoglobin. Calculations of the heat of oxygenation have shown that the different oxygenation steps are associated with different enthalpy changes (e.g. Mayo and Chien 1980; Jensen and Weber 1987). For example, the average heat of oxygenation (ΔH/mol oxygen) for carp haemoglobin at pH 7.3 and a temperature range from 10° to 25°C is –40 kJ mol^{-1}, and the heats of oxygenation for the consecutive oxygenation steps, K_1-K_4, are –6.3, –6.7, –47.7 and –97.4 kJ mol^{-1} respectively (Mayo and Chien 1980).

Within the physiological pH range a decrease in pH generally diminishes the effect of temperature on the oxygen affinity. This is also seen as a diminished heat of oxygenation (see e.g. Jensen and Weber 1987). Furthermore, as first shown by Benesch et al. (1969), organic phosphates also diminish the effect of temperature on the oxygen affinity. In both cases the effect is due to the fact that the liberation of the allosteric effectors (protons, organic phosphates) is an endothermic reaction, whereby the apparent overall heat of oxygenation is reduced. Again, however, there are exceptions. For example, the pH- and organic phosphate-independent haemoglobin of trout has a very low heat of oxygenation, which, furthermore, is completely independent of pH (see e.g. Brunori 1975).

The oxygen affinity of the haemoglobins of tunas (e.g. Carey and Gibson 1977), some sharks (e.g. Carey and Teal 1969) and sea turtles (Friedman et al. 1985) is little affected by temperature. Carey and Gibson (1977) and Ikeda-Saito et al. (1983) have examined the haemoglobin-oxygen equilibria of tuna (*Thunnus thynnus*) in detail. The oxygen affinity at low saturations decreases with increasing temperature, whereas the oxygen affinity at high saturation increases. The P_{50} value is hardly affected by temperature. Ikeda-Saito et al. (1983) give the apparent heats of oxygenation for the consecutive oxygenation steps: at pH 7.5 and at a temperature range from 10° to 30° they were −40, −18, 65 and 50 kJ mol^{-1} for the first to fourth oxygenation step respectively, giving an average heat of oxygenation of 14.2 kJ mol^{-1}. The value for n decreases markedly with decreasing temperature. The exceptional temperature dependence of tuna haemoglobins can be correlated to the partial homeothermy of these fish: they are able to maintain their core temperature 10–15 °C above ambient using a countercurrent heat exchange system in small arteries and veins (e.g. Carey 1973). If the temperature dependence of haemoglobins were normal, the heat exchanger would cause a marked decrease in the arterial oxygen saturation before the blood reached the sites of oxygen consumption. The peculiar temperature dependence of tuna haemoglobins, however, would be expected to cause a decrease in the arterial oxygen tension upon warming, whereby the oxygen diffusion from the arteries to the veins in the heat exchanger is reduced, and oxygen saturation of arterial blood maintained (see Fig. 10.11).

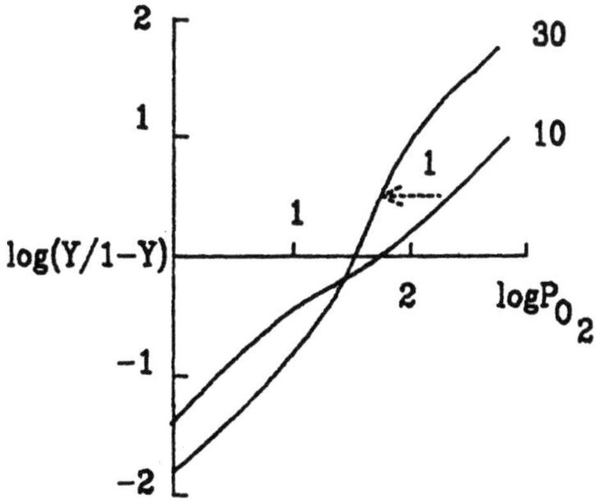

Fig. 10.11. A schematic representation of the effect of temperature on the oxygen equilibrium curves of the tuna (*Thunnus thynnus*; based on Ikeda-Saito et al. 1983). When arterial blood warms up in the heat exchanger upon passage from the surface to the core of the animal, the blood oxygen affinity increases. As a result, the oxygen tension of arterial blood drops, and the diffusion gradient from arterial to venous blood is reduced, reducing the loss of oxygen to the cooling, venous blood in the heat exchanger

10.2 Methods for Determining Blood/Haemoglobin Oxygen Content and Oxygen Equilibrium Curves

10.2.1 Preparation of Blood or Haemoglobin Solutions for Determinations of Oxygen Equilibrium Curves

Meaningful whole blood oxygen equilibrium curves can only be obtained if the temperature of the measurement, the carbon dioxide tension and pH of the incubation can be controlled. In addition, it would be advantageous if the red cell pH could be measured. This is important if the oxygen equilibrium curves of whole blood and haemoglobin solutions are to be compared. In haemoglobin solutions, the pH of the environment of the haemoglobin is generally kept constant. The oxygen equilibrium curves of whole blood are generally constructed at constant extracellular pH. Because the red cell pH of oxygenated blood is lower than that of deoxygenated blood, the oxygen equilibrium curve for whole blood will be left-shifted at low saturations and right-shifted at high saturations. As a result, the apparent Hill coefficient is lower for whole blood oxygen equilibrium curves than for oxygen equilibrium curves of haemoglobin solutions. The reduction of the apparent Hill coefficient depends on the magnitude of the Bohr effect: the greater the Bohr effect, the greater the reduction (see Jensen 1988).

In order to get meaningful data on the haemoglobin-oxygen equilibrium, it is necessary to initially produce cofactor-free haemoglobin, and thereafter to carefully control the composition of the haemoglobin solution. Organic phosphates and other cofactors can be removed from haemolysates either by column cromatography (ion exchange resins) or by dialysis (see e.g. Berman et al. 1971; Weber et al. 1977). The haemolysate must be kept alkaline, because at pH values below 7 considerable amounts of organic phosphates remain bound to the haemoglobin, regardless of the separation method used (Berman et al. 1971).

The presence of multiple haemoglobin components can be checked electrophoretically. After haemolysis, the haemoglobin solution (which may contain some methaemoglobin) is normally reduced to ferrous form using sodium dithionite, and thereafter liganded with carbon monoxide. CO-haemoglobin is more stable than the native haemoglobin. The haemoglobin components can be separated electrophoretically, using e.g. polyacrylamide gels, normally at relatively high pH (e.g. pH 8.9; Rodewald et al. 1984).

The haemoglobin components, later used for oxygen equilibrium studies, can be isolated by isoelectric focusing (see e.g. Weber and De Wilde 1976) in preparative columns containing 0.8–1% ampholines giving a pH range of, e.g. 5 to 9 (Weber et al. 1977). After the separation of components, the column contents are pumped out and the haemoglobin components obtained dialyzed against a buffer (e.g. EDTA-Tris buffer).

10.2.2 Methods for Determining Oxygen Equilibrium Curves

Methods for determining the oxygen equilibrium curves have been reviewed, e.g. by Torrance and Lenfant (1970), Imai (1981b), and Asakura and Reilly (1984). The methods can be divided into two major categories: dicontinuous and continuous. In discontinuous methods the oxygen equilibrium curve is measured point by point. Either the percentage of oxygen saturation is measured at various predetermined oxygen tensions, or the oxygen tension of samples having a range of predetermined oxygen saturations is measured. In continuous methods, the change in the oxygen content of blood is measured continuously against its oxygen tension during oxygenation/deoxygenation.

In the mixing method (e.g. Edwards and Martin 1966), fully oxygenated and fully deoxygenated blood (or haemoglobin solution) are mixed in known proportions, giving the oxygen saturation, and the oxygen tension of the mixture measured using polarographic electrode. It is important that the tonometry is effective – if deoxygenation or oxygenation is not complete, the measured values will be erroneous. Preferentially, the oxygen content of oxygenated and deoxygenated samples should be determined by an independent method (see e.g. Torrance and Lenfant 1970). Also, the organic phosphate concentration of nucleated red cells rapidly decreases in deoxygenated samples during equilibration. Therefore, for every admixture of oxygenated and deoxygenated blood, the organic phosphate concentration will be different.

In Tucker's (1967) method, blood oxygen content is measured by placing a blood sample in a stirred chamber filled with potassium ferricyanide. Ferricyanide liberates all haemoglobin-bound oxygen into solution. The oxygen content of the blood sample can then be calculated from the change in oxygen tension of the chamber, the volume of the chamber and the solubility of oxygen in the ferricyanide solution. The blood oxygen content can also be measured using the amperometric method (Lex-O_2-Con). In this method, the oxygen is released from the blood by a carbon monoxide-containing gas. The gas (and the oxygen liberated from blood) then flows over an oxygen-sensitive cell that gives an electrical output proportional to the number of oxygen molecules (see e.g. Hughes et al. 1982a). A third possibility of measuring the oxygen content (or oxygen saturation) of whole blood, using the polarographic electrode, has been described by Samaja and Rovida (1983). In their method, the measurement is done in an anaerobic measuring cuvette. The oxygen tensions of (1) air-equilibrated buffer before the addition of the blood sample, (2) after the addition of the sample and (3) after the addition of an oxidant are recorded. The blood oxygen content can then be calculated from the three oxygen tensions, the volumes of the reagents and the solubility coefficient of oxygen. The oxygen equilibrium curve can be obtained by varying the oxygen tension of the equilibration gas in each of the methods. The blood oxygen equilibrium curve can be measured with a very small sample size (e.g. 100–150 μl).

In spectrophotometric methods, the haemoglobin-oxygen saturation is determined from the (Soret) absorption spectra of deoxy- and oxyhaemoglobin, using dual wavelength spectrophotometry. In discontinuous spectrophotometric methods, the chamber is filled with blood or haemoglobin solution, equilibrated with a gas of known oxygen tension and the oxygen saturation measured. The gas composition is then changed in a stepwise fashion using gas mixing pumps (see e.g. Weber et al. 1976a). In continuous methods, blood or haemoglobin solution is placed in a chamber and fully deoxygenated (or oxygenated). Oxygen (nitrogen) is then allowed to flow in the chamber at a constant rate. Oxygen equilibrium curves can be plotted from simultaneous measurements of the oxygen saturation (spectrophotometrically) and oxygen tension (polarographically; see e.g. Reeves 1980; Imai 1981b; Gill 1981). Recently, the spectrophotometric methods have been modified to allow the analysis of ligand binding in single erythrocytes (Coletta et al. 1987).

Barnhart (1984) has described a dynamic method for the determination of oxygen equilibrium curves, which requires only the polarographic oxygen electrode. A stirred blood sample is oxygenated by equilibration across a gas-permeable membrane. The gas with high oxygen tension is then replaced with oxygen-free gas, which flows through the system at a known rate. Oxygen diffuses out of the sample through the gas-permeable membrane. The oxygen capacitance curve of the sample can be calculated from the change of the oxygen tension of the sample as a function of time, and then oxygen equilibrium curve integrated from the capacitance curve.

In each of the above methods, the pH can be regulated by controlling the carbon dioxide tension of the equilibration gas. Thus, the Bohr effects can be estimated by determining the oxygen equilibrium curves at different carbon dioxide tensions. The intrinsic alkaline Bohr effect of haemoglobin can also be measured by isoelectric focusing (Bunn and Riggs 1979; Poyart et al. 1981). Deoxygenated and carbon monoxide-treated haemoglobin samples are run through isoelectric focusing columns with a suitable pH gradient, and the pH values of the isoelectric points measured. The difference in the pI-values of the unliganded and liganded haemoglobin gives an estimate of the Bohr effect.

Recent developments have made it possible to measure the blood oxygen saturation in vivo. Steinke and Shepherd (1987) and Takatani et al. (1988a, b) have measured the oxygen saturation of blood in live animals, using reflection type oxygen sensors. Ellsworth et al. (1987) have measured the haemoglobin-oxygen saturation in red blood cells flowing through retractor muscle of the hamster cheek pouch, using a computer-aided videodensitometric method.

10.3 Molecular Aspects of Haemoglobin-Oxygen Binding

The molecular basis of haemoglobin function has been reviewed in detail, e.g. by Perutz (1978; 1979) and Dickerson and Geis (1983).

10.3.1 Binding of Oxygen to the Haem Group

The primary functional unit of haemoglobin is the haem group, i.e. ferrous ion surrounded by the porphyrin ring. Haem group is placed in a pocket formed by the characteristic folding of the globin chain. This arrangement is required in order to make the reaction between the ferrous ion and oxygen reversible.

Two factors influence the equilibrium constant between the ferrous ion and oxygen (see e.g. Winterhalter and De Iorio 1984):

1. The geometry of the haem pocket at the distal side (i.e. the ligand binding site). This influences the binding of oxygen to the ferrous ion by placing different degrees of steric hindrance between the oxygen molecule and the ferrous ion. The steric hindrance is largely caused by Histidine E7 and Valine E11 residues. Valine E11 of the β chains obstructs the ligand binding site of the haem in the T-form, whereby the oxygen affinity of β chains in the T-form is very much lower than that of the α chains (see Perutz 1979). Histidine E7 opposes the rotation of haem on ligand binding, thereby maintaining the restraint responsible for the low oxygen affinity of the T structure. Furthermore, it lowers the carbon monoxide affinity of the haem group and, finally, inhibits the haem iron from oxidation (for details, see Perutz 1979). Because of the steric hindrance on the distal side of the haem group, the oxygen – which binds to the iron by end-on geometry – is tilted away from the axis perpendicular to the porphyrin plane (see e.g. Bunn and Forget 1986).
2. The bond between the ferrous iron and the His(F8) residue. Increased strain (tilt of the iron-histidine bond away from a perpendicular to the haem plane) between the proximal base and the ferrous ion appears to decrease the oxygen affinity of haemoglobin. Increased strain between the ferrous iron of the haem and the proximal histidine may also cause dissociation of the haem group from the globin chain, and increase the rate of autoxidation of haemoglobin by a mechanism depicted in Fig. 10.12 (see Wilson and Knowles 1987).

In the unliganded T-state of human haemoglobin, the histidine-iron bond is tilted 7–8° from perpendicular to the haem plane, and the iron atom lies 0.06 nm out of the mean plane of the porphyrin ring. Such a structural arrangement is unfavourable for ligand binding, whereby the oxygen affinity of T-state is low. In the liganded R-state, the histidine-iron bond is straightened up, and the iron moves towards the plane of the porphyrin ring, whereby the ligand can bind to haem more easily. Binding of the ligand to the T-form cannot straighten the histidine-iron bond and allow the iron to move towards the porphyrin plane, because of steric hindrances characteristic of the T-state structure. The straightening of the histidine-iron bond and movement of iron towards the plane of the porphyrin ring are possible only upon the change of the quarternary structure to the R-state (for details, see Perutz 1979). Similarly, the steric hindrance placed by Valine E11 on the oxygen binding in the β chains can only be relieved after the change in the quarternary structure from T- to the R-state.

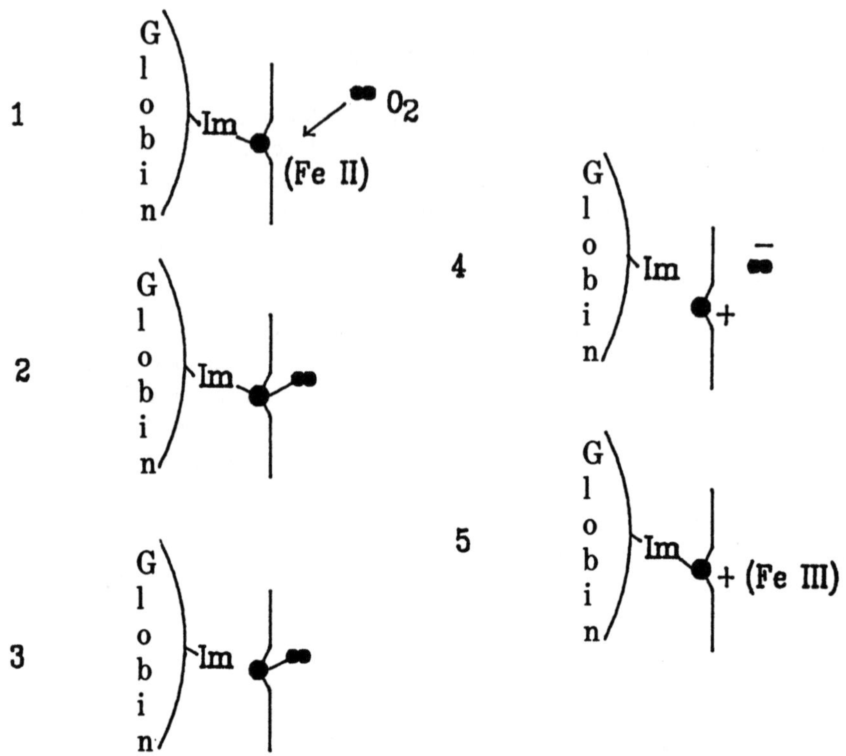

Fig. 10.12. A model for autoxidation of haemoglobin. *1* Oxygen binds to ferrous haemoglobin. *2* Because the iron-proximal histidine bond is bent and cannot be straightened upon ligand binding, *3* the ferrous haem-oxygen complex breaks free from proximal histidine. *4* Molecular oxygen oxidizes ferrous haem to ferric haem and becomes peroxide ion. *5* Ferric ion recombines with the proximal histidine, completing autoxidation (After Wilson and Knowles 1987)

10.3.2 Cooperativity

Cooperativity of oxygen binding results from the change of the low-affinity T-form of haemoglobin to the high-affinity R-form upon the binding of oxygen. Molecular aspects of cooperativity have been reviewed, e.g. by Perutz (1970, 1972, 1978, 1979) and Ten Eyck (1972). The functional unit behind cooperativity is the tetramer, as shown by the studies of Hewitt et al. (1972): the $\alpha\beta$ dimer did not show any cooperative behaviour. Furthermore, when the quarternary structures of the T- and R-form of haemoglobin are compared, the $\alpha_1\beta_1$ and $\alpha_2\beta_2$ dimers move as a unit – bonds are broken at the $\alpha_1\beta_2$ and $\alpha_2\beta_1$ contacts, whereby the two $\alpha\beta$ units move in relation to each other.

Although ligand binding to the α chains of the T-form does not cause any drastic changes in the haemoglobin structure, it causes a movement of the F-helix such that the bond between the penultimate tyrosine (Tyr HC2) and valine 98 (Val FG5) is weakened. Weakening this bond makes possible the breakage of the other bonds unique for deoxyformation.

The salt bridges at the $\alpha_1\beta_2$ interface of mammalian deoxy- and oxyhaemoglobin were given in Tables 3.4 and 3.5. The proximal histidine-iron bond can be straightened upon the breakage of the salt bridges and the iron can move towards porphyrin plane, whereby the oxygen affinity increases and the haemoglobin shows cooperative behaviour.

10.3.3 Bohr Effect

Recent detailed reviews on the molecular aspects of the Bohr effect have been written by Ho and Russu (1987) and Riggs (1988). The alkaline Bohr effect results from the stabilization of the salt bridges unique for the deoxyconformation. The following amino acid residues have been shown to be involved in the Bohr effect in man (see also Perutz et al. 1969; Kilmartin 1972):

1. *C-terminal histidine of the β chain (His HC3).* The imidazole of this histidine residue forms a salt bridge with aspartate (or glutamate) FG1 (β94) or with glutamate F6 in deoxyhaemoglobin (e.g. Perutz and Brunori 1982; Perutz et al. 1985). In the oxyconformation the carbonyl group of His HC3 forms a salt bridge with lysine HC1β. The C-terminal histidine residue accounts for 40–50% of the normal alkaline Bohr effect. This is shown by the observations that its removal halves the Bohr effect (Kilmartin and Wootton 1970), and that the Bohr effect of haemoglobin Cowtown, in which the His HC3 (146)β has been replaced by leucine, is halved as compared to normal human haemoglobin (Perutz et al. 1984). Russu et al. (1980; 1982) have suggested that the contribution of the C-terminal histidine depends on the solvent conditions being increased with increasing chloride concentration.

2. *N-terminal valine of the α chain.* In deoxyhaemoglobin Val1α_1 is in contact with the carbonyl group of the C-terminal arginine of the other α chain (Arg141α_2; see e.g. Kilmartin 1972), whereas in oxyhaemoglobin these residues are free. If the α-amino groups of α chains are blocked by cyanate, the Bohr effect is reduced by 25%, whereas blockade of the α-amino groups of the β chains had no effect on the Bohr effect (Kilmartin and Rossi-Bernardi 1969; Kilmartin 1972). Val1α contributes to the alkaline Bohr effect in chloride-containing media (see e.g. O'Donnell et al. 1979).

Additional Bohr groups. Generally, the contribution of the amino terminal valine of the α chain and carboxyterminal histidine of the β chain to the alkaline Bohr effect totals 75–85% of the total effect (see e.g. Kilmartin 1972). The groups contributing the additional 15–25% to the Bohr effect have, as yet, eluded definitive identification. One possible candidate is histidine 122 of the α chains (see e.g. Perutz 1970). However, this possibility appears unlikely in the light of the recent results of Di Cera et al. (1988). They have shown that His146β and Val1α account for 84% of the alkaline Bohr effect, and that the remaining 16% can be accounted for by a histidine group of the β chains.

The Bohr effect of mammalian haemoglobins is reversed below pH 6, i.e. oxygen affinity increases with decreasing pH. About half of this acid Bohr effect is caused by His(H21)β in human haemoglobin (Perutz et al. 1980).

In the absence of definitive X-ray crystallographic structures for the haemoglobins of nonmammalian vertebrates, the structural basis of alkaline Bohr effect is deduced supposing that the three-dimensional structure of haemoglobin is closely similar to that of the mammalian haemoglobins (see e.g. Perutz and Brunori 1982). The residues involved in the Bohr effect appear to be largely the same as in mammals. The C-terminal histidine of the β chain contributes ca. 50% to the alkaline Bohr effect of carp haemoglobin (Parkhurst et al. 1983). Whenever present, the α-amino group of the N-terminal amino acid of the α chain is expected to contribute to the Bohr effect as in mammals. However, the N-terminal amino acid of the α chain is acetylated in teleost and elasmobranch fish, and cannot play a role in the Bohr effect. In this instance it appears that Lys EF6 of the β chain might contribute to the stabilization of the T-form in the presence of protons (Perutz and Brunori 1982).

Perutz and Brunori (1982) have discussed the possible residues causing the extremely pronounced Bohr effect and the Root effect in teleost fish. They have suggested that the Root effect would result largely from a substitution of Cys F9 of the β chain by serine. The hydroxyl group of the serine would be able to form two hydrogen bonds with the C-terminal histidine of the β chain and stabilize the T-form of the haemoglobin. The role played by the C-terminal histidine is clear. Parkhurst et al. (1983) have shown that the Root effect is abolished in carp haemoglobins devoid of the C-terminal histidine of the β chain. Also, the C-terminal histidine has been replaced by phenylalanine in the pH-insensitive haemoglobins of teleost fish. However, the role played by the serine residue in position F9 of the β chain is less certain: although the haemoglobins of both the lungfish *Lepidosiren paradoxa* and the frog (*Rana*) have Ser F9β (see Sect. 3.5.1), they are devoid of the Root effect (Phelps et al. 1979; Wells and Weber 1985). In the toad *Xenopus*, the Root effect is absent from the red cells under physiological conditions (Bridges et al. 1983), but can be demonstrated by decreasing the pH of haemoglobin solutions in the presence of saturating concentrations of 2,3-diphosphoglycerate and inositolhexaphosphate (Brunori et al. 1987). Thus, in addition to the cysteine-serine substitution, other differences between mammalian and teleost haemoglobins must contribute to the Root effect. Possible substitutions important in this regard are the substitutions at positions HCl and F6 of the β chain. In the Root effect haemoglobins of teleosts, both these positions are occupied by amino acids which cannot form salt bridges with the C-terminal histidine of the β chain. Thus, only serine F9 is available for a salt bridge in the T-form, and no salt bridge is formed in the R-state. In contrast, in the amphibian and lungfish haemoglobins, which have serine F9β but do not show the Root effect, the amino acid at the position F6 is always a polar residue, whereby it can compete with Ser F9 for salt bridge formation with the C-terminal histidine (see Perutz and Brunori 1982). As a result, the bonds between the C-terminal histidine and Ser F9 may be weakened, whereby the Bohr effect could be diminished and the Root effect abolished. The amino acid residues at positions HClβ and F6β in different haemoglobins are listed in Table 10.1.

Table 10.1 Amino acid residues at positions HC1β and F6β of some vertebrates. Data are from Table 3.1

Species	HC1β	F6β
Lesser panda *Ailurus fulgens*	Lys	Glu
Pheasant *Phasianus colchinus*	Lys	Glu
Caiman *Caiman crocodylus*	Lys	Glu
Frog *Rana esculenta*	Ala	Glu
Goldfish *Carassius auratus*	Gln	Val
Dogfish *Squalus acanthias*	Gly	Lys

10.3.4 Binding of Organic Phosphates and Other Anions

Organic phosphates and other anions bind to the cavity between the β chains of haemoglobin tetramers. The following section describes the characteristics of the binding site in different vertebrates. Again, the stereochemical structure of the organic phosphate binding site in nonmammalian vertebrates assumes that the three-dimensional structure of all tetrameric haemoglobins is similar.

Mammals

In human haemoglobin, the organic phosphate binding site is lined with the α-amino terminus, His NA2, Lys EF6 and HisH21 of the β chains (see Arnone 1972). The binding site readily accepts 2,3-diphosphoglycerate with its five negative charges. In several mammalian haemoglobins, organic phosphate binding is diminished because of some structural alteration in the phosphate binding site. In the foetal haemoglobin of man, the His H21 (β143) has been replaced by a neutral residue (Ser). The small or nonexistent sensitivity of cat haemoglobins to organic phosphates can be correlated to a substitution of phenylalanine for histidine in the position NA2 (β2) in both components of haemoglobin components. In addition, the N-terminal amino acid is acetylated in one of the components (Taketa 1974). Similarly, in the llama (see e.g. Braunitzer 1980) and the elephant (Hiebl 1987) the His NA2 has been replaced by asparagine. The ruminant β chains also lack His NA2, whereby organic phosphate binding is markedly reduced (see e.g. Perutz 1970).

Birds

In the organic phosphate binding site of birds, His H21 has been replaced by arginine, which is also positively charged (e.g. Hiebl et al. 1987; Godovac-Zimmermann et al. 1988; Hiebl et al. 1988). Furthermore, two more positive charges may be present in the organic phosphate binding pocket of bird haemoglobins than in that of human haemoglobin: Arg 135β replaces Ala 135β and His 139β replaces Asn 139β (see Schnek et al. 1985). This may facilitate the binding of inositol pentaphosphate to the organic phosphate binding pocket.

Reptiles
Detailed structures for reptilian haemoglobins are known for crocodiles (e.g. Leclercq et al. 1981) and the primitive reptilian, *Sphenodon punctatus* (Abbasi et al. 1988). The anion-binding site of crocodilian haemoglobins does not accept organic phosphates. Instead, a binding site for bicarbonate is formed at each of the β chains by the residues serine (or acetylated alanine) 1β (NA1), lysine 82β (EF6) and glutamate 144β (H22; see Perutz et al. 1981; Leclercq et al. 1981). The organic phosphate binding site of the major haemoglobin of *Sphenodon* may resemble that of birds: the species has N-terminal valine, His NA2, His H17 (β139), and Arg H21 (β143) in the β chain. The minor haemoglobins lack Arg H21, but, instead, have the bird-like residue Arg at the position H13 (β135).

Amphibians
Since the β chains of frog haemoglobins lack of the first six amino acid residues (see Sect. 3.5.1), there is no His β2 lining the organic phosphate binding pocket. However, the other amino acids, corresponding to the amino acids found in the organic phosphate binding pocket of man, Lys EF6 and Lys H21, are both positively charged and capable of interacting with organic phosphates.

Fish
The organic phosphate binding site of a teleost fish, carp (*Cyprinus carpio*) has been examined by Gronenborn et al. (1984). The binding site is stereochemically compatible for the binding of both ATP and GTP. The amino acid residues forming bonds with NTPs appear to be Val NA1, Glu NA2, Lys EF6 and Arg H21. Binding of GTP to the deoxyhaemoglobin is stronger than that of ATP, since GTP is able to form one more hydrogen bond with the globin chain residues than ATP. In organic phosphate-sensitive rainbow trout (*Salmo gairdneri*) haemoglobin, ATP and GTP affect oxygen affinity equally. The only difference between the organic phosphate binding sites of rainbow trout and carp haemoglobins is that Glu NA2 of carp haemoglobin is replaced by aspartate in rainbow trout haemoglobin. This substitution may be adequate to prevent the formation of the additional hydrogen bond makes GTP a stronger effector than ATP in carp haemoglobin (see Gronenborn et al. 1984). The oxygen affinity of the cathodic haemoglobin of rainbow trout (Trout Hb1) is not effected by organic phosphates (see e.g. Brunori 1975). This difference in the behaviour between trout HbI and the other fish haemoglobins for which complete amino acid sequences are available may be due to the following amino acid substitutions: arginine H21 has been replaced by serine and lysine EF6 by leucine (see Barra et al. 1983). Thus, two of the positively charged amino acid residues, thought to be important in organic phosphate binding of carp haemoglobin, are replaced by nonpolar residues.

In the lungfish, *Lepidosiren paradoxa*, the organic phosphate binding site is the same as in birds – as compared to the amino acid residues of the human haemoglobin, His H21 has been replaced by arginine (Rodewald et al. 1984). In the elasmobranchs *Heterodontus portusjacksoni* and *Squalus acanthias*, the organic phosphate binding site is similar to that of man, except for the substitution of the positively charged lysine for histidine at the position H21 (Fisher et al. 1977; Aschauer et al. 1985).

10.4 Formation and Reduction of Methaemoglobin

Oxidation-reduction reactions between ferrous haemoglobin and ferric haemoglobin (methaemoglobin) have been reviewed, e.g. by Bunn and Forget (1986). Haemoglobin subunits that have been oxidized are not able to bind oxygen. Thus, methaemoglobin formation causes a reduction in the oxygen carrying capacity of blood. However, partially oxidized haemoglobins have a higher oxygen affinity than ferrous haemoglobin (see e.g. Bunn and Forget 1986).

The formation of methaemoglobin can result from autoxidation of haemoglobin, as depicted in Section 10.3.1. Autoxidation is facilitated by treatments that stabilize the T-form of haemoglobin. Oxidation of haemoglobin is also induced by several chemical agents. In physiological conditions the most important of these are hydrogen peroxide, thiols and nitrites.

The methaemoglobin concentration of mammalian red cells is low, generally less than 1% of total haemoglobin concentration (see Bunn and Forget 1986). Board et al. (1977) investigated the methaemoglobin concentration of some birds (*Gallus domesticus* and *Columba livia*) and several reptiles (the snakes *Pseudechis porphyriacus* and *Pseudonaja nuchalis*, the lizard *Tiliqua scincoides*, the tortoises *Emydura macquarii* and *Chelodina longicollis* and the crocodiles *Crocodylus johnstoni* and *C. porosus*). In all of these animals the proportion of methaemoglobin of the total haemoglobin was less than 3%. Pough (1969) reported much higher methaemoglobin concentrations (2-5%) in the blood of lizards. Sullivan and Riggs (1964) found very high and variable methaemoglobin concentrations (5-60%) in the blood of turtles. Similarly, Cameron (1971) observed that the methaemoglobin percentage of fish is high and variable (3-17%). Graham and Fletcher (1986) observed even higher methaemoglobin concentrations (7-27% of total haemoglobin) in the blood of temperate marine teleosts. Their data shows that the methaemoglobin concentration was lower in summer than in winter. Härdig and Höglund (1983) investigated the seasonal changes in the methaemoglobin concentration of the blood of the Baltic salmon (*Salmo salar*), and observed a tendency for lower concentrations in winter than in summer.

The proportion of methaemoglobin in the total haemoglobin depends on the balance between methaemoglobin formation and its reduction back to ferrous haemoglobin. In the red cells, methaemoglobin reduction proceeds mainly by the schemes described in Fig. 10.13. Both NADH and NADPH methaemoglobin reductase systems have been described for both mammalian and nucleated red

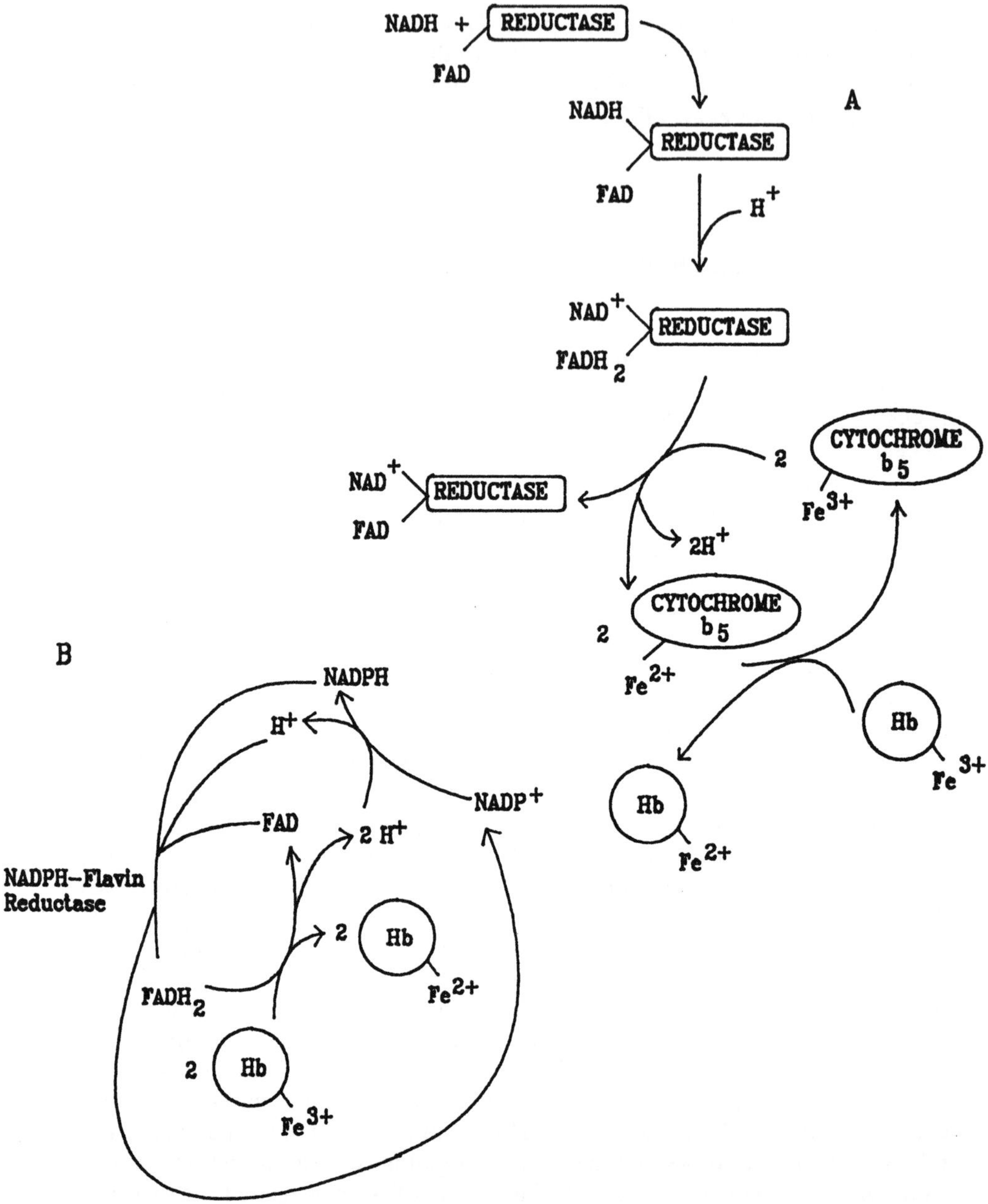

Fig. 10.13A,B. Reduction of methaemoglobin. *A* Cytochrome b_5 reductase pathway. *B* NADPH-flavin reductase pathway. Under physiological conditions methaemoglobin is mainly reduced by the cytochrome b_5 reductase pathway

cells (see e.g. Agar and Harley 1972; Board et al. 1977; Huey and Beitinger 1982). The NADH methaemoglobin reductase activity was greater in the reptilian red cells than in bird red cells (Board et al. 1977). Furthermore, whereas the enzyme in mammalian red cells is largely cytosolic, it is membrane-bound in nucleated red cells (Board et al. 1977).

10.5 Cellular and Molecular Adaptations of Haemoglobin Function to Variations in Respiratory Requirements

Reviews on the physiological adaptation of haemoglobin function to respiratory requirements include those of Johansen and Weber (1976); Wood and Lenfant (1979, 1987), Weber (1982); Baumann et al. (1987); and Jensen (1988). Many of the observed differences in the oxygen affinity of haemoglobin between different species, and adaptatory responses of haemoglobin-oxygen affinity, can be correlated either with differences in environmental oxygen availability or with differences in tissue oxygen consumption. However, it is important to remember that the haemoglobin-oxygen affinity is only a single component of the oxygen transport cascade. As pointed out by Wood and Lenfant (1987), changes in the oxygen affinity can be compensated for by changes in blood haemoglobin concentration, blood flow, tissue oxygen consumption, etc. Thus, roughly the same oxygen delivery to tissues can be achieved by mutant human haemoglobins with P_{50} values ranging from 12 to 70 mmHg.

Environmental oxygen availability is reduced at altitude, in burrows, in many freshwater environments either diurnally or seasonally and in the oxygen minimum zones of the oceans. Owing to the small capacitance coefficient of water for oxygen (see e.g. Dejours 1975), the oxygen concentration of small ponds and lakes is highly variable. In addition to the diurnally occurring hypoxic periods in eutrophic lakes and ponds, bouts of hyperoxia may occur during the day, owing to a pronounced photosynthetic activity of green plants. Since the oxygen concentration of air is much higher and much more stable than that of water, a pronounced change in environmental oxygen availability occurs upon transition from water- to air-breathing.

"Environmental" hypoxia also occurs during the ontogeny of most animals: the oxygen availability of embryos enclosed in eggs may be reduced, and the foetuses of viviparous animals must be able to extract oxygen from the maternal blood.

Marked changes in the demand for oxygen in the tissues occur seasonally in hibernating and aestivating animals, and in all poikilothermic animals upon changes in temperature.

In the following sections, examples are given of interspecific differences in haemoglobin-oxygen affinity between species inhabiting different environments, and of intraspecific adaptations to environmental changes.

The interspecies differences in the oxygen affinity of blood may result either from differences in the intrinsic properties of haemoglobin, from the sensitivity of haemoglobin to allosteric effectors, or from the concentrations of allosteric effectors within the cell. The responses of individual animals to environmental changes normally require changes in the red cell pH, changes in the concentration of organic phosphates (or other cofactors) within the cell, or changes in the red cell volume. Although changes in the haemoglobin pattern within individual animals have been described in response to environmental changes, their role in controlling the oxygen affinity of blood is not clear.

10.5.1 Ontogenetic Changes in Haemoglobin-Oxygen Affinity

Differences between the oxygen affinity of haemoglobin between the embryos, larvae, foetuses and adult animals have been described for all vertebrate groups.

Fish

The role of haemoglobin-oxygen affinity in the maternal-foetal oxygen transfer in viviparous fish has been reviewed by Ingermann and Terwilliger (1984). In addition, the oxygen affinity of embryonic/larval blood in oviparous fish may be different from that of the adult blood.

The ammonochaete larvae of the lampreys *Lampetra fluviatilis* and *Geotria australis* have much higher whole blood oxygen affinity than the adults (Bird et al. 1976; Macey and Potter 1982). The larvae of lampreys are relatively sedentary animals which live in burrows and, therefore, are continuously faced with a hypoxic environment. For these animals, a leftward shift of the oxygen dissociation curve is beneficial, since by these means the oxygen loading in gills can be maintained. Apparently, the fact that the oxygen affinity of larval blood is higher than that of adult blood is due to the presence of larval haemoglobins, which are different from adult ones (Potter and Brown 1975).

The blood oxygen affinity of the foetus of the ovoviviparous elasmobranch, *Squalus suckleyi*, is higher than that of the adult (Manwell 1958a). The higher oxygen affinity appears to be due to the presence of distinct foetal and adult haemoglobins. The "foetus" (9–27 g) and adult of oviparous skate, *Raja binoculata*, have similar blood oxygen affinities (Manwell 1958b). However, the "embryo" (1–1.6 g) of the same species has a higher oxygen affinity than the foetus or the adult. Again, the reason for the increased oxygen affinity may be the presence of embryonic haemoglobin, which is distinct from adult haemoglobin. Embryonic haemoglobin appears to be replaced by adult haemoglobins when the holes at the four corners of the egg case are opened and water can stream through the egg case (Manwell 1958b).

In addition to the different embryonic/foetal and adult haemoglobins, the fact that the organic phosphate concentrations of foetal elasmobranch red cells are lower than those of adult red cells (e.g. Bartlett 1982b) may contribute to the higher blood oxygen affinity of the foetal blood.

The viviparous teleosts *Zoarces viviparous* (see Weber and Hartvig 1984; Hartvig and Weber 1984) and *Embiotoca lateralis* (Ingermann and Terwilliger 1981b) have distinct foetal and adult haemoglobins. Again, the oxygen affinity of the foetal haemoglobin is higher than that of the adult haemoglobin. However, whereas the foetal-maternal difference in the blood oxygen affinity of *Zoarces* can be fully accounted for by the different haemoglobins (e.g. Weber and Hartvig 1984), the effect on *Embiotoca* is partly due to the lower NTP concentration (Ingermann and Terwilliger 1981a) and lower mean cellular haemoglobin concentration (Ingermann and Terwilliger 1982a) of foetal than of adult red cells. A lowered mean cellular haemoglobin concentration will increase haemoglobin-oxygen affinity by decreasing the interaction between haemoglobin and organic phosphates. Also, the dilution of haemoglobin within the cell will tend to increase intracellular pH (see e.g. Soivio and Nikinmaa 1981). It is apparent that the increased oxygen affinity of foetal blood is a hypoxia adaptation. Webb and Brett (1972) have estimated that the oxygen tension of the fluid of the ovarian sac in *Embiotoca lateralis* is substantially lower than the oxygen tension of maternal blood.

Iuchi (1973) has shown that the oxygen affinity of the embryonic haemoglobin of rainbow trout (newly hatched larvae) is greater than the oxygen affinity of adult haemoglobins. Furthermore, its pH sensitivity is reduced, and it does not show the Root effect. This indicates an ontogenetic change in haemoglobin-oxygen affinity in egg-laying teleosts as well.

Amphibians

The larval haemoglobin of the salamander *Pleurodeles waltii* has a higher oxygen affinity than the adult haemoglobin (Flavin et al. 1983). Furthermore, the larval haemoglobins show a reverse pH dependence (oxygen affinity increases with decreasing pH), whereas the adult haemoglobin shows a normal Bohr effect. The difference in haemoglobin-oxygen binding properties may be related to the water-breathing of larvae vs air-breathing of the adult. The foetus of the viviparous caecilian (*Typhlonectes compressicauda*) has a higher blood oxygen affinity than the adult (Garlick et al. 1979). The difference in oxygen affinity between foetal and adult blood can be accounted for by the lower ATP content of the foetal than of maternal red cells – no differences were observed between foetal and maternal haemoglobins.

Reptiles

Both in the viviparous lizard, *Sphenomorphus quoyii* (Grigg and Harlow 1981), and ovoviviparous snake, *Agkistrodon piscivorus* (Birchard et al. 1984), the blood oxygen affinity of the foetus is much higher than that of the adult. In both cases the electrophoretic pattern of foetal and adult haemoglobins is the same. Therefore, the difference in blood oxygen affinity appears to result solely from the lower NTP concentration of foetal than of adult blood (Birchard et al. 1984).

Birds

The changes in blood oxygen affinity during embryonic development have been studied in detail in the chicken *Gallus domesticus* (see e.g. Isaacks and Harkness 1980; Baumann 1984; Reeves 1984). There is a drop in blood oxygen affinity from day 4 to day 7–8 of embryonic development (Baumann 1984; Reeves 1984), which appears to correlate with the replacement of primitive erythrocytes by definitive erythrocytes. After day 8 of development, the oxygen affinity increases throughout the embryonic phase until pipping. This change is mainly due to the decrease in red cell ATP concentration, occurring throughout this period in all bird embryos studied (see Sect. 6.7.2.1 and e.g. Baumann 1984; Reeves 1984). In addition, the red cell pH of embryos with a low oxygen affinity (10-day embryo) is ca. 0.2 units lower than the red cell pH of embryos with a high oxygen affinity (14-day embryo; see Baumann 1984). After hatching, the red cell IPP concentration increases, and blood oxygen affinity decreases to adult levels. The changes in red cell organic phosphate concentration and red cell pH appear to be the predominant factors behind the increase in oxygen affinity from 8-day embryos to hatching, although the ratio of the haemoglobin components A and D also changes during development (see e.g. Baumann 1984).

Reeves (1984) has discussed the evidence for oxygen limitation of embryonic growth. Apparently, oxygen is not a limiting factor in the early embryonic development (up to day 12–14 of development). Afterwards, the embryos in normoxic environment become increasingly hypoxic, since the growth of the embryo can be speeded up in hyperoxic environments. Furthermore, the increase in oxygen affinity and drop in red cell ATP concentration can be accentuated by incubating the embryos in hypoxic conditions (Baumann et al. 1983; Ingermann et al. 1983; Baumann 1984). These observations suggest that the changes in oxygen affinity are an adaptation to a hypoxic environment within the egg.

Mammals

During mammalian development, three sets of haemoglobins are often found (see Sect. 3.5.2). Primitive erythrocytes, produced in the yolk sac early during gestation, contain the embryonic haemoglobins. Foetal haemoglobins are produced in definitive erythroid cells of the foetus and are normally replaced by adult haemoglobins upon birth. However, embryonic haemoglobins are directly replaced by adult haemoglobins in many mammals.

Information on the oxygen binding properties of the embryonic, foetal and adult haemoglobins, and on the whole blood oxygen affinities of the embryos, foetuses and adults are available for a few mammalian species. In man, the oxygen affinity of blood containing more than 30% of embryonic haemoglobins was similar to that of blood containing only foetal haemoglobin (Huehns and Farooqui 1975). The intrinsic oxygen affinity of foetal and adult haemoglobin is the same, and the concentration of 2,3-DPG within the foetal and adult red cells is the same. However, because the interaction between foetal haemoglobin and 2,3-DPG is reduced (see Sect. 10.3.4), the blood oxygen affinity is higher in the foetus than in the adult.

Haemoglobin function in the pig embryo has been studied in detail by Weber et al. (1987b). The pig has four embryonic haemoglobins consisting of two different α-like chains; the adult α-chain and the embryonic ξ-chain, and two different embryonic β-like chains, ε-chain and ϑ-chain. The haemoglobins are expressed sequentially during development. The haemoglobins Gower I (ξ_2 ε_2) and Heide I (ξ_2 ϑ_2) are the predominant haemoglobins during the early part of development. Throughout the early development the haemoglobins Gower II (α_2 ε_2) and Heide II (α_2 ϑ_2) are also present, although less abundantly. The embryonic haemoglobins are gradually replaced by adult haemoglobin. Pigs lack specific foetal haemoglobins. The oxygen affinity of the Gower I and Heide I haemoglobins is higher than that of Gower II and Heide II, which again have higher oxygen affinities than the adult haemoglobin. Furthermore, the oxygen binding of haemoglobins Gower I and Heide I is practically insensitive to pH, the respective Bohr factors being -0.12 and -0.06 in the absence of organic phosphates, and have low cooperativity (n_{50} is 1.4–1.5, in comparison to n_{50} of 2–2.7 in the adult haemoglobin). The reason for the higher oxygen affinity of Gower I and Heide I than that of Gower II and Heide II is due to substitution of Ala(130)α by Thr (see Weber et al. 1987b). The reason for the reduced Bohr effect of these haemoglobins is due to acetylation of the amino terminal of the α-like chain. In conclusion, the increased oxygen affinity of embryonic blood during the early parts of gestation in the pig is due to the presence of high-affinity embryonic haemoglobins.

The foetal blood has higher oxygen affinity (P_{50} 23.0 mmHg, pH 7.4, 37°C) than the adult blood (P_{50} 32–35 mmHg) also late in gestation, at which time the embryonic haemoglobins have been replaced by adult haemoglobins (see Baumann et al. 1973). This is due to the much lower 2,3-DPG concentration of foetal (2.4 mmol l^{-1} red blood cells) than adult (8–10 mmol l^{-1} red blood cells) red cells.

The regulation of red cell oxygen affinity during rabbit ontogeny has been reviewed by Jelkmann and Bauer (1980a). Altogether, six embryonic haemoglobins are present. As in pigs, one group of embryonic haemoglobins consists of embryonic α-type and embryonic β-type chains (Hb E I–III), and the other group consists of adult α and embryonic β-type chains (Hb L I–III). Both in the presence and absence of 2,3-DPG, the oxygen affinity of haemoglobins E I-E III is higher than that of haemoglobins L I-L III, which, again, have a higher oxygen affinity than adult haemoglobins. The specific effect of 2,3-DPG appears to be the same in all the haemoglobin types.

The embryonic haemoglobins are gradually replaced by adult haemoglobin, beginning on day 15 of the 31-day gestation period. Thus, the intrinsic oxygen affinity of the haemoglobins decreases from this time onward. However, the oxygen affinity of whole blood actually increases. This is due to a marked decrease in the red cell 2,3-DPG concentration from 1.4 mol 2,3-DPG mol^{-1} Hb in a 14-day embryo to 0.04 mol 2,3-DPG mol^{-1} Hb in a 30-day foetus (Jelkmann and Bauer 1977). The mechanism of the regulation of red cell 2,3-DPG concentration in rabbit red cells has been reviewed in Section 6.7.2.1. After birth, the oxygen affinity of blood, decreases once again, because of the marked increase in the red cell 2,3-DPG concentration.

A similar picture emerges from the ontogenetic changes in the oxygen affinity of the blood of the mouse. The blood oxygen affinity of the early embryo is high (e.g. Wells 1979), because the embryonic haemoglobin has a higher oxygen affinity and lower Bohr effect than adult haemoglobin (Bauer et al. 1975b). Significantly, the Bohr effect was reversed at pH values below 7.2. Late in gestation, the foetus has adult haemoglobin, but the whole blood oxygen affinity is higher than that of the adult, because the 2,3-DPG level of foetal red cells is lower than that of adult cells (Wells 1979).

The embryonic blood of sheep has a high oxygen affinity, and very little cooperativity of oxygen binding (Wells and Brittain 1983). Two embryonic haemoglobins are present before day 26 of gestation, after which they are replaced by foetal haemoglobin, which also has a higher oxygen affinity than that of the adult (see e.g. Isaacks and Harkness 1983). The difference between oxygen affinity in foetal and adult blood must be mainly due to the differences in the intrinsic oxygen affinities of the respective haemoglobins, since organic phosphates do not affect the oxygen binding of either haemoglobin.

The difference in the oxygen affinity of foetal and adult blood of the yak (*Bos grunniens*) and cattle (*Bos taurus*) is also due to the higher intrinsic oxygen affinity of foetal than adult haemoglobins (see e.g. Isaacks and Harkness 1983; Weber et al. 1988b). Foetal yak red cells contain two foetal haemoglobins, which have higher oxygen affinities than the four adult haemoglobins, both in the absence and the presence of organic phosphates (Weber et al. 1988b).

General Conclusions

As a generalization, vertebrate embryos and foetuses have a higher blood oxygen affinity than the adult. This can be correlated with a decreased oxygen availability in oviparous, ovoviviparous and viviparous animals. Three different strategies are apparent: (1) the intrinsic oxygen affinity of the embryonic/foetal haemoglobins is higher than that of the adult haemoglobins. (2) The intrinsic oxygen affinities are the same, but the sensitivity of foetal haemoglobins to organic phosphates is reduced. (3) There is no specific foetal haemoglobin, but the red cell organic phosphate concentration is lower in the foetus than in the adult animal. The foetus of the cat is an exception to the general rule: foetal blood of cats has the same oxygen affinity as adult blood (e.g. Dhindsa and Metcalfe 1974).

The increased oxygen affinity appears to be the major difference in the oxygen transport properties between embryonic/foetal and adult blood. Data on the blood haemoglobin concentration during the ontogeny of various vertebrates do not show any consistent differences between the embryo/foetus and the adult. The haemoglobin concentration of the pig foetus is almost the same as that of adult (Baumann et al. 1973). The embryos of the *Galliformes* birds have slightly lower blood haemoglobin concentrations than the adults (Isaacks et al. 1976). The oxygen capacity of the foetal blood of the ovoviviparous snake, *Agkistrodon piscivorus*, is almost the same as that of the adult (Birchard et al. 1984). The haemoglobin concentration of the foetal blood of viviparous teleosts *Zoarces viviparous* (Hartvig and Weber 1984) and *Embiotoca lateralis* (Ingermann and Terwilliger 1982a) does not differ significantly from that of the adult blood.

10.5.2 Responses of Air-Breathers to High Altitude

10.5.2.1 The Blood Oxygen Affinity in High-Altitude vs Low-Altitude Species

When comparisons between species are made, it is important that the species compared are closely similar in all respects apart from the oxygen availability of the habitat. Such comparisons can be made between the yak, which lives at high altitudes, and cattle (Weber et al. 1988b); between the llama, which lives at high altitudes, and the camel (see Braunitzer 1980; Isaacks and Harkness 1983); and between the bar-headed goose, which lives at high altitudes, and the Canada and greylag goose (Petchow et al. 1977). In all these cases, the species adapted to hypoxic environments have a higher blood oxygen affinity than species of the same group that live in well-oxygenated environments. In the vulture *Gyps rueppellii*, which soars to altitudes above 11 000 m, the oxygen affinity of haemolysate in the absence of organic phosphates is relatively low, with a P_{50} value of 16.4 mmHg (pH 7.5, 38 °C Weber et al. 1988a). This value is much higher than that of sea-level geese (see Petschow et al. 1977). It is possible that the vulture, in fact, does not consume great amounts of oxygen when soaring at high altitude. If this were the case, there would be no need for a high haemoglobin-oxygen affinity.

The bar-headed goose (*Anser indicus*) which lives at high altitudes has a higher blood oxygen affinity than its lowland relatives, the Canada goose (*Branta canadensis*), and the greylag goose (*Anser anser*), both as an embryo (Snyder et al. 1982) and as an adult (Petschow et al. 1977). Since the red cell inositol pentaphosphate concentrations are similar in all species, the whole blood oxygen affinity difference must be due to a higher intrinsic oxygen affinity of haemoglobin in the bar-headed goose than in the Canada and greylag goose. The P_{50} values in the absence of inositol pentaphosphate (at pH 7.2, 37 °C) were 4.6, 5.3 and 6.0 mmHg (Petchow et al. 1977). In addition, the interaction between organic phosphates and haemoglobin may be reduced (Petschow et al. 1977). Similarly, in the yak, as compared to cattle, the increased oxygen affinity is mainly due to a higher intrinsic oxygen affinity of haemoglobin (Weber et al. 1988).

The Andean guanaco (*Llama guanicoe;* Petchow et al. 1977), which is a high-altitude species, has a reduced organic phosphate effect on haemoglobin-oxygen affinity, because of the substitution of Asn for His at the position NA2 of the β chain (see Braunitzer 1980). Thus, the oxygen affinity of blood (in the presence of normal concentration of intracellular 2,3-DPG) is higher than in the camel, the haemoglobin of which has normal organic phosphate sensitivity (Braunitzer 1980).

10.5.2.2 Intraspecific Responses of Sea-Level Animals to High Altitude

The responses of sea-level humans to high altitude have been most intensively studied. It is generally considered that a rightward shift of the blood oxygen affinity is beneficial at altitudes below 4 300 m, and a leftward shift at altitudes above this (see e.g. Wood and Lenfant 1987). At moderate altitudes, the rightward shift of the oxygen equilibrium curve has a relatively small effect on the oxygen loading in lungs. However, via the reduction of oxygen affinity, the unloading partial pressure of oxygen can be maintained at a high level. As a result, the diffusion of oxygen from tissue capillaries to the oxygen-requiring structures proceeds faster than if the oxygen equilibrium curve were shifted to the left. In contrast, extreme altitudes would seriously reduce oxygen loading if the oxygen equilibrium curve were not shifted to the left. This is shown by the studies of Eaton (1974) and Eaton et al. (1974). The haemoglobin-oxygen affinity of rats was pharmacologically increased, and the animals were then subjected to extreme hypoxia (barometric pressure 233 mmHg) for 90 min. During this period of time, 8 of 10 control animals (with a P_{50} value of 37.3 mmHg at pH 7.4) died, whereas none of the experimental group (with a P_{50} value of 21.0) died. The authors calculated that the arterial oxygen content of the exeperimental group was, in this situation, approximately 14 vol%, whereas that of the control group was only 7 vol%.

The initial response to a moderately high altitude is an increase in ventilation and a respiratory alkalosis. In the short term, this increases the blood oxygen affinity (see e.g. Wood and Lenfant 1979). The increase in pH, however, increases the red cell 2,3-DPG concentration by a mechanism described in Section 6.3, whereby the blood oxygen affinity is decreased back to or below the sea-level value.

It should be noted that the changes in blood oxygen affinity are generally only a minor factor in the responses of animals to high altitude. For example, the decrease in arterial oxygen saturation is compensated for by an increase in red cell number, such that the arterial oxygen content can be maintained up to an altitude of 5500 m (see e.g. Cerretelli 1987).

10.5.3 Blood Oxygen Affinity in Burrowing Air-Breathers

Burrowing air-breathers may be exposed to hypoxic-hypercapnic conditions in their burrows. For example, concentrations of oxygen as low as 6% and carbon dioxide as high a 3.8% have been recorded in the burrows of different fossorial rodents (see Arieli et al. 1977). Thus, the animals could benefit from a high haemoglobin-oxygen affinity and from a reduced sensitivity of haemoglobin to carbon dioxide. This seems, indeed, to be the case. Hall (1965) observed that, among the *Sciuridae*, the oxygen affinity of the blood of the burrowing prairie dog is much higher that the blood oxygen affinity of the flying squirrel. Bartels et al. (1969) observed that, among insectivores, the oxygen affinity of the burrowing mole *Talpa europea* was higher and the Bohr effect smaller than in the shrew *Crocidura russula*. Also, the whole blood oxygen affinity of the armadillo (*Dasypus novemcinctus*; Dhindsa et al. 1971), the pocket gophers (*Thomomys bottae* and *T. umbricus melanotis;* Lechner 1976), the echnida (*Tachyglossus anatinus*; Isaacks et al. 1984) and the Chinese pangolin (*Manis pentadactyla;* Weber et al. 1986) is higher than that of similarly sized nonburrowing mammals. Jelkmann et al. (1981) also observed that carbon dioxide did not affect the haemoglobin-oxygen affinity of moles at constant pH in the presence of organic phosphates.

Jelkmann et al. (1981) examined the reason for the high haemoglobin-oxygen affinity of the mole, *Talpa europea*, and observed that it results from a decreased sensitivity of the haemoglobin to 2,3-DPG. Weber et al. (1986) have investigated the reason for the high blood oxygen affinity of the Chinese pangolin, and observed that it is due to a low 2,3-DPG/Hb ratio whithin the red cell. The 2,3-DPG/Hb ratio was lower than in non-burrowing animals. The oxygen affinity of stripped haemoglobin, and its sensitivity to organic phosphates were the same as in other similarly sized mammals.

In contrast to the blood oxygen affinity, the blood oxygen capacity of burrowing animals appears to be similar to that of nonburrowing animals: the haemoglobin concentration of the echidna is similar to that of the Tasmanian devil (*Sarcophilus harrissii*) and the wallaby (*Thylogale billardierii*; Isaacks et al. 1984). The haemoglobin concentration of burrowing squirrels is similar to that of nonburrowing squirrels (Hall 1965), and the oxygen capacities of mole and shrew blood are similar (Bartels et al. 1969).

10.5.4 Blood Oxygen Transport in Diving Animals

Respiratory adaptations of mammals to diving have been reviewed by Snyder (1983). He compared the blood oxygen affinities of diving mammals to those of terrestrial mammals, and observed that large caetaceans and pinnipeds had lower blood oxygen affinities, whereas small caetaceans and rodents had blood oxygen affinities higher than those of terrestrial mammals of similar weight. The major difference between the species with low-affinity blood and species with high-affinity blood appears to be that in the former the lungs are not used as an oxygen store during the dive (the lungs of large caetaceans collapse during the dive, and pinnipeds dive on expiration), whereas in the latter the air in the lungs is a potential oxygen store. Also, the large caetaceans and pinnipeds dive voluntarily for a longer period than the small caetaceans and rodents.

The Bohr effect of diving mammals tends to be greater than that of terrestrial mammals. The Bohr factors of porpoises and dolphins, weighing 30–150 kg, are around -0.55 (Horvath et al. 1968), somewhat greater than the values for man (-0.48). Similarly, the Bohr factor of the sea lion (*Zalophus californianus*); -0.61; Horvath et al. 1968) is greater than that of the dog (see e.g. Hilpert et al. 1963). The Bohr constant of the killer whale (*Orcinus orca*) is -0.6–0.7 (Lenfant et al. 1968; Dhindsa et al. 1974) far greater than that of the elephant, -0.36 (Hilpert et al. 1963). The Bohr constants of seals, usually above -0.5 (see Lenfant et al. 1969, 1970), are generally greater than those of similarly sized terrestrial mammals. Also, the Bohr constants of diving rodents (-0.54 to -0.66; Clausen and Ersland 1968; Rothstein et al. 1984) tend to be higher than those of terrestrial rodents (-0.43-0.55; Hilpert et al. 1963; Lechner 1976).

In the large caetaceans and pinnipeds, the low blood oxygen affinity and strong Bohr effect may help maintain the partial pressure gradient for oxygen between tissue capillaries and oxygen-consuming sites. In contrast, the high oxygen affinity of small caetaceans and rodents helps the animals to extract oxygen from the lung during the dive (see Snyder 1983). A possible reason for the pronounced Bohr effect is that it limits the decrease in venous pH and enhances carbon dioxide uptake in tissues, as outlined in Section 10.1.2.3. The metabolic acids and CO_2 can be effectively buffered, since the large Bohr effect implies a marked proton uptake by haemoglobin upon deoxygenation.

Again, the observed differences in the respiratory properties of red cells between diving and terrestrial mammals are not the only respiratory adaptations to diving. The blood oxygen capacity and the myoglobin concentration of muscles of divers especially is greater than that of terrestrial mammals (see Snyder 1983). The muscle myoglobin concentration is directly proportional to the duration of the dive.

The fragmentary data on diving birds suggest that their blood oxygen affinity is higher and the Bohr effect greater than the corresponding values in nondiving birds. Milsom et al. (1973) studied the blood oxygen equilibrium curves in Antarctic penguins (*Pygoscelis adeliae, P. papua* and *P. antarctica*), in the giant fulmar (*Macronectes giganteus*) and in the Antarctic skua (*Catharacta skua*). The P_{50} values for the penguins were 30–34 mmHg at pH 7.4 and 38 °C, and those for the nondiving birds 42.5 mmHg. The Bohr factors were -0.5–0.6 and -0.35 respectively. Also, the blood haemoglobin concentration was higher in the penguins than in the fulmar and the skua. Giardina et al. (1985) observed that in the presence of organic phosphates the oxygen affinity of the haemoglobin of the diving birds *Podiceps nigricollis* and *Phalacrocorax carbo sinensis* was higher than that of the similarly sized terrestrial birds. In these species, the intrinsic haemoglobin-oxygen affinity is similar to that of the terrestrial birds, but the effect of inositol hexaphosphate on the haemoglobin-oxygen affinity is markedly reduced. The blood haemoglobin concentration also correlates with the diving habit – the naturally diving tafted duck (*Aythya fuligula*), which dives for 15–40 s, has a higher blood haemoglobin concentration than the dabbling mallard duck (*Anas platyrhynchos*; Keijer and Butler 1982).

Several reptiles, especially turtles, can remain submerged for considerable periods of time: when water temperatures approach 0 °C, e.g. the turtles *Chrysemus picta bellii* can be submerged for several months (Maginniss et al. 1983). At summer temperatures (15 °–25 °C) the periods of submergence for turtles, e.g. *Pseudemys scripta*, vary from minutes to hours (Burggren and Shelton 1979). The active dives appear to be largely aerobic, the lung acting as an oxygen store (Burgren and Shelton 1979). Since the alveolar oxygen tensions decrease only moderately, e.g. from 120 mmHg to 80 mmHg, during these apnoeic periods, no specific requirements are placed on the red cell function. Indeed, no "obvious" adaptations to hypoxia are observed in the blood of active *Pseudemys*. The blood oxygen affinity is relatively high, with P_{50} values of 13.5 mmHg at 15 °C and pH 7.74, 19.3 mmHg at 25 °C and pH 7.55, and 32.8 mmHg at 35 °C and pH 7.44 (Maginniss et al. 1980). The Bohr factor increases with oxygen saturation of blood from -0.15 at low saturations to -0.4 at high saturations at 15 °C.

In contrast to the aerobic "summer" dives, the long periods of submergence at cold temperatures depend on the marked anoxic tolerance of turtles (see e.g. Jackson 1986). These long periods of hypometabolism can be considered as hibernation. Red cell function may play a role in the adaptation of these animals to long periods of submergence. Data are available on the blood oxygen affinity of *Chrysemus picta bellii* (Maginniss et al. 1983), submerged for 4–12 weeks at 3 °C. The carbon dioxide Bohr effect in the blood of *Chrysemus* with access to air and maintained at 24° is approximately -0.4 (Maginniss et al. 1983). Submergence in cold markedly reduces the CO_2 Bohr effect to ca. -0.2. In addition, despite the marked reduction in extracellular pH is submerged animals (the pH values measured at 3 °C were 7.44 for submerged animals and 8.09 for animals with free access to air), the oxygen affinity of blood at low saturations was similar in both groups (see Maginniss et al. 1983). This is possibly caused by impermeability of red cells to protons; Maginniss and Hitzig (1987) observed that the intracellular proton concentration of *Chrysemus* erythrocytes was little affected by submergence-induced acidification in cold. The relatively high blood oxygen affinity at low saturations and its insensitivity to pH may serve to reduce oxygen unloading in tissues during the submergence, whereby the tissues can be rendered hypometabolic, as suggested by Weber (1982) for aestivating lungfish.

The oxygen affinity of crocodilian (alligator) blood in different physiological situations has been studied by Weber and White (1986). Crocodiles dive voluntarily for 30–60 min, and can be made to dive for several hours at 22–27 °C (see e.g. Andersen 1961). During the early parts of the dive, oxygen is extracted from the lungs, as shown by the fact that one-half of the oxygen in the lung air may be consumed during the first 20 min of a two-hout dive (Andersen 1961). Afterwards, the oxygen extraction from the lungs diminishes. This is associated with right-to-left shunting of blood in the heart (i.e. pulmonary circulation is bypassed; e.g. White 1969).

Crocodilian haemoglobins are characterized by a small fixed acid Bohr effect, but a marked sensitivity to bicarbonate (see Sect. 10.1.2.6). During the dive, bicarbonate will be taken up by haemoglobin, whereby the increase in blood carbon dioxide tension remains relatively small (see Andersen 1961). Bicarbonate uptake will cause a decrease in the haemoglobin-oxygen affinity, whereby the blood will give up oxygen in the tissues at relatively high oxygen tensions, even when the pulmonary circulation is bypassed. When the animal surfaces, the blood carbon dioxide tension (and bicarbonate concentration) rapidly drops, and oxygen affinity increases, whereby oxygen loading in lungs is effective. Because the haemoglobin-oxygen affinity is relatively insensitive to fixed acid, effective oxygen loading is possible even if metabolic acids are produced during the dive.

The oxygen affinity of the aquatic snake, *Acrochordus javanicus*, is much higher, and the Bohr effect greater, than in the terrestrial snake *Boa constrictor* (see Johansen and Lenfant 1972). Thus, this species responds to the hypoxic conditions caused by diving in much the same way as mammals and birds. In contrast to the mammals and birds, however, the haemoglobin concentration of the blood of the aquatic snake is much lower than that of the terrestrial snake. This difference is probably related to the lower activity of the aquatic species.

10.5.5 Red Cell Function in Hypoxic Water-Breathers

Fish species living in hypoxic environments tend to have a higher blood oxygen affinity than the species living in well oxygenated waters (see e.g. Johansen and Lenfant 1972; Powers et al. 1979a; Powers 1980). Powers et al. (1979a) compared the blood oxygen affinities of a wide variety of Amazonian fish, and observed that the lotic species in the slow zone of the river, where oxygen tension was often

Table 10.2 P_{50} values for teleost whole blood at extracellular pH of 7.7 at 10°C. The fish were taken from a brackish-water environment (salinity ca. 0.6%)[a]

Species	P_{50} (mmHg)
Pike (*Esox lucius*)	6.5±0.7 (N=4)
Flounder (*Platichthys flesus*)	10.1±1.0 (N=6)
Rainbow trout (*Salmo gairdneri*)	14.4±0.8 (N=8)
Perch (*Perca fluviatilis*)	14.8±1.1 (N=8)
Cod (*Gadus morhua*)	19.4±1.0 (N=6)

[a] Data are from Nikinmaa (1983a). Means ± SEM are given.

reduced, had higher blood oxygen affinities than the lotic species in the fast-flowing zone of the river, where oxygen tension was high. The higher oxygen affinity of the hypoxia-tolerant species is also illustrated by the data in Table 10.2. The blood oxygen affinity of the pelagic cod is lower than the oxygen affinity of the bottom-dwelling flounder. Similarly, the blood oxygen affinity of pike, living in eutrophic lakes, is higher than that of the rainbow trout, which requires well oxygenated environment. The higher oxygen affinity of the hypoxia-tolerant species as compared to that of the species requiring well-oxygenated waters, appears to result mainly from the higher intrinsic oxygen affinity of the haemoglobins. This is illustrated by the data of Weber et al. (1976b) on rainbow trout, and the data of Jensen and Weber (1982) on tench. At pH 7.4 and 15°C, the P_{50} for the isolated haemoglobin components of rainbow trout were 14–18 mmHg, whereas those for the isolated haemoglobins of tench were 1.5–1.8 mmHg. However, there are notable exceptions to the general rule about the high blood oxygen affinity of hypoxia-tolerant fish. The fish living in the oxygen minimum zone of the Pacific Ocean have a low blood oxygen affinity (Douglas et al. 1975). They migrate between oxygen-rich and oxygen-poor waters during the day, and it is likely that these animals markedly reduce their oxygen uptake when staying in the hypoxic water.

The responses of the blood respiratory properties in teleosts encountering hypoxic environments have been studied, using both hypoxia-tolerant species; the eel, *Anguilla anguilla* (Wood and Johansen 1972, 1973a, b; Weber et al. 1976a), the carp (Weber and Lykkeboe 1978; Lykkeboe and Weber 1978; Nikinmaa et al. 1987a); the killifish *Fundulus heteroclitus* (Greaney and Powers 1978) and the tench (Jensen and Weber 1982; 1985b, c), and a species requiring well-oxygenated waters; the rainbow trout (Soivio et al. 1980; Tetens and Lykkeboe 1981, 1985; Nikinmaa and Soivio 1982; Tetens and Christensen 1987; Claireaux et al. 1988). Regardless of the species, the pattern is similar. The blood

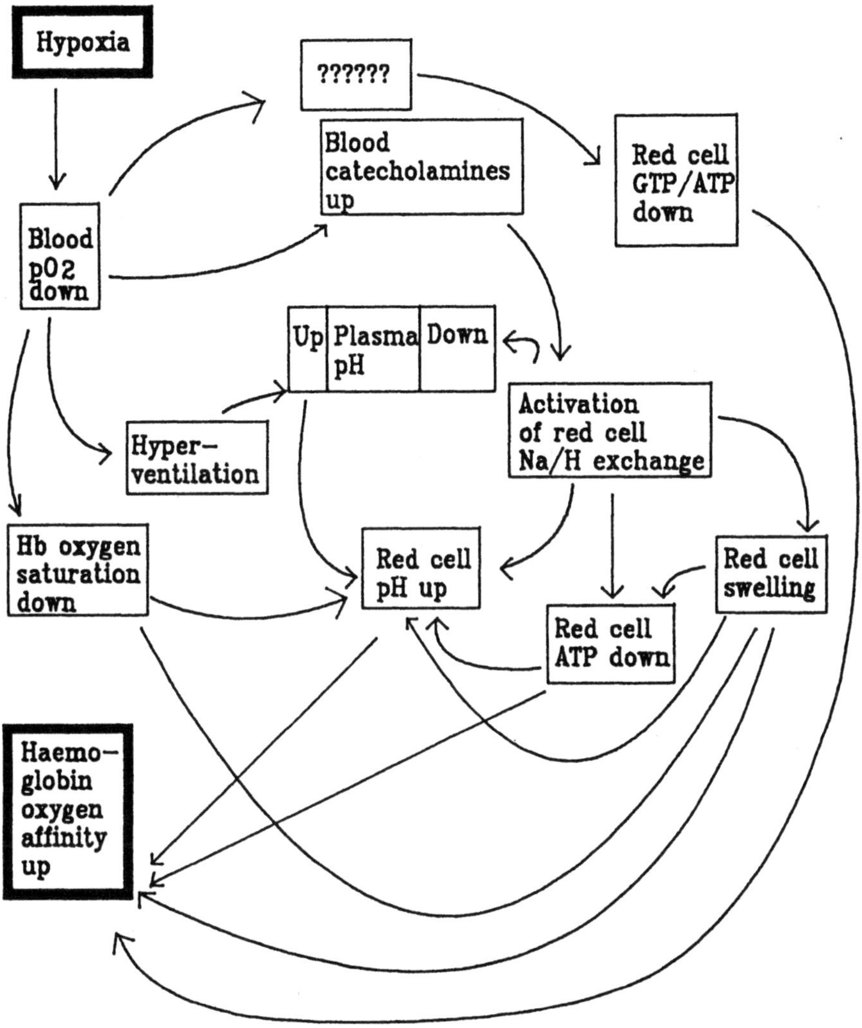

Fig. 10.14. Factors increasing the haemoglobin-oxygen affinity in acutely hypoxia-exposed teleost fish

oxygen affinity increases, whereby oxygen loading in the gills can be maintained in the face of reduced oxygen availability. Two major factors affect the red cell oxygen affinity (see also Fig. 10.14).

Catecholamines are liberated into the circulation of teleosts when they encounter hypoxia (Tetens and Christensen 1987). As a result, the sodium/proton exchange across the red cell membrane is activated, and the red cell pH increases (Tetens and Christensen 1987; Nikinmaa et al. 1987a; Fievet et al. 1988). This effect may coincide with the respiratory alkalosis occurring at the initial stages of hypoxia (e.g. Lykkeboe and Weber 1978). The blood oxygen affinity increases markedly following the adrenergic response (e.g. Tetens and

Christensen 1987). In addition to the adrenergic response, the red cell pH will increase in hypoxia, because the proportion of oxyhaemoglobin decreases and deoxyhaemoglobin increases. The pK value of oxyhaemoglobin is much lower than that of deoxyhaemoglobin (see Jensen 1986 for saturation-dependent changes in red cell pH at a constant extracellular pH).

Upon the onset of hypoxia, the red cell organic phosphate concentrations also start to decrease, as first shown by Wood and Johansen (1972; for discussion see Sect. 6.7.2.3). The decrease in cellular NTP concentration will increase the haemoglobin-oxygen affinity directly, and indirectly by increasing intracellular pH (see Wood and Johansen 1973b; Nikinmaa 1986b). The pH effect is similar to that caused by adrenergic stimulation of the sodium/proton exchange. To date the relative importance of these two factors during the course of hypoxia acclimation has not been elucidated. However, it is likely that the adrenergic response is important in acute hypoxia, and the metabolic drop of NTP concentration takes over in longer-term hypoxia (e.g. Nikinmaa 1983a).

In addition to the increased blood oxygen affinity in hypoxia, blood haemoglobin concentration generally increases in hypoxic fish, as compared to similarly treated normoxic animals (e.g. Wood and Johansen 1972; Soivio et al. 1980).

Hypoxic acclimation in the cyclostome, *Lampetra fluviatilis*, involves an increase in the red cell volume and an increase in the red cell pH (Nikinmaa and Weber 1984). Both these responses increase the haemoglobin-oxygen affinity and decrease the Bohr effect.

10.5.6 Blood Oxygen Affinity in Transition from Water- to Air-Breathing

The oxygen concentration of air is much higher than that of water. Furthermore, there are practically no fluctuations in the oxygen concentration at a given altitude. The density and viscosity of air are, furthermore, much lower than those of water, and the diffusion of oxygen through air is much faster than through water. Adoption of air-breathing has thus markedly contributed to the high oxygen consumption of the present-day mammals and birds. Because of the markedly different requirements of the oxygen transport system in water- and air-breathers, it can be expected that the blood oxygen affinity also changes upon the transition from water- to air-breathing. These changes can be investigated using fishes and amphibians, since in both groups both water-breathing and air-breathing species are found. Furthermore, among amphibians, the larvae breathe water, whereas the adults can be terrestrial. Several teleosts are facultative air-breathers, breathing water in normoxia, but resorting to air-breathing when the water becomes hypoxic.

Johansen and Lenfant (1972) reviewed the data on the blood oxygen affinities of water- and air-breathing amphibians and fishes, and concluded that upon transition from water- to air-breathing the blood oxygen affinity markedly decreased. The oxygen affinity of the blood of the gill-breathing amphibian, *Necturus*, was much higher than that of the predominantly lung-breathing *Rana catesbeiana* (Lenfant and Johansen 1967). The difference in oxygen affinity between these species probably results from differences in the intrinsic oxygen affinities of haemoglobin. This conclusion is based on the observation that the organic phosphate concentrations of the water-breather were higher than those of the air-breather. Since an increase in the organic phosphate concentration of the red cells normally decreases haemoglobin-oxygen affinity, a lower blood oxygen affinity would have been observed if the differences in oxygen affinity were caused by organic phosphates. The oxygen affinity of the axolotle, *Ambystoma mexicanum*, decreases upon triiodothyronine-induced metamorphosis (Gahlenbeck and Bartels 1970; Johansen and Lenfant 1972). Also, as reviewed earlier (Sect. 10.5.1), the blood oxygen affinity of the larval salamander, *P. waltii*, is higher than that of the adult.

When closely related water- and air-breathing teleosts from the Amazonian basin (*Arapaima gigas* vs *Osteoglossum bicirrhosum*, *Hoplerythrinus unitaeniatus* vs *Hoplias malabaricos* and *Pterygoplicthys* sp vs *Pseudoplatystoma* sp) were compared, the water-breathers had a higher blood oxygen affinity, lower haemoglobin concentrations and lower red cell organic phosphate concentrations than the air-breathers (Isaacks et al. 1978b; Johansen et al. 1978). Johansen et al. (1978) showed that the difference in the oxygen affinity between *Arapaima* and *Osteoglossum* blood was due to a difference in the intrinsic oxygen affinities of the haemoglobins. The different oxygen affinities of water- and air-breathers are only observed if closely related species are investigated; when water-breathers and air-breathers from various genera were compared (Powers et al. 1979a), no differences in the blood oxygen affinities between the water-breathers and air-breathers were observed.

Several facultative air-breathers breathe water at high oxygen tensions, but air when the oxygen tension of water is reduced. In this case, the responses of blood oxygen binding properties appear to depend on the breathing behaviour. *Hypostomus* sp. and *Pterygoplichthys* sp. gulp air, after which they remain submerged for considerable periods of time (e.g. Weber et al. 1979). In these species, hypoxia acclimation causes responses in the blood oxygen affinity similar to those in the temperate teleosts, which lack bimodal respiration ability. Thus, the blood oxygen affinity increases and the Bohr effect decreases, mainly because of a drop in the red cell GTP concentration (Weber et al. 1979). These changes in oxygen affinity may be adaptive during air-breathing in these fish for various reasons (see Weber et al. 1979). First, they will favour the depletion of oxygen stores obtained by one gulp, thereby allowing greater time intervals between the gulps. Second, the increased blood oxygen affinity may reduce the loss of oxygen to the surrounding hypoxic water. Third, the systemic arterial vessels are perfused with mixed arterial/venous blood. An increased oxygen affinity would reduce the efficiency loss in oxygen transport, caused by the mixed condition of the arterial blood. Alternatively, the observed changes in oxygen affinity may show that these fish are poorly adapted to air-breathing – the blood oxygen tensions during air-breathing may be lower than those during air-breathing. In this case, reduced internal oxygen tensions would be the reason for the increased oxygen affinity. In contrast to the armoured catfish (*Hypostomus* and *Pterygoplicthys*), the arterial oxygen saturations of the Amazonian eel

(*Synbranchus marmorata*) during air-breathing exceed those during water-breathing (Johansen 1966). Thus, this species is a more effective air-breather than the armoured catfish. Interestingly, no changes in blood oxygen affinity are observed in *Synbranchus* when it is acclimated to hypoxic water (Weber et al. 1979).

10.5.7 Effects of Temperature on Blood Oxygen Affinity in Poikilotherms

The effects of temperature on the haemoglobin-oxygen reaction were derived in Section 10.1.4. Adaptation of red cell function to temperature has been reviewed, e.g. by Wood (1980). Since the increase in temperature increases the oxygen consumption of poikilothermic animals, the rightward shift of the oxygen equilibrium curve with increasing temperature in beneficial in many instances: the delivery of oxygen in the tissues is favoured to fulfill the increased oxygen demand.

It is often thought that two factors contribute to the decrease in the oxygen affinity of the blood occurring with increasing temperature: first, the exothermic nature of the haemoglobin-oxygen reaction as such, and second, the decrease in blood pH which occurs when the temperature is increased. However, the pH as such is not important, but rather the ionization state of the haemoglobin molecule, which largely governs the relative stabilities of the T- and R-form. The most important ionizable group of haemoglobin within the physiological pH range is the imidazole group of histidine residues. The pK value of imidazole group decreases by 0.015–0.017 pH units per °C. Thus, the ionization state of the haemoglobin molecule remains constant if the red cell pH decreases by 0.015–0.017 pH units per °C. The reported temperature-induced changes in blood pH per °C are similar or smaller than the change in the pK value of the imidazole group (see Heisler 1986). Because the ionization state of haemoglobin molecule remains constant with increasing temperature, it is likely that the decrease in pH, which is observed with increasing temperature, does not contribute to the temperature-induced decrease in blood oxygen affinity.

Depending on the temperature range, the organic phosphate cofactors may either augment or attenuate the temperature-induced decrease in the blood oxygen affinity, caused by the exothermic nature of haemoglobin-oxygen reaction. At the lower end of the temperature scale, the organic phosphate concentration increases with increasing temperature (see Sect. 6.7.2.5), decreasing haemoglobin-oxygen affinity, and facilitating oxygen delivery. At the higher end of the temperature scale for a given species, the red cell NTP concentration drops and blood oxygen affinity increases (as compared to cold acclimated animals, if the oxygen affinities are measured at the same temperature, see e.g. Grigg 1969; Nikinmaa 1981; Laursen et al. 1985). The reason for the increased oxygen affinity may be that in the absence of the response, oxygen loading in gills would be reduced more than oxygen delivery would be augmented. As a result, tissue oxygen availability would be reduced. Nikinmaa (1981) observed that when environmental temperature was increased, the decrease in arterial oxygen saturation and increase in blood oxygen affinity coincided.

The temperature sensitivity of haemoglobins is reduced in "heterothermic" species – the tuna, some sharks, and some turtles (see Sect. 10.1.4). All these species maintain their core temperature considerably above the ambient temperature. In contrast, the thermal environment of the animals has little effect on the temperature sensitivity of haemoglobin-oxygen affinity. Powers et al. (1979b) investigated teleost species from a wide variety of thermal environments, and did not observe any general pattern between the thermal environment and the temperature sensitivity of haemoglobin-oxygen affinity. Pough (1980) did not find any differences in the heats of haemoglobin-oxygen reaction in stenothermal and eurythermal reptiles.

Although changes in the ionic microenvironment of haemoglobin have been reported during thermal acclimation (e.g. Houston and Smeda 1979; Houston and Koss 1984), their role in the regulation of haemoglobin-oxygen affinity is still unclear. However, the increase in Mg/NTP ratio with increasing temperature (Houston and Koss 1984) suggests that this effect may contribute to the high-temperature increase in blood oxygen affinity by complexing with NTPs and thereby decreasing their effect on the haemoglobin-oxygen affinity. Similarly, although the proportions of different haemoglobins change with thermal acclimation (e.g. Houston and Cyr 1974; Houston et al. 1976; Weber et al. 1976b), the effects of these changes on the blood oxygen affinity remain unclear. Also, the increase in the total blood haemoglobin concentration, described for teleosts (see e.g. De Wilde and Houston 1967) during acclimation to increased temperatures in some studies, has not been observed in other investigations (e.g. Houston and Smeda 1979; Nikinmaa et al. 1980). Obviously, other factors in addition to the temperature per se influence the number of circulating red cells.

10.5.8 Blood Oxygen Transport in Relation to Activity and Physical Disturbances

Schmidt-Nielsen and Larimer (1958) found that the oxygen affinity of the blood of mammals correlated with body weight – the greater the body weight the higher the oxygen affinity. Nakashima et al. (1985) found that the body weight/oxygen affinity correlation of mammalian bloods was only observed in intact erythrocytes; no such correlation was observed when the oxygen affinities of phosphate-free haemoglobins were compared. The most obvious difference between mammals of different size is the difference in their oxygen consumption. When the P_{50} values of mammalian whole blood are plotted against the oxygen consumption of the animal, a significant correlation is seen (Fig. 10.15): the higher the oxygen consumption, the greater the P_{50} value. It should be noted, however, that this correlation refers to constant carbon dioxide tension, pH and temperature – adequate data are not available to make comparisons using the in vivo values for different mammals. The correlation between body mass and blood oxygen affinity is also observed in poikilotherms, e.g. reptiles (see Wood and Lenfant 1976). Thus, high activity appears to be supported by a low oxygen

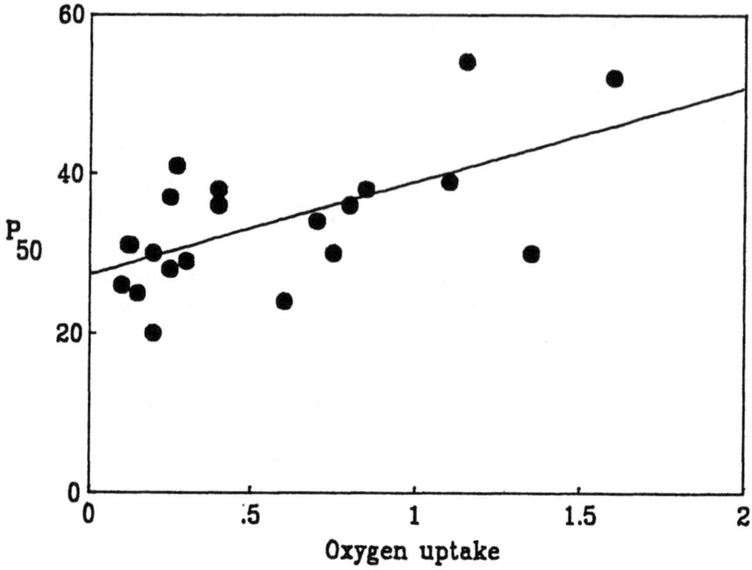

Fig. 10.15. The correlation between the P_{50} value and oxygen consumption in mammals. Data are from Altman and Dittmer (1961). The P_{50} data used were obtained at pH 7.4 and 37–38 °C. N = 21, the correlation coefficient = 0.64, $P < 0.001$

affinity, which makes effective oxygen unloading possible at high capillary oxygen tensions. As a result, the partial pressure gradient for oxygen between blood and the sites of oxygen consumption can be maintained.

In mammals, the oxygen delivery to the exercising tissue can be augmented by the hyperthermia and lactic acidosis occurring in the tissues during heavy exercise. Both these factors decrease the blood oxygen affinity, and, as a result, the required blood flow to the tissues is less than if there had been no shift in the oxygen equilibrium curve.

As exemplified by the temperature-induced changes in the blood oxygen affinity, a rightward shift of the oxygen equilibrium curve is also associated with an increase in activity in poikilotherms. In poikilotherms as well, the acidification of exercising tissues at a constant temperature will augment oxygen delivery by decreasing the blood oxygen affinity via the Bohr effect.

This effect is beneficial only as long as the oxygen loading in the gas exchange organ is not seriously affected. The pH of arterial blood is drastically decreased in exhaustively exercised (or physically disturbed) teleost fish (see e.g. Jensen et al. 1983; Holeton et al. 1983). Since most teleost fish have haemoglobins with a Root effect, oxygen loading in gills would be seriously affected if this drop in the arterial pH also occurred within the red cells. The oxygen loading in gills can be maintained owing to the adrenergic stimulation of red cells. Nikinmaa (1982a, b, 1983b) showed that β-adrenergic stimulation of the red cells of a teleost, rainbow trout, increased the blood oxygen affinity and reduced the Root effect. The improved oxygen affinity and reduction of the Root effect were mainly due to an increase in the red cell pH, although a decrease in the red cell ATP concentration has also been observed (Nikinmaa 1983b). Several studies have later shown that the blockade of β-adrenergic responses causes a considerable reduction in the red cell pH of heavily exercised teleosts (e.g. Nikinmaa et al. 1984; Primmett et al. 1986; Milligan and Wood 1987) and a consecutive drop in the arterial oxygen content.

As shown by Butler et al. (1986) and Ristori and Laurent (1985), significant elevations of plasma catecholamine concentrations in rainbow trout do not occur during sustained (aerobic) exercise, but only during exhaustive exercise or as a response to physical disturbance (e.g. tail grabbing). Notably, the blood pH can be maintained up to 90% of the critical swimming speed (Kiceniuk and Jones 1977), whereby the arterial oxygen loading is not likely to be affected by aerobic exercise.

The species (rainbow trout, striped bass) in which the adrenergic augmentation of blood oxygen transport in exhaustive exercise has been described are characterized by high arterial oxygen tensions at rest. In the tench, *Tinca tinca,* normoxic exercise does not elicit the adrenergic red cell responses (Jensen et al. 1983; Jensen 1987). In contrast, the arterial oxygen tension is markedly elevated and will, as such, cause an increase in arterial oxygen saturation.

10.5.9 Blood Oxygen Affinity in Hibernation, Aestivation and Torpor

Hibernation, aestivation and torpor represent hypometabolic states in which oxygen consumption is markedly reduced. In mammals, the body temperature decreases from ca. 35 °C in the active state to ca. 5 °C in hibernation (e.g. Snapp and Heller 1981). During hibernation the animals are acidotic. These two factors contribute to the reduced oxygen consumption. The decrease in body temperature causes a marked increase in blood oxygen affinity owing to the exothermic reaction between haemoglobin and oxygen. In addition, the 2,3-DPG concentration of red cells decreases in hibernation in the ground squirrel, *Citellus tridecemlineatus* (Musacchia and Volkert 1971), in the hamster, *Mesocricetus auratus* (Tempel and Musacchia 1975), in the hedgehog, *Erinaceus europaeus* (Kramm et al. 1975; Tähti et al. 1981) and in the woodchuck, *Marmota monax* (Harkness et al. 1974). The drop in red cell organic phosphate concentration further increases the blood oxygen affinity. Interestingly, the changes in blood oxygen affinity of poikilothermic animals with decreasing temperature closely resemble those observed for hibernators (see Sect. 10.5.7). Thus, the cold acclimation in poikilotherms may also be considered a form of "hibernation". The increase in blood oxygen affinity with decreasing temperature in both hibernators and poikilotherms serves to decrease the oxygen tension of capillary blood, whereby the diffusion of oxygen to the tissues is slowed down. This may be one of the factors which reduce aerobic metabolism in hibernation (see e.g. Leggio and Morpurgo 1968), although temperature-induced changes in metabolic reaction velocities are probably more important.

The increase in blood oxygen affinity also occurs in aestivation. Johansen et al. (1976) observed that the blood oxygen affinity of an aestivating lungfish (*Protopterus amphibius*) was significantly higher than that of an active animal.

This was obviously due to a reduced red cell NTP concentration. Jokumsen and Weber (1980) showed that the blood oxygen affinity of an aestivating amphibian *Xenopus laevis* was higher than that of an active animal in fresh water. In this case, red cell organic phosphates were not affected. However, a significant increase in red cell urea concentration was the probable cause of the decreased P_{50} value (Jokumsen and Weber 1980).

References

Abbasi A, Weber RE, Braunitzer G, Göltenboth R (1987) Molecular basis for ATP/2,3-bisphosphoglycerate control switch-over (poikilotherm/homeotherm). An intermediate amino-acid sequence in the hemoglobin of the great Indian Rhinoceros (*Rhinoceros unicornis*, Perissodactyla). Biol Chem Hoppe-Seyler 368:323–332

Abbasi A, Wells RMG, Brittain T, Braunitzer G (1988) Primary structure of the hemoglobins from Sphenodon (*Sphenodon punctatus*, Tuatara, Rynchocephalia). Biol Chem Hoppe-Seyler 369:755–764

Abramson S, Miller RG, Phillips RA (1977) The identification in adult bone marrow of pluripotent and restricted stem cells of the myeloid and lymphoid systems. J Exp Med 145:1567–1579

Ackers GK (1979) Linked functions in allosteric proteins: An exact theory for the effect of organic phosphates on oxygen affinity of hemoglobin. Biochemistry 18:3372–3380

Ackers GK, Smith FR (1985) Effects of site-specific amino acid modification on protein interactions and biological function. Annu Rev Biochem 54:597–629

Acquaye C, Walker EC, Schechter AN (1987) The development of a filtration system for evaluating flow characteristics of erythrocytes. Microvasc Res 33:1–14

Adair GS (1925) The hemoglobin system. IV. The oxygen dissociation curve of hemoglobin. J Biol Chem 63:529–545

Adams RLP, Burdon RH (1985) Molecular biology of DNA methylation. Springer, Berlin Heidelberg New York, 247 pp

Adorante JS, Cala PM (1987) Activation of electroneutral K flux in *Amphiuma* red blood cells by N-ethylmaleimide. Distinction between K/H exchange and KCl cotransport. J Gen Physiol 90:209–227

Agar NS (1979) Red cell enzymes-V. Enzyme activities in the red blood cells of Saanen and Angora goats. Comp Biochem Physiol 64B:239–240

Agar NS, Board PG (1983) Red cell metabolism. In: Agar NS, Board PG (eds) Red blood cells of domestic mammals. Elsevier, Amsterdam, pp 227–251

Agar NS, Harley JD (1972) Erythrocytic methaemoglobin reductases of various animal species. Experientia (Basel) 28:1248–1249

Agar NS, Smith JE (1972/73) Glutathione regeneration related to enzyme activities in erythrocytes of sheep. Enzyme (Basel) 14:82–86

Agar NS, Smith JE (1974) Enzymes and glycolytic intermediates in the rabbit erythrocyte. Enzyme (Basel) 17:205–209

Ahkong QF, Fisher D, Tampion W, Lucy JA (1975) Mechanisms of cell fusion. Nature 253:194–195

Aisen P (1982) Current concepts in iron metabolism. Clin Haematol 11:241–257

Aisen P, Listowsky I (1980) Iron transport and storage proteins. Annu Rev Biochem 49:357–393

Al-Badry KS, Nuzhy S (1983) Hematological and biochemical parameters in active and hibernating sand vipers. Comp Biochem Physiol 74A:137–141

Albers C (1985) Gas transport properties of fish blood. In: Gilles R (ed) Circulation, respiration, and metabolism. Current comparative approaches. Springer, Berlin Heidelberg New York, pp 82–90

Albers C, Goetz K-H, Hughes GM (1983) Effect of acclimation temperature on intraerythrocytic acid-base balance and nucleoside triphosphates in the carp, *Cyprinus carpio*. Respir Physiol 54:145–159

Al-Jobore A, Minocherhomjee AM, Villalobo A, Roufogalis BD (1984) Active calcium transport in normal and abnormal human erythrocytes. In: Kruckeberg WC, Eaton JW, Aster J, Brewer GJ (eds) Erythrocyte membranes: recent clinical and experimental advances, vol 3. Alan R Liss, New York, pp 243–292

Allan D, Thomas P (1981) Ca^{2+}-induced biochemical changes in human erythrocytes and their relation to microvesiculation. Biochem J 191:433–440
Allan D, Thomas P, Limbrick AR (1982) Microvesiculation and sphingomyelinase activation in chicken erythrocytes treated with ionophore A23187 (calcimycin) and calcium. Biochim Biophys Acta 693:53–67
Allard WJ, Lienhard GE (1985) Monoclonal antibodies to the glucose transporter from human erythrocytes. Identification of the transporter as a M_r = 55,000 protein. J Biol Chem 260:8668–8675
Alper SL, Palfrey HC, DeRiemer SA, Greengard P (1980) Hormonal control of protein phosphorylation in turkey erythrocytes. Phosphorylation by cAMP-dependent and Ca^{2+}-dependent protein kinases of distinct sites in goblin, a high molecular weight protein of the plasma membrane. J Biol Chem 255:11029–11039
Altland PD, Parker M (1955) Effects of hypoxia on the box turtle. Am J Physiol 180:421–427
Altman PL, Dittmer DS (1961) Biological Handbooks, Blood and other body fluids. Fed Am Soc Exp Biol, Washington, DC 540 pp
Alvarez J, Garcia-Sancho J (1987) An estimate of the number of Ca^{2+}-dependent K^+ channels in the human red cell. Biochim Biophys Acta 903:543–546
Andersen HT (1961) Physiological adjustments to prolonged diving in the American alligator, *Alligator mississippiensis.* Acta Physiol Scand 53:23–45
Anderson RA, Marchesi VT (1985) Regulation of the association of membrane skeletal protein 4.1 with glycophorin by a polyphosphoinositide. Nature 318:295–298
Andrew W (1965) Comparative hematology. Grune & Stratton, New York, 188 pp
Arad Z, Marder J, Eylath U (1983) Serum electrolyte and enzyme responses to heat stress and dehydration in the fowl (*Gallus domesticus*). Comp Biochem Physiol 74A:449–454
Arieli R, Ar A, Shkolnik A (1977) Metabolic responses of a fossorial rodent (*Spalax ehrenbergi*) to simulated burrow conditions. Physiol Zool 50:61–75
Arlot-Bonnemains Y, Fouchereau-Peron M, Moukhtar MS, Benson AA, Milhaud G (1985) Calcium-regulating hormones modulate carbonic anhydrase II in the human erythrocyte. Proc Natl Acad Sci USA 82:8832–8834
Arnone A (1972) X-ray diffraction study of binding of 2,3-diphosphoglycerate to human deoxyhemoglobin. Nature 237:146–149
Aronson PS (1985) Kinetic properties of the plasma membrane Na^+-H^+ exchanger. Annu Rev Physiol 47:545–560
Asai H, Takagi H, Tsunoda S (1976) Some characteristics of erythrocyte membrane and its ATPase from lamprey, *Entosphenus japonicus.* Comp Biochem Physiol 55B:69–75
Asakura T, Reilly MP (1984) Methods for the measurement of oxygen equilibrium curves of red cell suspensions and hemoglobin solutions. In: Nicolau C (ed) Oxygen transport in red blood cells. Pergamon, Oxford, pp 57–75
Aschauer H, Weber RE, Braunitzer G (1985) The primary structure of the hemoglobin of the dogfish shark (*Squalus acanthias*). Biol Chem Hoppe-Seyler 366:589–599
Astrup J (1974) Sodium and potassium in human red cells: variations among centrifuged cells. Scand J Clin Lab Invest 33:231–237
Atha DH, Riggs AF (1982) Conformational transitions in carp hemoglobin at low pH. In: Ho C (ed) Hemoglobin and oxygen binding, Elsevier, Amsterdam, pp 269–275
Atkin NB, Mattinson G, Becak W, Ohno S (1965) The comparative DNA content of 19 species of placental mammals, reptiles, and birds. Chromosoma (Berl) 17:1–10
Audit I, Deparis P, Flavin M, Rosa R (1976) Erythrocyte enzyme activities in diploid and triploid salamanders (*Pleurodeles waltlii*) of both sexes. Biochem Genet 14:759–769
Axelrad AA, McLeod DL, Shreeve MM, Heath DS (1974) Properties of cells that produce erythrocytic colonies in vitro. In: Robinson WA (ed) Hemopoiesis in culture. US Government Printing Office, Washington DC, pp 226–234
Axelrod D (1983) Lateral motion of membrane proteins and biological function. J Membr Biol 75:1–10
Bachand L, Leray C (1975) Erythrocyte metabolism in the yellow perch (*Perca flavescens* Mitchill) – I. Glycolytic enzymes. Comp Biochem Physiol 50 B:567–570
Backman L (1986) Shape control in the human red cell. J Cell Sci 80:281–298

Bagge U, Brånemark PI (1981) Red cell shapes in capillaries. Scand J Clin Lab Invest 41, Suppl 156:59–61

Baker RF (1981) Membrane deformability of metabolically depleted human red cells. Blood Cells 7:551–558

Baldini P, Incerpi S, Pascale E, Rinaldi C, Verna R, Luly P (1986) Insulin effects on human red blood cells. Mol Cell Endocr 46:93–102

Ballas SK, Kliman HJ, Smith ED (1985) Glyceraldehyde-3-phosphate dehydrogenase of rat erythrocytes has no membrane component. Biochim Biophys Acta 831:142–149

Bannai S, Tateishi N (1986) Role of membrane transport in metabolism and function of glutathione in mammals. J Membr Biol 89:1–8

Barnabas J, Goodman M, Moore GW (1971) Evolution of hemoglobin in primates and other therian mammals. Comp Biochem Physiol 39B:455–482

Barnhart MC (1984) Micromethod for dynamic determination of O_2 dissociation curves using a PO_2 electrode. J Appl Physiol 56:795–797

Barni S, Gerzeli G (1985) Comparative aspects of circulating erythrocytes in the trophic and reproductive phases of European eel. Ultrastructure and cytochemistry. Biol Cell 54: 261–270

Baroin A, Garcia-Romeu F, Lamarre T, Motais R (1984a) A transient sodium-hydrogen exchange system induced by catecholamines in erythrocytes of rainbow trout, *Salmo gairdneri*. J Physiol (Lond) 356:21–31

Baroin A, Garcia-Romeu F, Lamarre T, Motais R (1984b) Hormone-induced co-transport with specific pharmacological properties in erythrocytes of rainbow trout, *Salmo gairdneri*. J Physiol (Lond) 350:137–157

Barra D, Petruzzelli R, Bossa F, Brunori M (1983) Primary structure of hemoglobin from trout (*Salmo irideus*) amino acid sequence of the β chain of trout Hb I. Biochim Biophys Acta 742:72–77

Barrett LA, Dawson RB (1974) Avian erythrocyte development: microtubules and the formation of the disk shape. Dev Biol 36:72–81

Bartels H (1972) The biological significance of the Bohr effect. In: Rorth M, Astrup P (eds) Oxygen affinity of hemoglobin and red cell acid-base status, Alfred Benzon Symposium IV. Munksgaard, Copenhagen, pp 717–735

Bartels H, Schmelzle R, Ulrich S (1969) Comparative studies of the respiratory function of mammalian blood. V. Insectivora: shrew, mole and nonhibernating and hibernating hedgehog. Respir Physiol 7:278–286

Bartlett GR (1976) Phosphate compounds in red cells of reptiles, amphibians and fish. Comp Biochem Physiol 55A: 211–214

Bartlett GR (1978) Water-soluble phosphates of fish red cells. Can J Zool 56:870–877

Bartlett GR (1980) Phosphate compounds in vertebrate red blood cells. Am Zool 20:103–114

Bartlett GR (1982a) Phosphates in red cells of a hagfish and a lamprey. Comp Biochem Physiol 73A:141–145

Bartlett GR (1982b) Phosphate compounds in red cells of two dogfish sharks: *Squalus acanthias* and *Mustelus canis*. Comp Biochem Physiol 73A:135–140

Bartlett GR (1982c) Developmental changes of phosphates in red cells of the emu and the rhea. Comp Biochem Physiol 73A:129–134

Bartlett GR, Borgese TA (1976) Phosphate compounds in red cells of the chicken and duck embryo and hatchling. Comp Biochem Physiol 55A:207–210

Bauer C (1974) On the respiratory function of haemoglobin. Rev Physiol Biochem Pharmacol 70:1–31

Bauer C, Jelkmann W (1977) Carbon dioxide governs the oxygen affinity of crocodile blood. Nature 269:825–827

Bauer C, Engels U, Paleus S (1975a) Oxygen binding to haemoglobins of the primitive vertebrate *Myxine glutinosa* L. Nature 256:66–68

Bauer C, Tamm R, Petschow D, Bartels R, Bartels H (1975b) Oxygen affinity and allosteric effects of embryonic mouse haemoglobins. Nature 257:333–334

Bauer C, Forster M, Gros G, Mosca A, Perrella M, Rollema HS, Vogel D (1981) Analysis of bicarbonate binding to crocodilian haemoglobin. J Biol Chem 256:8429–8435

Baumann R (1984) Regulation of oxygen affinity of embryonic blood during hypoxic incubation. In: Seymour RS (ed) Respiration and metabolism of embryonic vertebrates. Junk, Dordrecht, pp 221–230
Baumann R, Engelke M (1987) The Cl^-/OH^- exchanger tributyltin depolarizes primitive red cells from chick embryo. In: Transport in cells and epithelia, 9th ESCPB Conference, Copenhagen, Denmark, Book of Abstracts, p 44
Baumann R, Teischel F, Zoch R, Bartels H (1973) Changes in red cell 2,3-diphosphoglycerate concentration as cause of the postnatal decrease of pig blood oxygen affinity. Respir Physiol 19:153–161
Baumann R, Padeken S, Haller E-A, Brilmayer T (1983) Effects of hypoxia on oxygen affinity, hemoglobin pattern, and blood volume of early chicken embryos. Am J Physiol 244:R733–R741
Baumann R, Bartels H, Bauer C (1987) Blood oxygen transport. In: Farhi LE, Tenney SM (eds) Handbook of physiology, section 3: The respiratory system, vol IV. Gas exchange. American Physiological Society, Bethesda, Maryland, pp 147–172
Bayliss LE (1962) The rheology of blood. In: Hamilton WF, Dow P (eds) Handbook of Physiology, section 2, Circulation, vol I. American Physiological Society, Washington DC, pp 137–150
Beaupain D (1985) Line-restricted hemoglobin synthesis in chick embryonic erythrocytes. Cell Differ 16:101–107
Behnke O (1970) A comparative study of microtubules of disk-shaped blood cells. J Ultrastruct Res 31:61–75
Benesch RE, Benesch R (1967) The effect of organic phosphates from the human erythrocyte on the allosteric properties of hemoglobin. Biochem Biophys Res Commun 26:162–167
Benesch RE, Benesch R (1974) The mechanism of interaction of red cell organic phosphates with hemoglobin. Adv Protein Chem 28:211–237
Benesch RE, Benesch R, Yu CI (1969) The oxygenation of hemoglobin in the presence of 2,3-diphosphoglycerate. Effect of temperature, pH, ionic strength, and hemoglobin concentration. Biochemistry 8:2567–2571
Bennett V, Stenbuck PJ (1979) The membrane attachment protein for spectrin is associated with band 3 in human erythrocytes. Nature 280:468–473
Bennett V, Stenbuck PJ (1980) Human erythrocyte ankyrin. J Biol Chem 255:2540–2548
Benovic JL, Mayor F Jr, Staniszewski C, Lefkowitz RJ, Caron MG (1987) Purification and characterization of the β-adrenergic receptor kinase. J Biol Chem 262:9026–9032
Benovic JL, Bouvier M, Caron MG, Lefkowitz RJ (1988) Regulation of adenylyl-cyclase coupled β-adrenergic receptors. Annu Rev Cell Biol 4:405–428
Bergenhem N, Carlsson U, Strid L (1986) The existence of glutathione and cysteine disulfide-linked to erythrocyte carbonic anhydrase from tiger shark. Biochim Biophys Acta 871:55–60
Berkowitz LR, Orringer EP (1987) Cell volume regulation in hemoglobin CC and AA erythrocytes. Am J Physiol 252:C300–C306
Berkowitz LR, Walstad D, Orringer EP (1987) Effect of N-ethylmaleimide on K transport in density-separated human red blood cells. Am J Physiol 253:C7–C12
Berlin RD, Oliver JM (1975) Membrane transport of purine and pyrimidine bases and nucleosides in animals cells. Int Rev Cytol 42:287–336
Berman M, Benesch R, Benesh RE (1971) The removal of organic phosphates from hemoglobin. Arch Biochem Biophys 145:236–239
Bernstein RE (1954) Potassium and sodium balance in mammalian red cells. Science 120: 459–460
Bernstein RS, Schraer R (1972) Purification and properties of an avian carbonic anhydrase from the erythrocytes of *Gallus domesticus*. J Biol Chem 247:1306–1322
Berridge MJ (1987) Inositol trisphosphate and diacylglycerol: two interacting second messengers. Annu Rev Biochem 56:159–193
Berridge MJ, Irvine RF (1984) Inositol trisphosphate, a novel second messenger in cellular signal transduction. Nature 312:315–321
Bessis M (1973) Red cell shapes. An illustrated classification and its rationale. In: Bessis M, Weed RI, Leblond PF (eds) Red cell shape, Springer, Berlin Heidelberg New York, pp 1–25
Bessis M, Mohandas N (1975) A diffractometric method for the measurement of cellular deformability. Blood Cells 1:307–313

Bethlenfalvay NC, Lima JE, Waldrup T (1984) Studies on the energy metabolism of opossum (*Didelphis virginiana*) erythrocytes. I. Utilization of carbohydrates and purine nucleosides. J Cell Physiol 120:69–74

Beutler E (1983) Active transport of glutathione disulfide from erythrocytes. In: Larsson A (ed) Functions of glutathione: biochemical, physiological, toxicological, and clinical aspects. Raven, New York, pp 65–74

Bidani A, Crandall ED (1988) Velocity of CO_2 exchanges in the lung. Annu Rev Physiol 50:639–652

Bidani A, Mathew SJ, Crandall ED (1983) Pulmonary vascular carbonic anhydrase activity. J Appl Physiol 55:75–83

Bihler I, Charles P, Sawh PC (1982a) Role of calcium in the regulation of sugar transport in the avian erythocyte: effects of the calcium ionophore, A23187. Cell Calcium 3:243–262

Bihler I, Charles P, Sawh PC (1982b) Sugar transport regulation in avian red blood cells: role of Ca^{2+} in the stimulatory effects of anoxia, adrenalin, and ascorbic acid. Can J Physiol Pharmacol 60:615–621

Birchard GF, Black CP, Schuett GW, Black V (1984) Foetal-maternal blood respiratory properties of an ovoviviparous snake the cottonmouth, *Agkistrodon piscivorus*. J Exp Biol 108:247–255

Bird DJ, Lutz PL, Potter IC (1976) Oxygen dissociation curves of the blood of larval and adult lampreys (*Lampetra fluviatilis*). J Exp Biol 65:449–458

Bishop C (1964) Overall red cell metabolism. In: Bishop C, Surgenor DM (eds) The red blood cell. Academic Press, New York pp 148–188

Bishop WR, Bell RM (1985) Assembly of the endoplasmic reticulum phospholipid bilayer: the phosphatidylcholine transport. Cell 42:51–60

Bishop WR, Bell RM (1988) Assembly of phospholipids into cellular membranes: biosynthesis, transmembrane movement and intracellular translocation. Annu Rev Cell Biol 4:579–610

Blikstad I, Nelson WJ, Moon RT, Lazarides E (1983) Synthesis and assembly of spectrin during avian erythropoiesis: stoichiometric assembly but unequal synthesis of α and β spectrin. Cell 32:1081–1091

Bly JE, Clem LW (1988) Temperature mediated processes in teleost immunity: homeoviscous adaptation by channel catfish peripheral blood cells. Comp Biochem Physiol 91A:481–485

Bly JE, Buttke TM, Meydrech EF, Clem LW (1986) The effects of in vivo acclimation temperature on the fatty acid composition of channel catfish (*Ictalurus punctatus*) peripheral blood cells. Comp Biochem Physiol 83B:791–795

Board PG, Agar NS (1983) Glutathione metabolism in erythrocytes. In: Agar NS, Board PG (eds) Red blood cells of domestic mammals. Elsevier, Amsterdam pp 253–269

Board PG, Agar NS, Gruca M, Shine R (1977) Methaemoglobin and its reduction in nucleated erythrocytes from reptiles and birds. Comp Biochem Physiol 57B:265–267

Boer de E, Antoniou M, Mignotte V, Wall L, Grosveld F (1988) The human β-globin promoter; nuclear protein factors and erythroid specific induction of transcription. EMBO J 7:4203–4212

Boivin P, Galand C, Bertrand O (1986) Properties of a membrane-bound tyrosine kinase phosphorylating the cytosolic fragment of the red cell membrane band 3 protein. Biochim Biophys Acta 860:243–252

Bolis L (1973) Comparative transport of sugars across red blood cells. In: Bolis L, Schmidt-Nielsen K, Maddrell SHP (eds) Comparative physiology. Elsevier, Amsterdam pp 583–590

Bolis L, Luly P, Baroncelli V (1971) D(+)-glucose permeability in brown trout *Salmo trutta* L. erythrocytes. J Fish Biol 3:273–275

Bonaventura C, Sullivan B, Bonaventura J, Bourne S (1977) Anion modulation of the negative Bohr effect of haemoglobin from a primitive amphibian. Nature 265:474–476

Borgese F, Garcia-Romeu F, Motais R (1986) Catecholamine-induced transport systems in trout erythrocyte. J Gen Physiol 87:551–556

Borgese F, Garcia-Romeu F, Motais R (1987) Ion movements and volume changes induced by catecholamines in erythrocytes of rainbow trout: effect of pH. J Physiol (Lond) 382:145–157

Bossa F, Barra D, Petruzzelli R, Martini F, Brunori M (1978) Primary structure of hemoglobin from trout (*Salmo irideus*). Amino acid sequence of α chain of Hb trout I. Biochim Biophys Acta 536:298–305

Bossak ET, Gordon AS, Charipper HA (1948) Influence of endocrine factors on hemopoiesis in the adult frog, *Rana pipiens*. J Exp Zool 109:13–32

Botta JA, De Mendoza D, Morero RD, Farias N (1983) High affinity L-triiodothyronine binding sites on washed rat erythrocyte membranes. J Biol Chem 258:6690–6692
Bottomley SS (1968) Characterization and measurement of heme synthetase in normal human bone marrow. Blood 31:314–322
Bourne PK, Cossins AR (1981) The effects of thermal acclimation upon ion transport in erythrocytes. J Therm Biol 6:179–181
Bourne PK, Cossins AR (1982) On the instability of K^+ influx in erythrocytes of the rainbow trout, *Salmo gairdneri*, and the role of catecholamine hormones in maintaining in vivo influx activity. J Exp Biol 101:93–104
Bourne PK, Cossins AR (1984) Sodium and potassium transport in trout (*Salmo gairdneri*) erythrocytes. J Physiol (Lond) 347:361–375
Boutilier RG, Iwama GK, Randall DJ (1986) The promotion of catecholamine release in rainbow trout, *Salmo gairdneri*, by acute acidosis: interactions between red cell pH and haemoglobin oxygen-carrying capacity. J Exp Biol 128:145–157
Boyd TA, Cha CJ, Forster RP, Goldstein L (1977) Free amino acids in tissues of the skate *Raja erinacea* and the stingray *Dasyatis sabina*: effects of environmental dilution. J Exp Zool 199:435–442
Brahm J (1977) Temperature-dependent changes of chloride transport kinetics in human red cells. J Gen Physiol 70:283–306
Branton D, Cohen CM, Tyler J (1981) Interaction of cytoskeletal proteins on the human erythrocyte membrane. Cell 24:2533–2541
Braunitzer G (1980) Phosphat-Hämoglobin-Wechselwirkung: Zur Atmung des adulten Menschen, des menschlichen Foetus, des Lamas und des Kamels. Klin Wochenschr 58:701–708
Braunitzer G, Godovac J (1982) The amino acid sequence of pheasant (*Phasanius colchinus colchinus*) hemoglobins. Hoppe-Seyler's Z Physiol Chem 363:229–238
Braunitzer G, Rodewald K (1980) Die Sequenz der α- und β-Ketten des Hämoglobins des Goldfisches (*Carassius auratus*). Hoppe-Seyler's Z Physiol Chem 361:587–590
Bridges CR, Hlastala MP, Riepl G, Scheid P (1983) Root effect induced by CO_2 and by fixed acid in the blood of the eel *Anguilla anguilla*. Respir Physiol 51:275–286
Bridges CR, Pelster B, Scheid P (1985) Oxygen binding in blood of *Xenopus laevis* (Amphibia) and evidence against Root effect. Respir Physiol 61:125–136
Briehl RW (1963) The relation between the oxygen equilibrium and aggregation of subunits in lamprey hemoglobin. J Biol Chem 238:2361–2366
Brittain T (1987) The Root effect. Comp Biochem Physiol 86B:473–481
Brock MA (1960) Production and life span of erythrocytes during hibernation in the golden hamster. Am J Physiol 198:1181–1186
Brock MA (1964) Hibernation and temperature effects on the ageing of red blood cells. Ann Acad Sci Fenn 71:53–63
Broyles RH, Johnson GM, Maples PB, Kindell GR (1981) Two erythropoietic microenvironments and two larval red cell lines in bullfrog tadpoles. Dev Biol 81:299–314
Brugnara C, Tosteson DC (1987) Cell volume, K transport, and cell density in human erythrocytes. Am J Physiol 252:C269–C276
Brugnara C, Bunn HF, Tosteson DC (1986) Regulation of erythrocyte cation and water content in sickle cell anemia. Science 232:388–390
Brunner A Jr, Coiro JRR, Menezes H, Mitsutani CY, Carvalho Dos Santos MAS (1975) Vesicles carrying nuclear material in mature *Cyprinus carpio* erythrocytes. Experientia (Basel) 31: 531–532
Brunori M (1975) Molecular adaptation to physiological requirements: the hemoglobin system of trout. Curr Top Cell Regul 9:1–39
Brunori M, Giardina B, Chiancone E, Spagnuolo C, Binotti I, Antonini E (1973) Studies on the properties of fish hemoglobins. Molecular properties and interaction with third components of the isolated hemoglobins from trout (*Salmo irideus*). Eur J Biochem 39:563–570
Brunori M, Bonaventura J, Focesi A, Galdames-Portus MI, Wilson MT (1979) Separation and characterization of the hemoglobin components of *Pterygoplichthys pardalis*, the Acaribodo. Comp Biochem Physiol 62A:173–177
Brunori M, Bellelli A, Giardina B, Condo S, Perutz MF (1987) Is there a Root effect in *Xenopus* hemoglobin? FEBS Lett 221:161–166

Buckley JA (1982) Hemoglobin-glutathione relationships in trout erythrocytes treated with monochloramine. Bull Environ Contam Toxicol 29:637–644

Bundy HF, Cheng B (1976) Amphibian carbonic anhydrase: purification and partial characterization of the enzyme from erythrocytes of *Rana catesbeiana*. Comp Biochem Physiol 55B;265–271

Bunn HF (1987) Subunit assembly of hemoglobin: an important determinant of hematologic phenotype. Blood 69:1–6

Bunn HF, Forget BG (1986) Hemoglobin: molecular, genetic and clinical aspects. Saunders, Philadelphia 690 pp

Bunn HF, McDonald MJ (1983) Electrostatic interactions in the assembly of haemoglobin. Nature 306:498–500

Bunn HF, Riggs A (1979) The measurement of the Bohr effect of fish hemoglobins by gel electrofocusing. Comp Biochem Physiol 62A:95–99

Bunn HF, Ransil BJ, Chao A (1971) The interaction between erythrocyte organic phosphates, magnesium ion and hemoglobin. J Biol Chem 246:5273–5279

Burggren WW, Shelton G (1979) Gas exchange and transport during intermittent breathing in chelonian species. J Exp Biol 82:75–92

Burwell EL, Brickley BA, Finch CA (1953) Erythrocyte life span in small animals. Comparison of two methods employing radioiron. Am J Physiol 172:718–724

Bushnell PG, Nikinmaa M, Oikari A (1985) Metabolic effects of dehydroabietic acid on rainbow trout erythrocytes. Comp Biochem Physiol 81C:391–394

Butler PJ, Metcalfe JD, Ginley SA (1986) Plasma catecholamines in the lesser spotted dogfish and in rainbow trout at rest and during different levels of exercise. J Exp Biol 123:409–421

Cabantchik ZI, Rothstein A (1972) The nature of membrane sites controlling anion permeability of human red blood cells as determined by studies with disulfonic stilbene derivatives. J Membr Biol 10:311–330

Cala PM (1977) Volume regulation by flounder red blood cells in anisotonic media. J Gen Physiol 69:537–552

Cala PM (1980) Volume regulation by *Amphiuma* red blood cells: the membrane potential and its implications regarding the nature of ion-flux pathways. J Gen Physiol 76:683–708

Cala PM (1983a) Cell volume regulation by *Amphiuma* red blood cells. The role of Ca^{2+} as a modulator of alkali metal/H^+ exchange. J Gen Physiol 82:761–784

Cala PM (1983b) Volume regulation by red blood cells: mechanisms of ion transport. Mol Physiol 4:33–52

Cala PM (1985a) Volume regulation by *Amphiuma* red blood cells: strategies for identifying alkali metal/H^+ transport. Fed Proc 44:2500–2507

Cala PM (1985b) Volume regulation by *Amphiuma* red blood cells: characteristics of volume-sensitive K/H and Na/H exchange. Mol Physiol 8:199–214

Cala PM (1986a) Volume-sensitive ion fluxes in *Amphiuma* red bood cells: general principles governing Na-H and K-H exchange transport and Cl-HCO_3 exchange coupling. Curr Top Membr Transp 27:193–218

Cala PM (1986b) Volume-sensitive alkali metal-H transport in *Amphiuma* red blood cells. Curr Top Membr Transp 26:79–99

Cala PM, Anderson SE, Gragoe EJ Jr (1988) Na-H exchange-dependent cell volume and pH regulation and disturbances. Comp Biochem Physiol 90A:551–555

Cameron JN (1971) Methemoglobin in erythrocytes of rainbow trout. Comp Biochem Physiol 40A:743–749

Canessa M, Spalvins A, Nagel RL (1986) Volume-dependent and NEM-stimulated K^+,Cl^- transport is elevated in oxygenated SS,SC and CC human red cells. FEBS Lett 200:197–202

Cantley LC (1981) Structure and mechanism of the Na,K-ATPase. Curr Top Bioenerg 11:201–307

Carafoli E, Zurini M (1982) The Ca^{2+}-pumping ATPase of plasma membranes. Purification, reconstitution and properties. Biochim Biophys Acta 683:279–301

Card RT, Valberg LS (1967) Characteristics of shortened survival of stress erythrocytes in the rabbit. Am J Physiol 213:566–572

Carey FG (1973) Fishes with warm bodies. Sci Am 228:36–44

Carey FG, Gibson QH (1977) Reverse temperature dependence of tuna hemoglobin oxygenation. Biochem Biophys Res Commun 78:1376–1382

Carey FG, Teal JM (1969) Mako and porbeagle: warm-bodied sharks. Comp Biochem Physiol 28:199-204

Carlsson U, Kjellström B, Antonsson B (1980) Purification and properties of cyclostome carbonic anhydrase from erythrocytes of hagfish. Biochim Biophys Acta 612:160-170

Carruthers A, Melchior DL (1986) How bilayer lipids affect membrane protein activity. Trends Biochem Sci 11:331-335

Carter MJ (1972) Carbonic anhydrase: isozymes, properties, distribution and functional significance. Biol Rev Camb Philos Soc 47:465-513

Cavieres JD (1977) The sodium pump in human red cells. In: Ellory JC, Lew VL (eds) Membrane transport in red cells. Academic Press, London, pp 1-37

Cavieres JD (1984) Calmodulin and the target size of the (Ca^{2+} + Mg^{2+})-ATPase of human red cell ghosts. Biochim Biophys Acta 771:241-246

Centonze VE, Ruben GC, Sloboda RD (1986) Structure and composition of the cytoskeleton of nucleated erythrocytes: III. Organization of the cytoskeleton of *Bufo marinus* erythrocytes as revealed by freeze-dried platinum-carbon replicas and immunofluorescence microscopy. Cell Motil Cytoskel 6:376-388

Cerione RA, Stanizewski C, Caron MG, Lefkowitz RJ, Codina J, Birnbaumer L (1985) A role for N_i in the hormonal stimulation of adenylate cyclase. Nature 318:293-295

Cerretelli P (1987) Extreme hypoxia in air breathers: some problems. In: Dejours P (ed), Comparative physiology of environmental adaptation, vol 2. Adaptations to extreme environments. Karger, Basel, pp 137-150

Chabanel A, Reinhart W, Chien S (1987) Increased resistance to membrane deformation of shaped-transforned human red blood cells. Blood 69:739-743

Chan L-NL (1977) Changes in the composition of plasma membrane proteins during differentiation of embryonic chick erythroid cell. Proc Natl Acad Sci USA 74:1062-1066

Chang C-H, Takeuchi H, Ito T, Machida K, Ohnishi S-I (1981) Lateral mobility of erythrocyte membrane proteins studied by fluorescence photobleaching recovery technique. J Biochem 90:997-1004

Chanutin A, Curnish RR (1967) Effect of organic and inorganic phosphates on the oxygen equilibrium of human erythrocytes. Arch Biohem Biophys 121:96-102

Charnay P, Maniatis T (1983) Transcriptional regulation of globin gene expression in the human erythroid cell line K562. Science 220:1281-1282

Chasis JA, Mohandas N (1986) Erythrocyte membrane deformability and stability: two distinct membrane properties that are independently regulated by skeletal protein associations. J Cell Biol 103:343-350

Chasis JA, Shohet SB (1987) Red cell biochemical anatomy and membrane properties. Annu Rev Physiol 49:237-248

Chasis JA, Mohandas N, Shohet SB (1985) Erythrocyte membrane rigidity induced by glycophorin A-ligand interaction. Evidence for a ligand-induced association between glycophorin A and skeletal proteins. J Clin Invest 75:1919-1926

Chauvet J-P, Acher R (1972) Phylogeny of hemoglobins. β chain of frog (*Rana esculenta*) hemoglobin. Biochemistry 11:916-926

Cherksey BD, Zadunaisky JA, Murphy RB (1980) Cytoskeletal constraint of the β-adrenergic receptor in frog erythrocyte membranes. Proc Natl Acad Sci USA 77:6401-6405

Chetrite G, Cassoly R (1985) Affinity of hemoglobin for the cytoplasmic fragment of human erythrocyte membrane band 3. J Mol Biol 185:639-644

Chien S (1970) Shear dependence of effective cell volume as a determinant of blood viscosity. Science 168:977-979

Chien S (1977) Principles and techniques for assessing erythrocyte deformability. Blood Cells 3:71-99

Chien S (1985) Role of blood cells in microcirculatory regulation. Microvasc Res 29:129-151

Chien S (1987) Red cell deformability and its relevance to blood flow. Annu Rev Physiol 49: 177-192

Chien S, Jan K-M (1973) Ultrastructural basis of the mechanism of rouleaux formation. Microvasc Res 5:155-166

Chien S, Usami S, Dellenback RJ, Bryant CA (1971) Comparative hemorheology – hematological implications of species differences in blood viscosity. Biorheology 8:35-57

Childs RA, Feizi T, Fukuda M, Hakomori S-I (1978) Blood-group-I activity associated with band 3, the major intrinsic membrane protein of human erythrocytes. Biochem J 173:333–336
Chipperfield AR (1980) An effect of chloride on (Na + K) co-transport in human red blood cells. Nature 286:281–282
Choi O-RB, Engel JD (1988) Developmental regulation of β-globin gene switching. Cell 55:17–26
Christiansen J, Douglas CG, Haldane JS (1914) The absorption and dissociation curve of carbon dioxide by human blood. J Physiol (Lond) 48:244–271
Christiansson A, Kuypers FA, Roelofsen B, Kamp Op Den JAF, Deenen Van LLM (1985) Lipid molecular shape affects erythrocyte morphology: a study involving replacement of native phosphatidylcholine with different species followed by treatment of cells with sphingomyelinase C or phospholipase A_2. J Cell Biol 101:1455–1462
Chu AH, Turner BW, Ackers GK (1984) Effects of protons on the oxygenation-linked subunit assembly in human hemoglobin. Biochemistry 23:604–617
Chudzik J, Houston AH (1983) Temperature and erythropoiesis in goldfish. Can J Zool 61:1322–1325
Claireaux G, Thomas S, Fievet B, Motais R (1988) Adaptive respiratory responses of trout to acute hypoxia. II Blood oxygen carrying properties during hypoxia. Respir Physiol 74:91–98
Clark MR (1988) Senescence of red blood cells: progress and problems. Physiol Rev 68:503–554
Clark MR, Mohandas N, Feo C, Jacobs MS, Shohet SB (1981) Separate mechanisms of deformability loss in ATP-depleted and Ca-loaded erythrocytes. J Clin Invest 67:531–539
Clark MR, Mohandas N, Shohet SB (1983) Osmotic gradient ektacytometry: comprehensive characterization of red cell volume and surface maintenance. Blood 61:899–910
Clausen G, Ersland A (1968) The respiratory properties of the blood of two diving rodents, the beaver and the water vole. Respir Physiol 5:350–359
Cline MJ, Golde DW (1979) Cellular interactions in haematopoiesis. Nature 277:177–181
Codina J, Hildebrandt JD, Sekura RD, Birnbaumer M, Bryan J, Manclark CR, Iyengar R, Birnbaumer L (1984) Ns and Ni, the stimulatory and inhibitory regulatory components of adenylyl cyclases. J Biol Chem 259:5871–5886
Cohen CM, Langley RC Jr (1984) Functional characterization of human erythrocyte spectrin α and β chains: association with actin and erythrocyte protein 4.1. Biochemistry 23:4488–4495
Cohen CM, Langley RC Jr, Foley SF, Korsgren C (1984) Functional associations of band 4.1 in the erythrocyte membrane skeleton and their role in inherited membrane skeletal abnormalities. In: Kruckeberg WC, Eaton JW, Aster J, Brewer GJ (eds) Erythrocyte membranes: recent clinical and experimental advances, vol 3. Alan R Liss, New York, pp 13–29
Cohen NS, Ekholm JE, Luthra MG, Hanahan DJ (1976) Biochemical characterization of density-separated human erythrocytes. Biochim Biophys Acta 419:229–242
Cohen WD (1978a) On erythrocyte morphology. Blood Cells 4:449–451
Cohen WD (1978b) Observations on the marginal band system of nucleated erythrocytes. J Cell Biol 78:260–273
Cohen WD, Bartelt D, Jaeger R, Langford G, Nemhauser I (1982) The cytoskeletal system of nucleated erythrocytes. I. Composition and function of major elements. J Cell Biol 93:828–838
Coleman JE (1984) Carbonic anhydrase: zinc and the mechanism of catalysis. Ann NY Acad Sci 429:26–48
Coletta M, Brumen M, Giardina B, Benedetti PA, Brunori M (1987) A microspectroscopic analysis of ligand binding in single erythrocyte. Biomed Biochim Acta 46:108–112
Coll J, Ingram VM (1978) The stimulation of heme accumulation and erythroid colony formation in cultures of chick bone marrow cells by chick plasma. J Cell Biol 76:184–190
Collins FS, Weissman SM (1984) The molecular genetics of human hemoglobin. Progr Nucl Acid Res Molec Biol 31:315–462
Comi P, Giglioni B, Pozzoli ML, Ottolenghi S, Gianni AM, Migliaccio AR, Migliaccio G, Lettieri F, Peschle C (1981) Biosynthesis of globin chains in fetal liver and adult marrow cultures. Comparative analysis of individual colonies derived from early, intermediate or late erythroid progenitors. Exp Cell Res 133:347–356
Connolly TJ, Carruthers A, Melchior DL (1985) Effects of bilayer cholesterol on human erythrocyte hexose transport protein activity in synthetic lecithin bilayers. Biochemistry 24:2865–2873
Cooper RA (1978) Influence of increased membrane cholesterol on membrane fluidity and cell function in human red blood cells. J Supramol Struct 8:413–430

Cossins AR (1976) Changes in muscle lipid composition and resistance adaptation to temperature in the freshwater crayfish *Australopotamobius pallipes*. Lipids 11:307–316

Cossins AR (1977) Adaptation of biological membranes to temperature-the effect of temperature acclimation of goldfish upon the viscosity of synaptosomal membranes. Biochim Biophys Acta 470:395–411

Cossins AR (1983) The adaptation of membrane structure and function to changes in temperature. In: Cossins AR, Sheterline P (eds) Cellular acclimation to environmental change, Cambridge University Press, Cambridge, pp 3–32

Cossins AR, Lee JAC (1985) The adaptation of membrane structure and lipid composition to cold. In: Gilles R (ed) Circulation, respiration, and metabolism. Current comparative approaches. Springer, Berlin Heidelberg New York, pp 543–552

Cossins AR, Richardson PA (1984) Na^+/H^+ exchange in fish erythrocytes. J Physiol (Lond) 351:43P

Cossins AR, Richardson PA (1985) Adrenalin-induced Na^+/H^+ exchange in trout erythrocytes and its effects upon oxygen-carrying capacity. J Exp Biol 118:229–246

Craik CS, Buchman SR, Beychok S (1980) Characterization of globin domains: Heme binding to the central exon product. Proc Natl Acad Sci USA 77:1384–1388

Craik CS, Buchman SR, Beychok S (1981) O_2 binding properties of the product of the central exon of β-globin gene. Nature 291:87–90

Crandall ED, O'Brasky JE (1978) Direct evidence for participation of rat lung carbonic anhydrase in CO_2 reactions. J Clin Invest 62:618–622

Crandall ED, Critz AM, Osher AS, Keljo DJ, Forster RE (1978) Influence of pH on elastic deformability of the human erythrocyte membrane. Am J Physiol 235:C269–C278

Czech MP, Massague J, Pilch PF (1981) The insulin receptor: structural features. Trends Biochem Sci 6:222–225

Dabrowski K (1982) Postprandial distribution of free amino acids between plasma and erythrocytes of common carp (*Cyprinus carpio* L.). Comp Biochem Physiol 72A:753–763

Dainiak N, Cohen CM (1985) Regulation of human erythroid proliferation in vitro by leukocyte surface components. Ann N Y Acad Sci 459:129–142

Daleke DL, Huestis WH (1985) Incorporation and translocation of aminophospholipids in human erythrocytes. Biochemistry 24:5406–5416

Dalmark M (1975) Chloride and water distribution in human red cells. J Physiol (Lond) 250:65–84

Dautry-Varsat A, Ciechanover A, Lodish HF (1983) pH and the recycling of transferrin during receptor mediated endcytosis. Proc Natl Acad Sci USA 80:2258–2262

Davies BN, Withrington PG (1973) The actions of drugs on the smooth muscle of the capsule and blood vessels of the spleen. Pharmacol Rev 25:373–413

Davoren PR, Sutherland EW (1963) The effect of l-epinephrine and other agents on the synthesis and release of adenosine 3′,5′-phosphate by whole pigeon erythrocytes. J Biol Chem 238:3009–3015

Dawson RMC, Elliott DC, Elliott WH, Jones KM (1987) Data for biochemical research, 3rd edn. Clarendon, Oxford, 580 pp

DeGroot H, Noll T (1987) Oxygen gradients: the problem of hypoxia. Biochem Soc Trans 15:363–365

Dejours P (1975) Principles of comparative respiratory physiology. Elsevier, Amsterdam, 253 pp

Dekowski SA, Rybicki A, Drickamer K (1983) A tyrosine kinase associated with the red cell membrane phosphorylates band 3. J Biol Chem 258:2750–2753

Denton JE, Yousef MK (1975) Seasonal changes in hematology of rainbow trout, *Salmo gairdneri*. Comp Biochem Physiol 51A:151–153

Denton MJ, Arnstein HRV (1973) Characterization of developing adult mammalian erythroid cells separated by velocity sedimentation. Br J Haematol 24:7–17

DePont JJHHM, Prooijen Van – Eeden Van A, Bonting SL (1978) Role of negatively charged phospholipids in highly purified Na,K-ATPase from rabbit kidney outer medulla. Biochim Biophys Acta 508:464–477

Deuticke B, Haest CWM (1987) Lipid modulation of transport proteins in vertebrate cell membranes. Annu Rev Physiol 49:221–235

De Wilde MA, Houston AH (1967) Hematological aspects of the thermoacclimatory process in the rainbow trout, *Salmo gairdneri*. J Fish Res Board Can 24:2267–2281

Dexter TM, Ponting ILO, Roberts RA, Spooncer E, Heyworth C, Gallagher JT (1988) Growth and differentiation of hematopoietic stem cells. In: Gunn RB, Parker JC (eds) Cell physiology of blood, Rockefeller University Press, New York pp 25–38

Dhindsa DS, Hoversland AS, Metcalfe J (1971) Comparative studies of the respiratory functions of mammalian blood. VII. Armadillo (*Dasypus novemcinctus*). Respir Physiol 13:198–208

Dhindsa DS, Metcalfe J (1974) Post-natal changes in oxygen affinity and the concentration of 2,3-diphosphoglycerate in cat blood. Respir Physiol 21:37–46

Dhindsa DS, Metcalfe J, Hoversland AS, Hartman RA (1974) Comparative studies of the respiratory functions of mammalian blood X. Killer whale (*Orcinus orca* Linnaeus) and Beluga whale (*Delphinapterus leucas*). Respir Physiol 20:93–103

Di Cera E, Doyle ML, Gill SJ (1988) Alkaline Bohr effect of human hemoglobin A_0. J Mol Biol 200:593–599

Di Cera E, Robert CH, Gill SJ (1987) Allosteric interpretation of the oxygen-binding reaction of human hemoglobin tetramers. Biochemistry 26:4003–4008

Dickerson RE, Geis I (1983) Hemoglobin: structure, function, evolution, and pathology. Cummings, Menlo Park, California 176 pp

Dierks P, Ooyen Van A, Mantei N, Weissman C (1981) DNA sequences preceding the rabbit β-globin gene are required for formation in mouse L cells of β-globin RNA with the correct 5′ terminus. Proc Natl Acad Sci USA 78:1411–1415

Dische Z (1964) The pentose phosphate metabolism in red cells. In: Bishop C, Surgenor DM (eds) The red blood cell. Academic Press, London pp 189–209

Dixon RAF, Sigal IS, Rands E, Register RB, Candelore MR, Blake AD, Stradel CD (1987) Ligand binding to the β-adrenergic receptor involves its rhodopsin-like core. Nature 326:73–77

Dobson GP, Baldwin J (1982) Regulation of blood oxygen affinity in the Australian blackfish *Gadopsis marmoratus* II. Thermal acclimation. J Exp Biol 99:245–254

Dockham PA, Vidaver GA (1987) Comparison of human and pigeon erythrocyte membrane proteins by one- and two-dimensional gel electrophoresis. Comp Biochem Physiol 87B: 171–177

Dohi Y, Sugita Y, Yoneyama Y (1973) The self-association and oxygen equilibrium of hemoglobin from the lamprey, *Entosphenus japonicus*. J Biol Chem 248:2354–2363

Dolan M, Sugarman BJ, Dodgson JB, Engel JD (1981) Chromosomal arrangement of the chicken β-type globin genes. Cell 24:669–677

Domin BA, Mahony WB, Zimmerman TP (1988) Purine nucleobase transport in human erythrocytes. Reinvestigation with a novel "inhibitor-stop" assay. J Biol Chem 263:9276–9284

Domm LV, Taber E, Davis DE (1943) Comparison of erythrocyte numbers in normal and hormone treated brown leghorn fowl. Proc Soc Exp Biol Med 52:49–50

Dorn AR, Broyles RH (1982) Erythrocyte differentiation during the metamorphic hemoglobin switch of *Rana catesbeiana*. Proc Natl Acad Sci USA 79:5592–5596

Douglas EL, Friedl WA, Pickwell GV (1976) Fishes in oxygen-minimum zones: blood oxygenation characteristics. Science 191:957–959

Drickamer LK (1975) The red cell membrane contains three different adenosine triphosphatases. J Biol Chem 250:1952–1954

Dubinsky WP Jr, Frizzell RA (1983) A novel effect of amiloride on H^+-dependent Na^+ transport. Am J Physiol 245:C157–C159

Duhm J (1972) The effect of 2,3-DPG and other organic phosphates on the Donnan equilibrium and the oxygen affinity of human blood. In: Rorth M, Astrup P (eds) Oxygen affinity of hemoglobin and red cell acid-base status. Alfred Benzon Symposium IV. Munksgaard, Copenhagen, pp 583–594

Duhm J (1974) Inosine permeability and purine nucleoside phosphorylase activity as limiting factors for the synthesis of 2,3-diphosphoglycerate from inosine, puryvate, and inorganic phosphate in erythrocytes of various mammalian species. Biochim Biophys Acta 343:89–100

Duhm J (1976) Influence of 2,3-diphosphoglycerate on the buffering properties of human blood. Role of the red cell membrane. Pflügers Arch 363:61–67

Duhm J (1987) Furosemide-sensitive K^+ (Rb^+) transport in human erythrocytes: modes of operation, dependence on extracellular and intracellular Na^+, kinetics, pH dependency and the effect of cell volume and N-ethylmaleimide. J Membr Biol 98:15–32

Duhm J, Göbel BO (1984) Role of the furosemide-sensitive Na^+/K^+ transport system in determining the steady-state Na^+ and K^+ content and volume of human erythrocytes in vitro and in vivo. J Membr Biol 77:243–254
Duling BR, Desjardins C (1987) Capillary hematocrit-what does it mean? News Physiol Sci 2:66–69
Dunham PB, Ellory JC (1981) Passive potassium transport in low potassium sheep red cells: dependence upon cell volume and chloride. J Physiol (Lond) 318:511–530
Dunham PB, Hoffman JF (1980) Na and K transport in red blood cells. In: Andreoli TE, Hoffman JF, Fanestil DD (eds) Membrane physiology. Plenum, New York, pp 255–272
Dunham PB, Stewart GW, Ellory JC (1980) Chloride-activated passive potassium transport in human erythrocytes. Proc Natl Sci USA 77:1711–1715
Dutta-Roy AK, Sinha AK (1985) Binding of prostaglandin E_1 to human erythrocyte membrane. Biochim Biophys Acta 812:671–678
Dutta-Roy AK, Ray TK, Sinha AK (1985) Control of erythrocyte membrane microviscosity by insulin. Biochim Biophys Acta 816:187–190
Eaton JW (1974) Oxygen affinity and environmental adaptation. Ann N Y Acad Sci 241:491–497
Eaton JW, Skelton TD, Berger E (1974) Survival at extreme altitude: protective effect of increased hemoglobin-oxygen affinity.Science 183:743–744
Eaton WA (1980) The relationship between coding sequences and function in haemoglobin. Nature 284:183–185
Eavenson E, Christensen HN (1967) Transport systems for neutral amino acids in the pigeon erythrocyte. J Biol Chem 242:5386–5396
Eddy FB (1977) Oxygen uptake by rainbow trout blood, *Salmo gairdneri*. J Fish Biol 10:87–90
Edwards MJ, Martin RJ (1966) Mixing technique for the oxygen-hemoglobin equilibrium and Bohr effect. J Appl Physiol 21:1898–1902
Effros RM, Chang RSY, Silverman P (1978) Acceleration of plasma bicarbonate conversion to carbon dioxide by pulmonary carbonic anhydrase. Science 199:427–429
Efstratiadis A, Posakony JW, Maniatis T, Lawn RM, O'Connell C, Spritz RA, DeRiel JK, Forget BG, Weissman SM, Slightom JL, Blechl AE, Smithies O, Baralle FE, Shoulders CC, Proudfoot NJ (1980) The structure and evolution of the human β-globin gene family. Cell 21:653–668
Elgsaeter A, Stokke BT, Mikkelsen A, Branton D (1986) The molecular basis of erythrocyte shape. Science 234:1217–1223
Ellory JC (1977) The sodium pump in ruminant cells. In: Ellory JC, Lew VL (eds) Membrane transport in red cells. Academic Press, London pp 363–382
Ellory JC (1982) Flux measurements. In: Techniques in cellular physiology, Pt II, Elsevier, Amsterdam P 129:1–11
Ellory JC, Hall AC (1987) Temperature effects on red cell membrane transport processes. In: Bowler K, Fuller BJ (eds) Temperature and animal cells. Company of Biologists Ltd, Cambridge, pp 53–66
Ellory JC, Hall AC (1988) Human red cell volume regulation in hypotonic media. Comp Biochem Physiol 90A:533–537
Ellory JC, Tucker EM (1983) Cation transport in red blood cells. In: Agar NS, Board PG (eds) Red blood cells of domestic mammals. Elsevier, Amsterdam, pp 291–313
Ellory JC, Jones SEM, Preston RL, Young JD (1981a) A high-affinity sodium-dependent transport system for glutamate in dog red cells. J Physiol (Lond) 320:79P
Ellory JD, Jones SEM, Young JD (1981b) Glycine transport in human erythrocytes. J Physiol (Lond) 320:403–422
Ellory JC, Dunham PB, Logue PJ, Steward GW (1982) Anion-dependent cation transport in erythrocytes. Phil Trans R Soc Lond B 299:483–495
Ellory JC, Flatman PW, Stewart GW (1983) Inhibition of human red cell sodium and potassium transport by divalent cations. J Physiol (Lond) 340:1–17
Ellory JC, Hall AC, Stewart GW (1985) Volume-sensitive cation fluxes in mammalian red cells. Mol Physiol 8:235–246
Ellory JC, Wolowyk MW, Young JD (1987) Hagfish (*Eptatretus stouti*) erythrocytes show minimal chloride transport activity. J Exp Biol 129:377–383
Ellsworth ML, Pittman RN, Ellis CG (1987) Measurement of hemoglobin oxygen saturation in capillaries. Am J Physiol 252:H1031–H1040

Escobales N, Canessa M (1986) Amiloride-sensitive Na^+ transport in human red cells: evidence for a Na/H exchange system. J Membr Biol 90:21–28
Escobales N, Rivera A (1987) Na^+ for H^+ exchange in rabbit erythrocytes. J Cell Physiol 132:73–80
Evans E, Leung A (1984) Adhesivity and rigidity of erythrocyte membrane in relation to wheat germ agglutinin binding. J Cell Biol 98:1201–1208
Evans T, Reitman M, Felsenfeld G (1988) An erythroid-specific DNA-binding factor recognizes a regulatory sequence common to all chicken globin genes. Proc Natl Acad Sci USA 85:5976–5980
Everaarts J (1978) The haemoglobin of the herring, *Clupea harengus*. Neth J Sea Res 12:1–57
Fagg B (1981) Is erythropoietin the only factor which regulates late erythroid differentiation? Nature 289:184–185
Fairbanks G, Steck TL, Wallach DFH (1971) Electrophoretic analysis of the major polypeptides of the human erythrocyte membrane. Biochemistry 10:2606–2617
Fairbanks MB, Hoffert JR, Fromm PO (1969) The dependence of the oxygen-concentrating mechanism of the teleost eye (*Salmo gairdneri*) on the enzyme carbonic anhydrase. J Gen Physiol 54:203–211
Fänge R (1966) Physiology of the swimbladder. Physiol Rev 46:299–322
Fänge R (1987) Lymphomyeloid system and blood cell morphology in elasmobranchs. Arch Biol 98:187–208
Fantoni A, Chapelle De La A, Rifkind RA, Marks PA (1968) Erythroid cell development in fetal mice: Synthetic capacity for different proteins. J Mol Biol 33:79–91
Farmer M, Fyhn HJ, Fyhn UEH, Noble RW (1979) Occurrence of Root effect hemoglobins in Amazonian fishes. Comp Biochem Physiol 62A:115–124
Ferguson RA, Boutilier RG (1988) Metabolic energy production during adrenergic pH regulation in red cells of the Atlantic salmon, *Salmo salar*. Respir Physiol 74:65–76
Fermi G, Perutz M (1981) Haemoglobin and myoglobin. In: Phillips DC, Richards FM (eds) Atlas of molecular structures in biology. Clarendon, Oxford, pp 22–24
Fermi G, Perutz MF, Shaanan B, Fourme R (1984) The crystal structure of human deoxyhaemoglobin at 1.74 Å resolution. J Mol Biol 175:159–174
Ferrell JE Jr, Huestis WH (1984) Phosphoinositide metabolism and the morphology of human erythrocytes. J Cell Biol 98:1992–1998
Ferrell JE Jr, Lee K-J, Huestis WH (1985) Membrane bilayer balance and erythrocyte shape: a quantitative assessment. Biochemistry 24:2849–2857
Fievet B, Claireaux G, Thomas S, Motais R (1988) Adaptive respiratory responses of trout to acute hypoxia. III. Ion movements and pH changes in the red blood cell. Respir Physiol 74:99–114
Fincham DA, Young JD (1983) Amino acid transport deficiency in horse erythrocytes. Biochem Soc Trans 11:776–777
Fincham DA, Willis JS, Young JD (1984) Red cell amino acid transport. Evidence for the presence of system Gly in guinea pig reticulocytes. Biochim Biophys Acta 777:147–150
Fincham DA, Wolowyk MW, Young JD (1986) Evidence for the presence of system asc in erythrocytes from the Pacific hagfish (*Eptatretus stouti*). J Physiol 382:140P
Fincham DA, Wolowyk MW, Young JD (1987) Volume-sensitive taurine transport in fish erythrocytes. J Membr Biol 96:45–56
Fincham DA, Mason DK, Young JD (1988) Dibasic amino acid interactions with Na^+-independent transport system asc in horse erythrocytes. Kinetic evidence of functional and structural homology with Na^+-dependent system ASC. Biochim Biophys Acta 937:184–194
Fischer TM (1978) A comparison of the flow behavior of disc shaped versus elliptic red blood cells (RBC). Blood Cells 4:453–461
Fischer TM, Haest CWM, Stöhr M, Kamp D, Deuticke B (1978a) Selective alteration of erythrocyte deformability by SH-reagents. Evidence for an involvement of spectrin in membrane shear elasticity. Biochim Biophys Acta 510:270–282
Fischer TM, Stöhr-Liesen M, Schmid-Schönbein H (1978b) The red cell as a fluid droplet: tank tread-like motion of the human erythrocyte membrane in shear flow. Science 202:894–896
Fisher TJ, Tucker EM, Young JD (1986) Relationship between cell age, glutathione and cation concentrations in sheep erythrocytes with a normal and detective transport system for amino acids. Biochim Biophys Acta 884:211–214

Fisher WK, Nash AR, Thompson EOP (1977) Haemoglobins of the shark, *Heterodontus portusjacksoni* III. Amino acid sequence of the β-chain. Aust J Biol Sci 30:487–506
Flatman PW (1983) Sodium and potassium transport in ferret red cells. J Physiol (Lond) 341:545–557
Flatman PW (1988) The effects of magnesium on potassium transport in ferret red cells. J Physiol (Lond) 397:471–487
Flatman PW, Andrews PLR (1983) Cation and ATP content of ferret red cells. Comp Biochem Physiol 74A:939–943
Flavin M, Blouquit Y, Rosa J (1978a) Biochemical studies of the hemoglobin switch during metamorphosis in the salamander *Pleurodeles waltli* – I. Partial characterization of the adult hemoglobin. Comp Biochem Physiol 61B:533–537
Flavin M, Blouquit Y, Duprat AM, Rosa J (1978b) Biochemical studies of the hemoglobin switch during metamorphosis in the salamander *Pleuronectes waltli* – II. Comparative studies of larval and adult hemoglobins. Comp Biochem Physiol 61B:539–544
Flavin M, Thillet J, Rosa J (1983) Oxygen equilibrium of larval and adult hemoglobins of the salamander, *Pleurodeles waltii.* Comp Biochem Physiol 75A:81–85
Fletcher GL, Haedrich RT (1987) Rheological properties of rainbow trout blood. Can J Zool 65:879–883
Forget BG (1983) Normal and abnormal human globin genes. In: Goldwasser E (ed) Regulation of hemoglobin biosynthesis. Elsevier, Amsterdam, pp 27–38
Fornaini G, Magnani M, Fazi A, Accorsi A, Stocchi V, Dacha M (1985) Regulatory properties of human erythrocyte hexokinase during cell ageing. Arch Biochem Biophys 239:352–358
Forster ME (1985) Blood oxygenation in shortfin eels during swimming and hypoxia: influence of the Root effect. New Zealand J Mar Freshwater Res 19:247–251
Forster RE (1972) The effect of dilution in saline on the oxygen affinity of human hemoglobin. In: Rorth M, Astrup P (eds) Oxygen affinity of hemoglobin and red cell acid-base status, IV Alfred Benzon Symposium. Munksgaard, Copenhagen, pp 162–165
Forster RE, Crandall ED (1975) Time course of exchanges between red cells and extracellular fluid during CO_2 uptake. J Appl Physiol 38:710–718
Fortes PAG (1977) Anion movements in red blood cells. In: Ellory JC, Lew VL (eds) Membrane transport in red cells. Academic Press, London, pp 175–195
Fossel ET, Solomon AK (1978) Ouabain-sensitive interaction between human red cell membrane and glycolytic enzyme complex in cytosol. Biochim Biophys Acta 510:99–111
Fossel ET, Solomon AK (1981) Relation between red cell membrane ($Na^+ + K^+$)-ATPase and band 3 protein. Biochim Biophys Acta 649:557–571
Fowler VM, Bennett V (1984) Tropomyosin: a new component of the erythrocyte membrane skeleton. In: Kruckeberg WC, Eaton JW, Aster J, Brewer GJ (eds) Erythrocyte membranes: recent clinical and experimental advances, vol 3. Alan R Liss, New York, pp 57–71
Frangioni G, Borgioli G (1988) Sites and trend of erythropoiesis in anemic, normal and splenectomized newts. J Exp Zool 247:244–250
Franzke R, Jelkmann W (1982) Characterization of the pyruvate kinase which induces the low 2,3-DPG level of fetal rabbit red cells. Pflügers Arch 394:21–25
Frazier J, Caskey J, Yoffe M, Seligman P (1982) Studies of the transferrin receptor on both human reticulocytes and nucleated human cells in culture: Comparison of factors regulating receptor density. J Clin Invest 69:853–865
Freedman JC, Hoffman JF (1979) Ionic and osmotic equilibria of human red blood cells treated with nystatin. J Gen Physiol 74:187–212
Fried W, Plzak LF, Jacobson LO, Goldwasser E (1957) Studies on erythropoiesis III. Factors controlling erythropoietin production. Proc Soc Exp Biol Med 94:237–241
Friedman JM (1985) Structure, dynamics, and reactivity in hemoglobin. Science 228:1273–1280
Friedman JM, Simon SR, Scott TW (1985) Structure and function in sea turtle hemoglobins. Copeia 1985(3):679–693
Fronticelli C, Bucci E, Orth C (1984) Solvent regulation of oxygen affinity in haemoglobin. Sensivity of bovine haemoglobin to chloride ions. J Biol Chem 259:10841–10844
Fucci L, Cirotto C, Tomei L, Geraci G (1983) Synthesis of globin chains in the erythropoietic sites of the early chick embryo. J Embryol Exp Morphol 77:153–165
Fucci L, Vitale E, Cirotto C, Geraci G (1987) Evidences that hemoglobin switch in the chick embryo depends on erythroid cell line substitution. Cell Differ 20:55–63

Fuchs DA, Albers C (1988) Effect of adrenaline and blood gas conditions on red cell volume and intraerythrocytic electrolytes in the carp, *Cyprinus carpio*. J Exp Biol 137:457–475
Fugelli K (1967) Regulation of cell volume in flounder (*Pleuronectes flesus*) erythrocytes accompanying a decrease in plasma osmolarity. Comp Biochem Physiol 22:253–260
Fugelli K, Reiersen LO (1978) Volume regulation in flounder erythrocytes. In: Jorgensen CB, Skadhauge E (eds) Osmotic and volume regulation. Alfred Benzon Symposium XI. Munksgaard, Copenhagen, pp 418–428
Fugelli K, Rohrs H (1980) The effect of Na^+ and osmolality on the influx and steady state distribution of taurine and gamma-aminobutyric acid in flounder (*Platichthys flesus*) erythrocytes. Comp Biochem Physiol 67A:545–551
Fugelli K, Thoroed SM (1986) Taurine transport associated with cell volume regulation in flounder erythrocytes under anisosmotic conditions. J Physiol (Lond) 374:245–261
Fugelli K, Zachariassen KE (1976) The distribution of taurine, gamma-aminobutyric acid and inorganic ions between plasma and erythrocytes in flounder (*Platichthys flesus*) at different plasma osmolalities. Comp Biochem Physiol 55A:173–177
Fujii H, Miwa S (1986) Red cell enzymes. In: Schmidt RM, Fairbanks VF (eds) CRC handbook series in clinical laboratory science, Section I: Hematology, vol 4. CRC Press, Boca Raton, Florida, pp 307–352
Fujii T, Sato T, Tamura A, Wakatsuki M, Kanaho Y (1979) Shape changes of human erythrocytes induced by various amphipathic drugs acting on the membrane of the intact cells. Biochem Pharmacol 28:613–620
Fujiki H, Braunitzer G, Rudloff V (1970) N-formylproline as N-terminal amino acid of lamprey hemoglobin. Hoppe-Seyler's Z Physiol Chem 351:901–902
Fujise H, Lauf PK (1987) Swelling, NEM, and A23187 activate Cl^--dependent K^+ transport in high-K^+ sheep red cells. Am J Physiol 252:C197–C204
Funder J, Parker JC, Wieth JO (1987) Further evidence for coupling of sodium and proton movements in dog red blood cells. Biochim Biophys Acta 899:311–312
Fyhn UEH, Sullivan B (1975) Elasmobranch hemoglobins: dimerization and polymerization in various species. Comp Biochem Physiol 50B:119–129
Gaehtgens P, Schmidt F, Will G (1981a) Comparative rheology of nucleated and non-nucleated red blood cells. I. Microrheology of avian erythrocytes during capillary flow. Pflügers Arch 390:278–282
Gaehtgens P, Will G, Schmidt F (1981b) Comparative rheology of nucleated and non-nucleated red blood cells. II. Rheological properties of avian red cell suspensions in narrow capillaries. Pflügers Arch 390:283–287
Gaetani GD, Parker JC, Kirkman HN (1974) Intracellular restraint: a new basis for the limitation in response to oxidative stress in human erythrocytes containing low-activity variants of glucose-6-phosphate dehydrogenase. Proc Natl Acad Sci USA 71:3584–3587
Gahlenbeck H, Bartels H (1970) Blood gas transport properties in gill and lung forms of the axolotl (*Ambystoma mexicanum*). Respir Physiol 9:175–182
Gambhir KK, Archer JA, Bradley CJ (1978) Characteristics of human erythrocyte insulin receptors. Diabetes 27:701–708
Garby L, Robert M, Zaar B (1972) Proton- and carbamino-linked oxygen affinity of normal human blood. Acta Physiol Scand 84:482–492
Garcia-Romeu F, Motais R, Borgese F (1988) Desentitization by external Na of the cyclic AMP-dependent Na^+/H^+ antiporter in trout red blood cells. J Gen Physiol 91:529–548
Gardner JD, Mensh RS, Kiino DR, Aurbach GD (1975) Effects of β-adrenergic catecholamines on potassium transport in turkey erythrocytes. J Biol Chem 250:1155–1163
Gardos G (1959) The role of calcium and potassium permeability of human erythrocytes. Acta Physiol Acad Sci Lond 15:121–125
Garlick RL, Davis BJ, Farmer M, Fyhn HJ, Fyhn UEH, Noble RW, Powers DA, Riggs A, Weber RE (1979) A fetal-maternal shift in the oxygen equilibrium of hemoglobin from the viviparous caecilian, *Typhlonectes compressicauda*. Comp Biochem Physiol 62A:239–244
Garrick MD (1983) Hemoglobin switching. In: Agar NS, Board, PG (eds) Red blood cells of domestic mammals. Elsevier, Amsterdam, pp 209–225
Garrick MD, Garrick LM (1983) Hemoglobins and globin genes. In: Agar NS, Board, PG (eds) Red blood cells of domestic mammals. Elsevier, Amsterdam, pp 165–207

Gasson JC, Bersch N, Golde DW (1985) Characterization of purified human erythroid-potentiating activity. In: Cronkite EP, Dainiak N, McCaffrey RP, Palek J, Quesenberry PJ (eds) Hematopoietic stem cell physiology. Alan R Liss, New York, pp 95–104

Geoghegan WD, Poluhowich JJ (1974) The major erythrocytic organic phosphates of the American eel, *Anguilla rostrata*. Comp Biochem Physiol 49B:281–290

Gianni AM, Comi P, Giglioni B, Ottolenghi S, Migliaccio AR, Migliaccio G, Lettieri F, Maguire YP, Peschle C (1980) Biosynthesis of Hb in individual fetal liver bursts. γ-chain production peaks earlier than β-chain in the erythropoietic pathway. Exp Cell Res 130:345–352

Giardina B, Corda M, Pellegrini MG, Condò SG, Brunori M (1985) Functional properties of the hemoglobin system of two diving birds (*Podiceps nigricollis* and *Phalacrocorax carbo sinensis*). Mol Physiol 7:281–292

Gil A, Proudfoot NJ (1984) A sequence downstream of AAUAAA is required for rabbit β-globin mRNA 3′-end formation. Nature 312:473–474

Gill SJ (1981) Measurement of oxygen binding by means of a thin-layer optical cell. Methods Enzymol 76:427–438

Gill SJ, Sköld R, Fall L, Shaeffer T, Spokane R, Wyman J (1978) Aggregation effects on oxygen binding of sickle cell hemoglobin. Science 201:362–364

Gillen RG, Riggs A (1971) The hemoglobins of a fresh-water teleost, *Cichlasoma cyanoguttatum* (Baird and Girard) — I. The effects of phosphorylated organic compounds upon the oxygen equilibria. Comp Biochem Physiol 38B:585–595

Gillen RG, Riggs A (1972) Structure and function of the hemoglobins of the carp, *Cyprinus carpio*. J Biol Chem 245:6039–6046

Gillen RG, Riggs A (1973) Structure and function of the isolated hemoglobins of the American eel. J Biol Chem 246:1961–1969

Ginsberg BH, Kahn CR, Roth J (1977) The insulin receptor of the turkey erythrocyte: similarity to mammalian insulin receptors. Endocrinology 100:82–90

Giraud F, M'Zali H, Chailley B, Mazet F (1984) Changes in morphology and in polyphosphoinositide turnover of human erythrocytes after cholesterol depletion. Biochim Biophys Acta 778:191–200

Glynn IM (1985) The Na^+,K^+-transporting adenosine triphosphatase. In: Martonosi AN (ed) The enzymes of biological membranes, 2nd Edn. Plenum, New York, pp 35–114

Glynn IM, Karlish SJD (1975) The sodium pump. Annu Rev Physiol 37:13–55

Godovac-Zimmermann J, Braunitzer G (1983) The amino acid sequence of northern mallard (*Anas platyrhynchos platyrhynchos*). Hoppe-Seyler's Z Physiol Chem 364:665–674

Godovac-Zimmermann J, Braunitzer G (1984) The amino-acid sequence of α^A- and β-chains from the major hemoglobin component of American flamingo (*Phoenicopterus ruber ruber*). Hoppe-Seyler's Z Physiol Chem 365:437–443

Godovac-Zimmermann J, Kösters J, Braunitzer G, Göltenboth R (1988) Structural adaptation of bird hemoglobins to high-altitude respiration and the primary sequences of black-headed gull (*Larus ridibundus*, Charadriiformes) α_A-and β/β′-chains. Biol Chem Hoppe-Seyler 369:341–348

Goldstein L, Boyd TA (1978) Regulation of β-alanine transport in skate (*Raja erinacea*) erythrocytes. Comp Biochem Physiol 60A:319–325

Goldwasser E (1975) Erythropoietin and the differentiation of red blood cells. Fed Proc 34: 2285–2292

Goodman SR, Krebs KE, Whitfield CF, Riederer BM, Zagon IS (1988) Spectrin and related molecules. CRC Crit Rev Biochem 23:171–234

Gordon AG (1960) Humoral influences on blood cell formation and release. In: Wolstenholme GEW, O'Connor M (eds) Haemopoiesis. Cell production and its regulation. Churchill, London, pp 325–362

Graham MS, Fletcher GL (1983) Blood and plasma viscosity of winter flounder: influence of temperature, red cell concentration, and shear rate. Can J Zool 61:2344–2350

Graham MS, Fletcher GL (1986) High concentrations of methemoglobin in five species of temperate marine teleosts. J Exp Zool 239:139–142

Granger BL, Lazarides E (1984) Membrane skeletal protein 4.1 of avian erythrocytes is composed of multiple variants that exhibit tissue-specific expression. Cell 37:595–607

Granick S, Urata G (1963) Increase in activity of δ-aminolevulinic acid synthetase in liver mitochondria induced by feeding of 3,5-dicarbethoxy-1,4- dihydrocollidine. J Biol Chem 238:821–827

Grant WS, Root WS (1952) Fundamental stimulus for erythropoiesis. Physiol Rev 32:449–498
Grayzel AI, Horchner P, London IM (1966) The stimulation of globin synthesis by heme. Proc Natl Acad Sci USA 55:650–655
Greaney GS, Powers DA (1978) Allosteric modifiers of fish hemoglobins: in vitro and in vivo studies of the effect of ambient oxygen and pH on erythrocyte ATP concentrations. J Exp Zool 203:339–350
Grebe R, Wolff H, Schmid-Schönbein H (1988) Influence of red cell surface charge on red cell membrane curvature. Pflügers Arch 413:77–82
Gregory CJ (1976) Erythropoietin sensitivity as a differentiation marker in the hemopoietic system: studies of three erythropoietic colony responses in culture. J Cell Physiol 89:289–302
Gregory CJ, Eaves AC (1977) Human marrow cells capable of erythropoietic differentiation in vitro: definition of three erythroid colony responses. Blood 49:855–864
Grigg GC (1969) Temperature induced changes in the oxygen equilibrium curve of the blood of the brown bullhead, *Ictalurus nebulosus*. Comp Biochem Physiol 28:1203–1223
Grigg GC, Harlow P (1981) A fetal-maternal shift of blood oxygen affinity in an Australian viviparous lizard, *Sphenomorphus quoyii* (Reptilia, Scincidae). J Comp Physiol B 142: 495–499
Grinstein S, Smith JD (1987) Asymmetry of the Na^+/H^+ antiport of dog red cell ghosts. Sidedness of inhibition by amiloride. J Biol Chem 262:9088–9092
Grinstein S, Cohen S, Goetz JD, Rothstein A, Mellors A, Gelfand EW (1986) Activation of the Na^+-H^+ antiport by changes in cell volume and by phorbol esters; possible role of protein kinase. Curr Top Membr Transp 26:115–134
Gronenborn AM, Clore GM, Brunori M, Giardina B, Falcioni G, Perutz MF (1984) Stereochemistry of ATP and GTP bound to fish haemoglobins. A transferred nuclear Overhauser enhancement. ^{31}P-nuclear magnetic resonance, oxygen equilibrium and molecular modelling study. J Mol Biol 178:731–742
Groom AC (1987) The microcirculatory society Eugene M. Landis award lecture. Microcirculation of the spleen: new concepts, new challenges. Microvasc Res 34:269–289
Gros G, Dodgson SJ (1988) Velocity of CO_2 exchange in muscle and liver. Annu Rev Physiol 50:669–694
Gros G, Forster RE, Lin L (1976) The carbamate reaction of glycylglycine, plasma, and tissue extracts evaluated by pH stopped flow apparatus. J Biol Chem 251:4398–4407
Gros G, Rollema HS, Forster RE (1981) The carbamate equilibrium of α- and ε-amino groups of human hemoglobin at 37°C. J Biol Chem 256:5471–5480
Gross DM, Fisher JW (1980) Erythropoietic effects of PGE_2 and 2 endoperoxide analogs. Experientia (Basel) 36:458–459
Gross M, Kaplansky DA (1983) Effect of hemin-controlled translational repressor and the double-stranded RNA activated inhibitor on polypeptide chain initiation in rabbit reticulocyte lysate. In: Goldwasser E (ed) Regulation of hemoglobin biosynthesis. Elsevier, Amsterdam, pp 211–227
Gross M, Rabinovitz M (1972) Control of globin synthesis in cell-free preparations of reticulocytes by formation of a translational repressor that is inactivated by hemin. Proc Natl Acad Sci USA 69:1565–1568
Grosweld GC, De Boer E, Shewmaker CK, Flavell RA (1982) DNA sequences necessary for transcription of the rabbit β-globin gene in vivo. Nature 295:120–126
Groudine M, Weintraub H (1981) Activation of globin genes during chicken development. Cell 24:393–401
Groudine M, Kohwi-Shigematsu T, Gelinas R, Stamatoyannopoulos G, Papayannopoulou T (1983) Human fetal to adult hemoglobin switching: Changes in chromatin structure of the β-globin gene locus. Proc Natl Acad Sci USA 80:7551–7555
Grujic-Injac B, Braunitzer G, Stangl A (1980) Hämoglobine, XXXV Die Sequenz der β_A- und β_B-ketten der hämoglobine des karpfens (*Cyprinus carpio* L.). Hoppe-Seyler's Z Physiol Chem 361:1629–1639
Grunze M, Forst B, Deuticke B (1980) Dual effect of membrane cholesterol on simple and mediated transport processes in human erythrocytes. Biochim Biophys Acta 600:860–869
Guesnon P, Poyart C, Bursaux E, Bohn B (1979) The binding of lactate and chloride ions to human adult hemoglobin. Respir Physiol 38:115–129

Gupta NK, Bagchi M, Das A, Ghosh-Dastidar P, Grace M, Nasrin N, Ralston R, Roy R (1983) Regulation of protein synthesis in reticulocyte lysate by eIF-2 and eIF-2-ancillary protein factors. In: Goldwasser E (ed) Regulation of hemoglobin biosynthesis. Elsevier, Amsterdam, pp 230–251

Guttman SI (1970a) An electrophoretic study of the hemoglobins of the sand lizards, *Callisaurus Cophosaurus, Holbrookia* and *Uma*. Comp Biochem Physiol 34:569–574

Guttman SI (1970b) Hemoglobin electrophoresis and relationships within the lizard genus *Sceloporus* (Sauria: Iguanidae). Comp Biochem Physiol 34:563–568

Haas M, McManus TJ (1985) Effect of norepinephrin on swelling-induced potassium transport in duck red cells. Evidence against a volume-regulatory decrease under physiological conditions. J Gen Physiol 85:649–667

Haas M, Schmidt WF III, McManus TJ (1982) Catecholamine-stimulated ion transport in duck red cells. Gradient effects in electrically neutral [Na + K + 2 Cl] cotransport. J Gen Physiol 80:125–147

Haest CWM (1982) Interactions between membrane skeleton proteins and the intrinsic domain of the erythrocyte membrane. Biochim Biophys Acta 694:331–352

Haest CWM, Fisher TM, Plasa G (1980) Stabilization of erythrocyte shape by chemical increase in membrane shear stiffness. Blood Cells 6:539–553

Hågå P, Kristiansen S (1981) Role of kidney in foetal erythropoiesis: erythropoiesis and erythropoietin levels in newborn mice with renal agenesis. J Embryol Exp Morphol 61:165–173

Haigh LS, Hellewell S, Roninson IB, Owens BB, Ingram VM (1982) Control of hemoglobin expression in chick embryonic development. In: Akoyunoglou G, Evangelopoulos AE, Goergatsos J, Palaiologos G, Trakatellis A, Tsiganos CP (eds) Cell function and differentiation, Part A. Alan R Liss, New York, pp 35–46

Hall AC, Ellory JC (1985) Measurement and stoichiometry of bumetanide-sensitive (2Na:1K:3Cl) cotransport in ferret red cells. J Membr Biol 85:205–213

Hall AC, Ellory JC (1986) Effects of high hydrostatic pressure on 'passive' monovalent cation transport in human red cells. J Membr Biol 94:1–17

Hall AC, Willis JS (1984) Differential effects of temperature on three components of passive permeability to potassium in rodent red cells. J Physiol (Lond) 348:629–643

Hall AC, Ellory JC, Klein RA (1982) Pressure and temperature effects on human red cell cation transport. J Membr Biol 68:47–56

Hall AC, Wolowyk MW, Wang LCH, Ellory JC (1987) The effects of temperature on Ca^{2+} transport in red cells from a hibernator (*Spermophilus richardsonii*). J Therm Biol 12:61–63

Hall FG (1965) Hemoglobin and oxygen: affinities in seven species of Sciuridae. Science 148:11350–11351

Hall GE, Schraer R (1979) Purification and partial characterization of high and low activity carbonic anhydrase isoenzymes from *Malaclemys terrapin centrata*. Comp Biochem Physiol 63B:561–567

Hall GE, Schraer R (1983) Characterization of a high affinity carbonic anhydrase isozyme purified from erythrocytes of *Salmo gairdneri*. Comp Biochem Physiol 75B:81–92

Hamasaki N, Kawano Y, Inoue H (1987) The active center of transport for phosphoenolpyruvate and inorganic phosphate in the human erythrocyte membrane. Biomed Biochem Acta 46:S51–S54

Hanss M (1983) Erythrocyte filtrability measurement by the initial flow rate method. Biorheology 20:199–211

Härdig J (1978) Maturation of circulating red blood cells in young Baltic salmon (*Salmo salar* L.). Acta Physiol Scand 102:290–300

Härdig J, Höglund LB (1983) Seasonal and ontogenetic effects on methaemoglobin and reduced glutathione contents in the blood of reared Baltic salmon. Comp Biochem Physiol 75A:27–34

Härdig J, Höglund LB (1984) Seasonal variation in blood components of reared Baltic salmon, *Salmo salar* L. J Fish Biol 24:565–579

Härdig J, Olsson LA, Höglund LB (1978) Autoradiography on erythrokinesis and multihemoglobins in juvenile *Salmo salar* L. at various respiratory gas regimes. Acta Physiol Scand 103:240–251

Harkness DR, Roth S, Goldman P (1974) Studies on the red cell oxygen affinity and 2,3-diphosphoglyceric acid in the hibernating woodchuck (*Marmota monax*). Comp Biochem Physiol 48A:591–599

Harrison PR, Conkie D, Affara N, Paul J (1974) In situ localization of globin messenger RNA formation. J Cell Biol 63:402–413
Hartvig M, Weber RE (1984) Blood adaptations for maternal-fetal oxygen transfer in the viviparous teleost, *Zoarces viviparus* L. In: Seymour RS (ed) Respiration and metabolism of embryonic vertebrates. Junk, Dordrecht, pp 17–30
Haswell MS, Zeidler R, Kim HD (1978) Chloride transport in red cells of the teleost, *Tilapia mossambica*. Comp Biochem Physiol 61A:217–220
Haugaard N, Haugaard ES, Stadie WC (1954) Combination of insulin with cells. J Biol Chem 211:289–295
Hazel JR (1979) The influence of thermal acclimation on membrane lipid composition of rainbow trout liver. Am J Physiol 236:R91–R101
Hazel JR (1984) Effects of temperature on the structure and metabolism of cell membranes in fish. Am J Physiol 246:R460–R470
Heasley LE, Brunton LL (1985) Prostaglandin A_1 metabolism and inhibition of cyclic AMP extrusion by avian erythrocytes. J Biol Chem 260:11514–11519
Heasley LE, Watson MJ, Brunton LL (1985) Putative inhibitor of cyclic AMP efflux: chromatography, amino acid composition, and identification as a prostaglandin A_1-glutathione adduct. J Biol Chem 260:11520–11523
Heinz E (1981) Electrical potentials in biological membrane transport. Springer, Berlin Heidelberg New York, 85 pp
Heinz E, Geck P, Pietrzyk C (1975) Driving forces of amino acid transport of animal cells. Ann NY Acad Sci 264:428–441
Heisler N (1986) Comparative aspects of acid-base regulation. In: Heisler N (ed) Acid-base regulation in animals. Elsevier, Amsterdam, pp 397–450
Heming TA (1984) The role of fish erythrocytes in transport and excretion of carbon dioxide. PhD Thesis, University of British Columbia, Vancouver, 177 pp
Heming TA, Randall DJ, Boutilier RG, Iwama GK, Primmett D (1986) Ionic equilibria in red blood cells of rainbow trout (*Salmo gairdneri*): Cl^-, HCO_3^-, and H^+. Respir Physiol 65:223-234
Heming TA, Randall DJ, Mazeaud MM (1987) Effects of adrenaline on ionic equilibria in red blood cells of rainbow trout (*Salmo gairdneri*). Fish Physiol Biochem 3:83–90
Henry RP, Dodgson SJ, Forster RE, Storey BT (1986) Rat carbonic anhydrase: activity, localization and isozymes. J Appl Physiol 60:638–645
Henry RP, Smatresk NJ, Cameron JN (1988) The distribution of branchial carbonic anhydrase and the effects of gill and erythrocyte carbonic anhydrase inhibition in the channel catfish *Ictalurus punctatus*. J Exp Biol 134:201–218
Herrmann A, Müller P (1986) Correlation of the internal microviscosity of human erythrocytes to cell volume and the viscosity of hemoglobin solutions. Biochem Biophys Acta 885:80–87
Hershko A, Razin A, Shoshani T, Mager J (1967) Turnover of purine nucleotides in rabbit erythrocytes II. Studies in vitro. Biochim Biophys Acta 149:59–73
Hesketh JE (1986) Insulin inhibits the phosphorylation of the membrane cytoskeletal protein spectrin in pig erythrocytes. Cell Biol Int Rep 10:623–629
Hespel P, Lijnen P, Fagard R, Hoof Van R, Goossens W, Amery A (1988) Effects of training on erythrocyte 2,3-diphosphoglycerate in normal men. Eur J Appl Physiol 57:456–461
Hevesy G, Lockner D, Sletten K (1964) Iron metabolism and erythrocyte formation in fish. Acta Physiol Scand 60:256–266
Hewitt JA, Kilmartin JV, Ten Eyck LF, Perutz MF (1972) Noncooperativity of the alpha-beta dimer in the reaction of hemoglobin with oxygen. Proc Natl Acad Sci USA 69:203–207
Hiebl I (1987) Hannibals alpenübergang aus molecularbiologischer sicht. Naturwissensch Rundsch 40:14–15
Hiebl I, Braunitzer G, Schneeganss D (1987) High altitude respiration of geese. The primary structures of the major and minor hemoglobin-components of adult andean goose (*Chloephaga melanoptera*, Anatidae): the mutation Leu – Ser in position 55 of the β-chains. Biol Chem Hoppe-Seyler 368:1559–1569
Hiebl I, Weber RE, Schneeganss D, Kösters J, Braunitzer G (1988) Structural adaptations in the major and minor hemoglobin components of adult Rüppell's griffon (*Gyps rueppellii*, Aegypiinae): a new molecular pattern for hypoxic tolerance. Biol Chem Hoppe-Seyler 369:217–232

Hildebrandt JD, Codina J, Risinger J, Birnbaumer L (1984) Identification of a subunit associated with the adenylyl cyclase regulatory components Ns and Ni. J Biol Chem 259:2039–2042

Hilpert P, Fleischmann RG, Kempe D, Bartels H (1963) The Bohr effect related to blood and erythrocyte pH. Am J Physiol 205:337–340

Hilse K, Braunitzer B (1968) Die aminosauresequenz der α-ketten der beiden hauptkomponenten des karpfenhämoglobins. Hoppe-Seyler's Z Physiol Chem 349:433–450

Hjorth JP (1974) Genetics of *Zoarces* populations. VII: Fetal and adult hemoglobins and a polymorphism common to both. Hereditas 78:69–72

Hladky SB (1977) A comment on the semantics of the "determination" of membrane potential. In: Ellory JC, Lew VL (eds) Membrane transport in red cells. Academic Press, London, pp 173–174

Hladky SB, Rink TJ (1977) pH equilibrium across the red cell membrane. In: Ellory JC, Lew VL (eds) Membrane transport in red cells. Academic Press, London, pp 115–135

Hlastala MP (1985) Interactions between O_2 and CO_2 in blood. In: Nicolau C (ed) Oxygen transport in red blood cells. Pergamon, Oxford, pp 95–103

Ho C, Russu IM (1987) How much do we know about the Bohr effect of hemoglobin? Biochemistry 26:6299–6305

Hochmuth RM, Waugh RE (1987) Erythrocyte membrane elasticity and viscosity. Annu Rev Physiol 49:209–219

Hochmuth RM, Worthy PR, Evans EA (1979) Red cell extensional recovery and the determination of membrane viscosity. Biophys J 26:101–114

Hochmuth RM, Buxbaum KL, Evans EA (1980) Temperature dependence of the viscoelastic recovery of red cell membrane. Biophys J 29:177–182

Hoffman JF, Laris PC (1974) Determination of membrane potential in human and *Amphiuma* red blood cells by means of a fluorescent probe. J Physiol (Lond) 239:519–552

Hoffmann EK (1983) Volume regulation by animal cells. In: Cossins AR, Sheterline P (eds) Cellular acclimatisation to environmental change. Cambridge University Press, Cambridge, pp 55–80

Hokin LE, Dahl JL, Deupree JD, Dixon JF, Hackney JF, Perdue JF (1973) Studies on the characterization of the sodium-potassium transport adenosine triphosphatase. X. Purification of the enzyme from the rectal gland of *Sgualus acanthias*. J Biol Chem 248:2593–2605

Holeton GF, Neumann P, Heisler N (1983) Branchial ion exchange and acid-base regulation after strenuous exercise in rainbow trout (*Salmo gairdneri*). Respir Physiol 51:303–318

Hombrados I, Rodewald K, Neuzil E, Braunitzer G (1983) Haemoglobins, LX. Primary structure of the major haemoglobin of the sea lamprey *Petromyzon marinus* (var. Garonne, Loire). Biochimie (Paris) 65:247–257

Hombrados I, Rodewald K, Allard M, Neuzil E, Braunitzer G (1987) Primary structure of the minor haemoglobins from the sea lamprey (*Petromyzon marinus*, Cyclostomata). Biol Chem Hoppe-Seyler 368:145–154

Honzatko RB, Hendrickson WA (1986) Molecular models for the putative dimer of sea lamprey hemoglobin. Proc Natl Acad Sci USA 83:8487–8491

Honzatko RB, Hendrickson WA, Love WE (1985) Refinement of a molecular model for lamprey hemoglobin from *Petromyzon marinus*. J Mol Biol 184:147–164

Hope MJ, Cullis PR (1987) Lipid asymmetry induced by transmembrane pH gradients in large unilamellar vesicles. J Biol Chem 262:4360–4366

Hopkins BE, Wagner HJ, Smith JW (1976) Sodium and potassium activated adenosine triphosphatase of the nasal salt gland of the duck (*Anas platyrhynchos*). J Biol Chem 251:4365–4371

Horvath SM, Chiodi H, Ridgway SH, Azar S, Jr (1968) Respiratory and electrophoretic characteristics of hemoglobin of porpoises and sea lion. Comp Biochem Physiol 24:1027–1033

Hosbach HA, Wyler T, Weber R (1983) The *Xenopus laevis* globin gene family: Chromosomal arrangement and gene structure. Cell 32:45–53

Houslay MD, Gordon LM (1983) The activity of adenylate cyclase is regulated by the nature of its lipid environment. Curr Top Membr Transp 18:179–231

Houston AH (1985) Erythrocytic magnesium in freshwater fishes. Magnesium 4:106–128

Houston AH (1988) Insulin affects ionic composition of rainbow trout erythrocytes. Regul Pept 22:199–204

Houston AH, Cyr D (1974) Thermoacclimatory variation in the haemoglobin system of goldfish (*Carassius auratus*) and rainbow trout (*Salmo gairdneri*). J Exp Biol 61:455–461

Houston AH, Koss TF (1984) Erythrocytic haemoglobin, magnesium and nucleoside triphosphate levels in rainbow trout exposed to progressive heat stress. J Therm Biol 9:159–164
Houston AH, Smeda JS (1979) Thermoacclimatory changes in the ionic microenvironment of haemoglobin in the stenothermal rainbow trout (*Salmo gairdneri*) and eurythermal carp (*Cyprinus carpio*). J Exp Biol 80:317–340
Houston AH, Mearow KM, Smeda JS (1976) Further observations upon the haemoglobin systems of thermally acclimated freshwater teleosts: pumpkinseed (*Lepomis gibbosus*), carp (*Cyprinus carpio*), goldfish (*Carassius auratus*) and carp goldfish hybrids. Comp Biochem Physiol 54A:267–273
Huebers HA, Finch CA (1987) The physiology of transferrin and transferrin receptors. Physiol Rev 67:520–582
Huehns ER, Farooqui AM (1975) Oxygen dissociation properties of human embryonic red cells. Nature 254:335–337
Huey DW, Beitinger TL (1982) A methemoglobin reductase system in channel catfish *Ictalurus punctatus*. Can J Zool 60:1511–1513
Hughes GM, Kikuchi Y (1984) Effect of in vivo and in vitro changes in pO_2 on the deformability of red blood cells of rainbow trout (*Salmo gairdneri* R.). J Exp Biol 111:253–257
Hughes GM, Horimoto M, Kikuchi Y, Kakiuchi Y, Koyama T (1981) Blood flow velocity in microvessels of the gill filaments of the goldfish (*Carassius auratus* L.). J Exp Biol 90:327–331
Hughes GM, Belaud A, Peyraud C, Adcock PJ (1982a) A comparison of two methods for measurement of O_2 content of small (20 μl) samples of fish blood. J Exp Biol 96:417–420
Hughes GM, Kikuchi Y, Watari H (1982b) A study of the deformability of red blood cells of a teleost fish, the yellowtail (*Seriola quinqueradiata*), and a comparison with human erythrocytes. J Exp Biol 96:209–220
Hunter AS, Hunter FR (1957) A comparative study of erythrocyte metabolism. J Cell Comp Physiol 49:479–502
Huot SJ, Cassel D, Igarashi P, Cragoe EJ Jr, Slayman CW, Aronson PS (1989) Identification and purification of a renal amiloride-binding protein with properties of the Na^+-H^+ exchanger. J Biol Chem 264:683–686
Hyde DA, Moon TW, Perry SF (1987) Physiological consequences of prolonged aerial exposure in the American eel, *Anguilla rostrata*: blood respiratory and acid base status. J Comp Physiol 157:635–642
Iacopetta BJ, Morgan EH, Yeoh GCT (1982) Transferrin receptors and iron uptake during erythroid cell development. Biochim Biophys Acta 687:204–210
Ibrahim NG, Gruenspecht NR, Freedman ML (1978) Hemin feedback inhibition at reticulocyte δ-aminolevulinic acid synthetase and δ-aminolevulinic acid dehydratase. Biochem Biophys Res Commun 80:722–728
Ibrahim NG, Friedland ML, Levere RD (1983) Heme metabolism in erythroid and hepatic cells. In: Brown EB (ed) Progress in hematology, vol 13. Grune & Stratton, New York, pp 75–129
Ikeda-Saito M, Yonetani T, Gibson QH (1983) Oxygen equilibrium studies on hemoglobin from the bluefin tuna (*Thunnus Thynnus*). J Mol Biol 168:673–686
Imai K (1973) Analyses of oxygen equilibria of native and chemically modified human adult hemoglobins on the basis of Adair's stepwise oxygenation theory and the allosteric model of Monod, Wyman and Changeux. Biochemistry 12:798–808
Imai K (1981a) Analysis of ligand binding equilibria. Methods Enzymol 76:470–486
Imai K (1981b) Measurement of accurate oxygen equilibrium curves by an automatic oxygenation apparatus. Methods Enzymol 76:438–449
Imai K, Yonetani T (1975) pH dependence of the Adair constants of human hemoglobin. Nonuniform contribution of successive oxygen bindings to the alkaline Bohr effect. J Biol Chem 250:2227–2231
Inaba M, Maeda Y (1988) A new major transmembrane glycoprotein, gp155, in goat erythrocytes. J Biol Chem 263:17763–17771
Ingermann RL (1982) Physiological significance of Root effect hemoglobins in trout. Respir Physiol 49:1–10
Ingermann RL, Terwilliger RC (1981a) Intraerythrocytic organic phosphates of fetal and adult seaperch (*Embiotoca lateralis*): their role in maternal-fetal oxygen transport. J Comp Physiol 144:253–259

Ingermann RL, Terwilliger RC (1981b) Oxygen affinities of fetal and maternal hemoglobins of the viviparous seaperch, *Embiotoca lateralis*. J Comp Physiol 142:523–531
Ingermann RL, Terwilliger RC (1982a) Blood parameters and facilitation of maternal-fetal oxygen transfer in a viviparous fish (*Embiotoca lateralis*). Comp Biochem Physiol 73A:497–501
Ingermann RL, Terwilliger RC (1982b) Presence and possible function of Root effect hemoglobins in fishes lacking functional swim bladders. J Exp Zool 220:171–177
Ingermann RL, Terwilliger RC (1984) Facilitation of maternal-fetal oxygen transfer in fishes:anatomical and molecular specializations. In: Seymour RS (ed) Respiration and metabolism of embryonic vertebrates. Junk, Dordrecht, pp 1–15
Ingermann RL, Stock MK, Metcalfe J, Shin T-B (1983) Effect of ambient oxygen on organic phosphate concentrations in erythrocytes of the chick embryo. Respir Physiol 51:141–152
Ingermann RL, Hall RE, Bissonotte JM, Terwilliger RC (1984) Monosaccharide transport into erythrocytes of the Pacific hagfish, *Eptatretus stouti*. Mol Physiol 6:311–320
Ingermann RL, Bissonnette JM, Hall RE (1985a) Sugar uptake by red blood cells. In: Circulation, Respiration, and Metabolism, Gilles R (ed), Springer, Berlin Heidelberg, pp 290–300
Ingermann RL, Stock MK, Metcalfe J, Bissonnette JM (1985b) Monosaccharide uptake by erythrocytes of the embryonic and adult chicken. Comp Biochem Physiol 80A:369–372
Ingram VM (1985) Erythropoiesis: cellular and molecular mechanisms. In: Gilles R (ed) Circulation, respiration and metabolism. Current comparative approaches. Springer, Berlin Heidelberg New York, pp 322–332
Isaacks RE, Harkness DR (1980) Erythrocyte organic phosphates and hemoglobin function in birds, reptiles, and fishes. Am Zool 20:115–129
Isaacks RE, Harkness DR (1983) Erythrocyte organic phosphates and hemoglobin function in domestic mammals. In: Agar NS, Board, PG (eds) Red blood cells of domestic mammals. Elsevier, Amsterdam, pp 315–337
Isaacks RE, Harkness DR, Sampsell RN, Adler JL, Kim CY, Goldman PH (1976) Studies on avian erythrocyte metabolism. IV. Relationship between the major phosphorylated metabolic intermediates and oxygen affinity of whole blood in adults and embryos in several Galliformes. Comp Biochem Physiol 55A:29–33
Isaacks RE, Harkness DR, Witham PR (1978a) Relationship between the major phosphorylated metabolic intermediates and oxygen affinity of whole blood in the loggerhead (*Caretta caretta*) and the green sea turtle (*Chelonia mydas*). Dev Biol 62:344–353
Isaacks RE, Kim HD, Harkness DR (1978b) Relationship between phosphorylated metabolic intermediates and whole blood oxygen affinity in some air-breathing and water-breathing teleosts. Can J Zool 56:887–890
Isaacks RE, Kim CY, Johnson AEJ, Goldman PH, Harkness DR (1982) Studies on avian erythrocyte metabolism. XII. The synthesis and degradation of inositol pentakis (dihydrogen phosphate). Poult Sci 61:2271–2281
Isaacks RE, Nicol S, Sallis J, Zeidler R, Kim HD (1984) Erythrocyte phosphates and hemoglobin function in monotremes and some marsupials. Am J Physiol 246:R236–R241
Iscove NN (1977) The role of erythropoietin in regulation of population size and cell cycle of early and late erythroid precursors in mouse bone marrow. Cell Tissue Kinet 10:323–334
Iscove NN, Guilbert LJ (1978) Erythropoietin-independence of early erythropoiesis and a two-regulator model of proliferative control in the hemopoietic system. In: Murphy MJ (ed) In vitro aspects of erythropoiesis. Springer, Berlin Heidelberg New York, pp 3–7
Iscove NN, Keller G, Roitsch C (1985) Factors required by pluripotential stem cells in culture. In: Cronkite EP, Dainiak N, McCaffrey RP, Palek J, Quesenberry PJ (eds) Hematopoietic stem cell physiology. Alan R Liss, New York, pp 105–115
Isomaa B, Hägerstrand H, Paatero G (1987) Shape transformations induced by amphiphiles in erythrocytes. Biochim Biophys Acta 899:93–103
Iuchi I (1973) Chemical and physiological properties of the larval and the adult hemoglobins in rainbow trout, *Salmo gairdnerii irideus*. Comp Biochem Physiol 44B:1087–1101
Iuchi I (1985) Cellular and molecular bases of the larval-adult shift of hemoglobins in fish. Zool Sci 2:11–23
Iuchi I, Yamamoto M (1983) Erythropoiesis in the developing rainbow trout (*Salmo gairdneri irideus*). Histochemical and immunochemical detection of erythropoietic organs. J Exp Zool 226:409–417

Jackson DC (1986) Acid-base regulation of reptiles. In: Heisler N (ed) Acid-base regulation in animals. Elsevier, Amsterdam, pp 235–263
Jackson P, Morgan DB (1982) The relation between the membrane cholesterol content and anion exchange in the erythrocytes of patients with cholestasis. Biochim Biophys Acta 693:99–104
Jacobs MH, Stewart DR (1942) The role of carbonic anhydrase in certain ionic exchanges involving the erythrocyte. J Gen Physiol 25:539–552
Jacobs MH, Stewart DR (1947) Osmotic properties of the erythrocyte XII. Ionic and osmotic equilibria with a complex external solution. J Cell Comp Physiol 30:79–103
Jacquez JA (1984) Red blood cell as glucose carrier: significance for placental and cerebral glucose transport. Am J Physiol 246:R289–R298
Jarrett HW, Penniston JT (1977) Partial purification of the (Ca^{2+} + Mg^{2+})-ATPase activator from human erythrocytes. Its similarity to the activator of 3′:5′-cyclic nucleotide phosphodiesterase. Biochem Biophys Res Comm 77:1210–1216
Jarvis SM, Young JD, Ansay M, Archibald AL, Harkness RA, Simmonds RJ (1980) Is inosine the physiological energy source of pig erythrocytes? Biochim Biophys Acta 597:183–188
Jay DG (1983) Characterization of the chicken erythrocyte anion exchange protein. J Biol Chem 258:9431–9436
Jay DG, Cantley L (1986) Structural aspects of the red cell anion exchange protein. Annu Rev Biochem 55:511–538
Jeffreys AJ, Wilson V, Wood D, Simons JP, Kay RM, Williams JG (1980) Linkage of adult α- and β-globin genes in X. laevis and gene duplication by tetraploidization. Cell 21:555–564
Jelkmann W (1986) Renal erythropoietin: properties and production. Rev Physiol Biochem Pharmacol 104:139–215
Jelkmann W, Bauer C (1977) Oxygen affinity and phosphate compounds of red blood cells during intrauterine development of rabbits. Pflügers Arch 372:149–156
Jelkmann W, Bauer C (1978) High pyruvate kinase activity causes low concentration of 2,3-diphosphoglycerate in fetal rabbit red cells. Pflügers Arch 375:189–195
Jelkmann W, Bauer C (1980a) The regulation of red cell oxygen affinity and 2,3-P_2-glycerate level during rabbit ontogeny. Adv Physiol Sci 6:115–123
Jelkmann W, Bauer C (1980b) 2,3-DPG levels in relation to red cell enzyme activities in rat fetuses and hypoxic newborns. Pflügers Arch 389:61–68
Jelkmann W, Oberthür W, Kleinschmidt T, Braunitzer G (1981) Adaptation of hemoglobin function to subterranean life in the mole, *Talpa europaea*. Respir Physiol 46:7–16
Jenkins DMG, Lew VL (1973) Ca uptake by ATP-depleted red cells from different species with and without associated increase in K permeability. J Physiol (Lond) 234:41P-42P
Jennings ML (1985) Kinetics and mechanism of anion transport in red blood cells. Annu Rev Physiol 47:519–533
Jennings ML, Adams-Lackey M (1982) A rabbit erythrocyte membrane protein associated with l-lactate transport. J Biol Chem 257:12866–12871
Jennings ML, Nicknish JS (1985) Localization of a site of intermolecular cross-linking in human red blood cell band 3 protein. J Biol Chem 260:5472–5479
Jennings ML, Anderson MP, Monaghan R (1986a) Monoclonal antibodies against human erythrocyte band 3 protein. J Biol Chem 261:9002–9010
Jennings ML, Douglas SM, McAndrew PE (1986b) Amiloride-sensitive sodium-hydrogen exchange in osmotically shrunken rabbit red blood cells. Am J Physiol 251:C32–C40
Jensen FB (1986) Pronounced influence of Hb-O_2 saturation on red cell pH in tench blood in vivo and in vitro. J Exp Zool 238:119–124
Jensen FB (1987) Influences of exercise-stress and adrenaline upon intra- and extracellular acid-base status, electrolyte composition and respiratory properties of blood in tench (*Tinca tinca*) at different seasons. J Comp Physiol B 157:51–60
Jensen FB (1988) Acid-base regulation and blood gas transport in freshwater teleosts. Environmental dependence and adaptation. Dr Scient Thesis, University of Odense, Denmark, 63 pp
Jensen FB, Weber RE (1982) Respiratory properties of tench blood and hemoglobin. Adaptation to hypoxic-hypercapnic water. Mol Physiol 2:235–250
Jensen FB, Weber RE (1985a) Proton and oxygen equilibria, their anion sensitivities and interrelationships in tench hemoglobin. Mol Physiol 7:41–50

Jensen FB, Weber RE (1985b) Kinetics of the acclimational responses of tench to combined hypoxia and hypercapnia. II. Extra- and intracellular acid-base status in the blood. J Comp Physiol B 156:205–211
Jensen FB, Weber RE (1985c) Kinetics of the acclimational responses of tench to combined hypoxia and hypercapnia. I. Respiratory responses. J Comp Physiol B 156:197–203
Jensen FB, Weber RE (1987) Thermodynamic analysis of precisely measured oxygen equilibria of tench (*Tinca tinca*) hemoglobin and their dependence on ATP and protons. J Comp Physiol B 157:137–143
Jensen FB, Nikinmaa M, Weber RE (1983) Effects of exercise stress on acid-base balance and respiratory function in blood of the teleost *Tinca tinca*. Respir Physiol 51:291–301
Jinbu Y, Sato S, Nakao M, Tsukita S, Ishikawa H (1984) The role of ankyrin in shape deformability change of human erythrocyte ghosts. Biochim Biophys Acta 773:237–245
Johansen K (1966) Airbreathing in the teleost *Synbranchus marmoratus*. Comp Biochem Physiol 18:383–395
Johansen K, Lenfant C (1972) A comparative approach to the adaptability of O_2-Hb affinity. In: Rorth M, Astrup P (eds) Oxygen affinity of hemoglobin and red cell acid base status, Alfred Benzon Symposium IV. Munksgaard, Copenhagen, pp 750–780
Johansen K, Weber RE (1976) On the adaptibility of haemoglobin function to environmental conditions. In: Davies S (ed) Perspectives in experimental biology, vol 1. Zoology, Pergamon, Oxford, pp 219–234
Johansen K, Lenfant C, Hanson D (1973) Gas exchange in the lamprey, *Entosphenus tridentatus*. Comp Biochem Physiol 44A:107–119
Johansen K, Lykkeboe G, Weber RE, Maloiy GMO (1976) Respiratory properties of blood in awake and estivating lungfish, *Protopterus amphibius*. Respir Physiol 27:335–345
Johansen K, Mangum CP, Weber RE (1978) Reduced blood O_2 affinity associated with air breathing in osteoglossid fishes. Can J Zool 56:891–897
Johansen K, Berger M, Bicudo JEPW, Ruschi A, De Almeida PJ (1987) Respiratory properties of blood and myoglobin in hummingbirds. Physiol Zool 60:269–278
Johnson PC (1971) Red cell separation in the mesenteric capillary network. Am J Physiol 221:99–104
Jokumsen A, Weber RE (1980) Haemoglobin-oxygen binding properties in the blood of *Xenopus laevis*, with special reference to the influences of aestivation and of temperature and salinity acclimation. J Exp Biol 86:19–37
Jones MN, Nickson JK (1981) Monosaccaride transport proteins of the human erythrocyte membrane. Biochim Biophys Acta 650:1–20
Jonsson BH, Steiner H, Lindskog S (1976) Participation of buffer in the catalytic mechanism of carbonic anhydrase. FEBS Lett 64:310–314
Jordan HE (1933) The evolution of blood-forming tissues. Q Rev Biol 8:58–76
Jordan HE (1938) Comparative hematology. In: Downey H (ed) Handbook of hematology. Hamish Hamilton, London, pp 699–862
Jordan HE, Speidel C (1930) Blood formation in cyclostomes. Am J Anat 46:355–391
Jorgensen PL (1974) Purification and characterization of Na,K-ATPase. III. Purification from outer medulla of mammalian kidney after selective removal of membrane components by sodium dodecylsulfate. Biochim Biophys Acta 356:36–52
Joseph-Silverstein J, Cohen WD (1984) The cytoskeletal system of nucleated erythrocytes. III. Marginal band function in mature cells. J Cell Biol 98:2118–2125
Kaji D, Kahn T (1985) Kinetics of Cl-dependent K influx in human erythrocytes with and without external Na: effect of NEM. Am J Physiol 249:C490–C496
Kaloyianni-Dimitriades M, Beis I (1984) Studies on the energy metabolism of *Rana ridibunda* erythrocytes. J Comp Physiol 155B:109–115
Kannan KK, Notstrand B, Fridborg K, Lövgren S, Ohlsson A, Petef M (1975) Crystal structure of human erythrocyte carbonic anhydrase B. Three-dimensional structure at a nominal 2.2-Å resolution. Proc Natl Acad Sci USA 72:51–55
Kapitza H-G, Sackmann E (1980) Local measurement of lateral motion in erythrocyte membranes by photobleaching technique. Biochim Biophys Acta 595:56–64
Kaplan BH (1970) The control of heme synthesis. In: Gordon AS (ed) Regulation of hematopoiesis, vol 1. Red cell production. Meredith, New York, pp 677–700
Kaplan JH (1985) Ion movements through the sodium pump. Annu Rev Physiol 47:535–544

Kaul RK, Murthy SNP, Reddy AG, Steck TL, Kohler H (1983) Amino acid sequence of the N^{α}-terminal 201 residues of human erythrocyte membrane band 3. J Biol Chem 258:7981–7990
Kay MMB, Goodman SR, Sorensen K, Whitfield CF, Wong P, Zaki L, Rudloff V (1983) Senescent cell antigen is immunologically related to band 3. Proc Natl Acad Sci USA 80:1631–1635
Kay MMB, Bosman GJCGM, Lawrence C (1988a) Functional topography of band 3: Specific structural alteration linked to functional aberrations in human erythrocytes. Proc Natl Acad Sci USA 85:492–496
Kay MMB, Bosman GJCGM, Johnson GJ, Beth AH (1988b) Band-3 polymers and aggregates, and hemoglobin precipitates in red cell aging. Blood Cells 14:275–289
Keijer E, Butler PJ (1982) Volumes of the respiratory and circulatory systems in tufted and mallard ducks. J Exp Biol 101:213–220
Kellett GL (1971) Dissociation of hemoglobin into subunits. Ligand-linked dissociation at neutral pH. J Mol Biol 59:401–424
Kendall MD, Ward P (1974) Erythropoiesis in avian thymus. Nature 249:366–367
Khalifah RG (1971) The carbon dioxide hydration activity of carbonic anhydrase: stop flow kinetic studies on the native human isozymes B and C. J Biol Chem 246:2561–2573
Khodadad JK, Weinstein RS (1983) The band 3-rich membrane of llama erythrocytes: studies on cell shape and the organization of membrane proteins. J Membr Biol 72:161–171
Kiceniuk JW, Jones DR (1977) The oxygen transport system in trout (*Salmo gairdneri*) during sustained exercise. J Exp Biol 69:247–260
Kikuchi Y, Koyama T (1984) Red blood cell deformability and protein adsorption on red blood cell surface. Am J Physiol 247:H739–H747
Kikuchi Y, Arai T, Koyama T (1983) Improved filtration method for red cell deformability measurement. Med Biol Eng Comput 21:270–276
Kikuchi Y, Hughes GM, Koyama T, Kakiuchi Y, Araiso T (1985) Effects of temperature and transfer from seawater to freshwater on blood microrheology in Pacific salmon. Jpn J Physiol 35:683–688
Kilmartin JV (1972) Molecular mechanism of the Bohr effect. In: Rorth M, Astrup P (eds) Oxygen affinity of hemoglobin and red cell acid-base status, Alfred Benzon Symposium IV. Munksgaard, Copenhagen, pp 93–99
Kilmartin JV, Rossi-Bernardi L (1969) Inhibition of CO_2 combination and reduction of the Bohr effect in haemoglobin chemically modified at its α-amino groups. Nature 222:1243–1246
Kilmartin JV, Rossi-Bernardi L (1973) Interaction of hemoglobin with hydrogen ions, carbon dioxide, and organic phosphates. Physiol Rev 53:836–890
Kilmartin JV, Wootton JF (1970) Inhibition of Bohr effect after removal of C-terminal histidines from haemoglobin β-chains. Nature 228:766–767
Kim HD (1983) Postnatal changes in energy metabolism of mammalian red blood cells. In: Agar NS, Board PG (eds) Red blood cells of domestic mammals. Elsevier, Amsterdam, pp 339–355
Kim HD (1985) ATP metabolism in mammalian red blood cells. In: Gilles R (ed) Circulation, respiration, and metabolism. Current comparative approaches. Springer, Berlin Heidelberg New York, pp 312–321
Kim HD, Isaacks RE (1978) The membrane permeability of nonelectrolytes and carbohydrate metabolism of Amazon fish red cells. Can J Zool 56:863–869
Kim HD, McManus TJ (1971a) Studies on the energy metabolism of pig red cells. I. The limiting role of membrane permeability in glycolysis. Biochim Biophys Acta 230:1–11
Kim HD, McManus TJ (1971b) Studies on the energy metabolism of pig red cells. II. Lactate formation from free ribose and deoxyribose with maintenance of ATP. Biochim Biophys Acta 230:12–19
Kim HD, Watts RP, Luthra MG, Schwalbe CR, Conner RT, Brendel K (1980) A symbiotic relationship of energy metabolism between a 'non-glycolytic' mammalian red cell and the liver. Biochim Biophys Acta 589:256–263
Kim J-S, Gay CV, Schraer R (1983) Purification and properties of carbonic anhydrase from salmon erythrocytes. Comp Biochem Physiol 76B:523–527
Kim S, Magendantz M, Katz W, Solomon F (1987) Development of a differentiated microtubule structure: formation of the chicken erythrocyte marginal band in vivo. J Cell Biol 104:51–59
Kimmich GA, Randles J, Restrepo D, Montrose M (1985) A new method for determination of relative ion permeabilities in isolated cells. Am J Physiol 248:C399–C405

Kirk RG, Lee P (1988) Anion transport during maturation of erythroblastic cells. J Membr Biol 101:173–178
Kister J, Poyart C, Edelstein SJ (1987) An expanded two-state allosteric model for interactions of human hemoglobin A with nonsaturating concentrations of 2,3-diphosphoglycerate. J Biol Chem 262:12085–12091
Kleinschmidt T, Rücknagel KP, Weber RE, Koop BF, Braunitzer G (1987) Primary structure and functional properties of the hemoglobin from the free-tailed bat *Tadarida brasiliensis* (Chiroptera). Small effect of carbon dioxide on oxygen affinity. Biol Chem Hoppe-Seyler 368:681–690
Kleyman TR, Cragoe EJ Jr (1988) Amiloride and its analogs as tools in the study of ion transport. J Membr Biol 105:1–21
Klocke RA (1978) Catalysis of CO_2 reactions by lung carbonic anhydrase. J Appl Physiol 44: 882–888
Klocke RA (1987) Carbon dioxide transport. In: Farhi LE, Tenney SM (eds) Handbook of physiology, section 3. The respiratory system, vol. 4. Gas exchange. American Physiological Society, Bethesda, Maryland, pp 173–197
Klocke RA (1988) Velocity of CO_2 exchange in blood. Annu Rev Physiol 50:625–637
Knauf PA, Fuhrmann GF, Rothstein S, Rothstein A (1977) The relationship between anion exchange and net anion flow across the human red blood cell membrane. J Gen Physiol 69:363–386
Kodicek M, Mircevova L, Marik T (1987) Energy requirements of erythrocytes under mechanical stress. Biomed Biochim Acta 46:103–107
Koike TI, Pryor LR, Neldon HL (1983) Plasma volume and eletrolytes during progressive water deprivation in chickens (*Gallus domesticus*). Comp Biochem Physiol 74A:83–87
Kopito RR, Lodish HF (1985) Primary structure and transmembrane orientation of the murine anion exchange protein. Nature 316:234–238
Korsgren C, Cohen CM (1988) Associations of human erythrocyte band 4.2. Binding to ankyrin and to the cytoplasmic domain of band 3. J Biol Chem 263:10212–10218
Koury ST, Repasky EA, Eckerts BS (1987) The cytoskeleton of isolated murine primitive erythrocytes. Cell Tissue Res 249:69–77
Koyama T (1985) Shear rate and orientation of erythrocytes in pulmonary microvessels of bullfrogs. Biorheology 22:379–384
Kracke GR, Anatra MA, Dunham PB (1988) Asymmetry of Na-K-Cl cotransport in human erythrocytes. Am J Physiol 254:C243–C250
Kramm C, Sattrup G, Baumann R, Bartels H (1975) Respiratory function of blood in hibernating and non-hibernating hedgehogs. Respir Physiol 25:311–318
Krantz SB, Jacobson LO (1970) Erythropoietin and the regulation of erythropoiesis. University of Chicago Press, Chicago 330 pp
Kregenow FM (1971) The response of duck erythrocytes to hypertonic media. J Gen Physiol 58:396–412
Kregenow FM (1973) The response of duck erythrocytes to norepinephrine and an elevated extracellular potassium. J Gen Physiol 61:509–527
Kregenow FM (1978) An assessment of the cotransport hypothesis as it applies to the norepinephrine and hypertonic responses. In: Jorgensen CB, Skadhauge E (eds) Osmotic and volume regulation. Alfred Benzon Symposium XI. Munksgaard, Copenhagen, pp 378–391
Kregenow FM (1981) Osmoregulatory salt transporting mechanisms: control of cell volume in anisotonic media. Annu Rev Physiol 43:493–505
Kregenow FM, Robbie DE, Orloff J (1976) Effect of norepinephrine and hypertonicity on K influx and cyclic AMP in duck erythrocytes. Am J Physiol 231:306–312
Kregenow FM, Caryk T, Siebens AW (1985) Further studies of the volume-regulatory response of *Amphiuma* red cells in hypertonic media. J Gen Physiol 86:565–584
Kurtz A, Jelkmann W, Bauer C (1982) A new candidate for the regulation of erythropoiesis. Insulin-like growth factor I. FEBS Lett 149:105–108
Kuypers FA, Roelofsen B, Berendsen W, Op Den Kamp JAF, Van Deenen LLM (1984) Shape changes in human erythrocytes induced by replacement of the native phosphatidylcholine with species containing various fatty acids. J Cell Biol 99:2260–2267

Lacombe C, Da Silva J-L, Bruneval P, Fournier J-G, Wendling F, Casadevall N, Camilleri J-P, Bariety J, Varet B, Tambourin P (1988) Peritubular cells are the site of erythropoietin synthesis in the murine hypoxic kidney. J Clin Invest 81:620–623
Lane HC (1979) Progressive changes in hematology and tissue water of sexually mature trout, *Salmo gairdneri* Richardson during the autumn and winter. J Fish Biol 15:425–436
Lane HC (1984) Nucleoside triphosphate changes during the peripheral life-span of erythrocytes of adult rainbow trout (*Salmo gairdneri*). J Exp Zool 231:57–62
Lange Y (1984) The dynamics of erythrocyte membrane cholesterol. In: Kruckeberg WC, Eaton JW, Aster J, Brewer GJ (eds) Erythrocyte membranes: recent clinical and experimental advances, vol 3. Alan R Liss, New York, pp 137–151
Lapennas GN (1983) The magnitude of the Bohr coefficient: optimal for oxygen delivery. Respir Physiol 54:161–172
Lapennas GN, Reeves RB (1983) Oxygen affinity of blood of adult domestic chicken and red jungle fowl. Respir Physiol 52:27–39
Larsen B (1968) Red cell life span in hibernating hedgehogs (*Erinaceus europaeus* L). Acta Univ Bergensis Ser Math Rer Nat 1967(6):1–8
Lassen UV (1977) Electrical potential and conductance of the red cell membrane. In: Ellory JC, Lew VL (eds) Membrane transport in red cells. Academic Press, London, pp 137–172
Lassen UV, Pape L, Vestergaard-Bogind B (1976) Effect of calcium on the membrane potential of *Amphiuma* red cells. J Membr Biol 26:51–70
Lauf PK (1982) Evidence for chloride dependent potassium and water transport induced by hyposmotic stress in erythrocytes of the marine teleost, *Opsanus tau*. J Comp Physiol 146:9–16
Lauf PK (1985a) Passive K^+-Cl^- fluxes in low K^+ sheep erythrocytes: modulation by A23187 and bivalent cations. Am J Physiol 249:C271–C278
Lauf PK (1985b) On the relationship between volume- and thiol-stimulated K^+Cl^- fluxes in red cell membranes. Mol Physiol 8:215–234
Lauf PK, Bauer J (1987) Direct evidence for chloride-dependent volume reduction in macrocytic sheep reticulocytes. Biochem Biophys Res Comm 144:849–855
Lauf PK, Theg BE (1980) A chloride dependent K flux induced by N-ethylmaleimide in genetically low K sheep and goat erythrocytes. Biochem Biophys Res Comm 92:1422–1428
Lauf PK, Valet G (1983) Na^+K^+ pump and passive K^+ transport in large and small red cell population of anemic high and low K^+ sheep. J Cell Physiol 116:35–44
Lauf PK, Perkins CM, Adragna NC (1985) Cell volume and metabolic dependence of NEM-activated K^+-Cl^- flux in human red blood cells. Am J Physiol 249:C124–C128
Lauf PK, McManus TJ, Haas M, Forbush B III, Duhm J, Flatman PW, Milton H, Saier MH Jr, Russell JM (1987) Physiology and biophysics of chloride and cation cotransport across cell membranes. Fed Proc 46:2377–2394
Laursen JS, Andersen NA, Lykkeboe G (1985) Temperature acclimation and oxygen binding properties of the European eel, *Anguilla anguilla*. Comp Biochem Physiol 81A:79–86
Leblond PF, Coulombe L (1979) The measurement of erythrocyte deformability using micropore membranes. J Lab Clin Med 94:133–143
Lechner AJ (1976) Respiratory adaptations in burrowing pocket gophers from sea level and high altitude. J Appl Physiol 41:168–173
Leclercq F, Schnek AG, Braunitzer G, Stangl A, Schrank B (1981) Direct reciprocal allosteric interaction of oxygen and hydrogen carbonate sequence of the haemoglobins of the caiman (*Caiman crocodylus*), the Nile crocodile (*Crocodylus niloticus*) and the Mississippi crocodile (*Alligator mississippiensis*). Hoppe-Seyler's Z Physiol Chem 362:1151–1158
Lee P, Miles PR (1972) Density distribution and cation composition of red blood cells in newborn puppies. J Cell Physiol 79:377–388
Lefkowitz RJ, Caron MG, Michel T, Stadel JM (1982) Mechanisms of hormone receptor-effector coupling: the beta-adrenergic receptor and adenylate cyclase. Fed Proc 41:2664–2670
Lefkowitz RJ, Stadel JM Caron MG (1983) Adenylate-cyclase coupled beta-adrenergic receptors. Annu Rev Biochem 52:159–186
Lefkowitz RJ, Cerione RA, Codina J, Birnbaumer L, Caron MG (1985) Reconstitution of the β-adrenergic receptor. J Membr Biol 87:1–12

Leggio T, Morpurgo G (1968) Dissociation curves of toad haemoglobin and a hypothesis for the cause of hibernation. Nature 219:493–494

Lehninger A (1975) Biochemistry, 2nd edition, Worth, New York. 1107 pp

Leite MV, Goldstein L (1987) Ca^{2+} ionophore and phorbol ester stimulate taurine efflux from skate erythrocytes. J Exp Zool 242:95–97

Lemez L (1971) The causes of the extremely short average life spans of the primitive erythrocytes (E I) in avian and mammalian embryos. In: Travnicek T, Neuwirt J (eds) The regulation of erythropoiesis and haemoglobin synthesis. Universita Karlova, Praha, pp 210–216

Lemke PR, Graf G (1974) Isolation and partial characterization of carbonic anhydrase from erythrocytes of *Meleagris gallopavo*. Mol Cell Biochem 4:141–147

Lenfant C, Johansen K (1967) Respiratory adaptations in selected amphibians. Respir Physiol 2:247–260

Lenfant C, Kenney DW, Aucutt C (1968) Respiratory function in the killer whale *Orcinus orca* (Linnaeus). Am J Physiol 215:1506–1511

Lenfant C, Elsner R, Kooyman GL, Drabek CM (1969) Respiratory function of blood of the adult and fetus Weddell seal *Leptonychotes weddellii*. Am J Physiol 216:1595–1597

Lenfant C, Johansen K, Torrance JD (1970) Gas transport and oxygen storage capacity in some pinnipeds and the sea otter. Respir Physiol 9:277–286

Lenfant C, Bellingham AJ, Detter JC (1972) Physiological factors influencing the haemoglobin affinity for oxygen. In: Rorth M, Astrup P (eds) Oxygen affinity of haemoglobin and red cell acid base status, Alfred Benzon Symposium IV. Munksgaard, Copenhagen, pp 736–747

Leray C, Bachand L (1975) Erythrocyte metabolism in the yellow perch (*Perca flavescens* Mitchill) – II. Intermediates, nucleotides and free energy changes in glycolytic reactions. Comp Biochem Physiol 51B:349–353

Lerner MH, Lowy BA (1974) The formation of adenosine in rabbit liver and its possible role as a direct precursor of erythrocyte adenine nucleotides. J Biol Chem 249:959–966

Lesley J, Hyman R, Schulte R, Trotter J (1983) Expression of transferrin receptor on murine hematopoietic progenitors. Cell Immunol 83:14–25

Levere RD, Kappas A, Granick S (1967) Stimulation of hemoglobin synthesis in chick blastoderms by certain 5β androstane and 5β pregnane steroids. Biochemistry 58:985–990

Levitzki A (1984) Receptors. A quantitative approach. Cummings, Menlo Park, 142 pp

Levitzki A (1986) β-adrenergic receptors and their mode of coupling to adenylate cyclase. Physiol Rev 66:819–854

Levitzki A (1988) From epinephrine to cyclic AMP. Science 241:800–805

Lew VL, Bookchin RM (1986) Volume, pH, and ion-content regulation in human red cells: analysis of transient behavior with an integrated model. J Membr Biol 92:57–74

Lew VL, Ferreira HG (1978) Calcium transport and the properties of a calcium-activated potassium channel in red cell membranes. Curr Top Membr Transp 10:217–271

Lewis RE, Czech MP (1988) Molecular biology of the insulin receptor. Prog Nucl Acid Res Mol Biol 35:157–171

Lijnen P, Hespel P, Fagard R, Lysens R, Eynde Van den E, Goris M, Goossens W, Amery A (1988) Erythrocyte 2,3-diphosphoglycerate concentration before and after a marathon in men. Eur J Appl Physiol 57:452–455

Liljas A, Kannan KK, Bergstén P-C, Waara I, Fridborg K, Strandberg B, Carlbom U, Järup L, Lövgren S, Petef M (1972) Crystal structure of human carbonic anhydrase C. Nature 235:131–137

Lim BC, Morgan EH (1984) Transferrin endocytosis and the mechanism of iron uptake by reticulocytes in the toad (*Bufo marinus*). Comp Biochem Physiol 79A:317–323

Lindskog S, Engberg P, Forsman C, Ibrahim SA, Jonsson B-H, Simonsson I, Tibell L (1984) Kinetics and mechanism of carbonic anhydrase isozymes. Ann N Y Acad Sci 429:61–75

Lingrel JB, Schon EA, Cleary ML, Shapiro SG (1983) Structure and evolution of the developmentally regulated globin genes of the goat. In: Goldwasser E (ed) Regulation of hemoglobin biosynthesis. Elsevier, Amsterdam, pp 89–104

Liu S-C, Palek J (1984) Hemoglobin enhances the self-association of spectrin heterodimers in human erythrocytes. J Biol Chem 259:11556–11562

London IM, Fagard R, Leroux A, Levin DH, Matts R, Petryshyn R (1983) The regulation of hemoglobin synthesis by heme and protein kinases. In: Goldwasser E (ed) Regulation of hemoglobin biosynthesis. Elsevier, Amsterdam pp 165–181

Lönnerholm G (1982) Pulmonary carbonic anhydrase in the human, monkey, and rat. J Appl Physiol 52:352–356
Lovrien RE, Anderson RA (1980) Stoichiometry of wheat germ agglutinin as a morphology controlling agent and as a morphology protecting agent for the human erythrocyte. J Cell Biol 85:534–548
Low PS (1986) Structure and function of the cytoplasmic domain of band 3: center of erythrocyte membrane-peripheral protein interactions. Biochim Biophys Acta 864:145–167
Low PS, Waugh SM, Zinke K, Drenckhahn D (1985) The role of hemoglobin denaturation and band 3 clustering in red blood cell aging. Science 227:531–533
Low PS, Allen DP, Zioncheck TF, Chari P, Willardson BM, Geahlen RL, Harrison ML (1987) Tyrosine phosphorylation of band 3 inhibits peripheral protein binding. J Biol Chem 262:4592–4596
Lowe AG, Lambert A (1983) Chloride-bicarbonate exchange and related transport processes. Biochim Biophys Acta 694:353–374
Lowe AG, Walmsley AR (1987) A single half-turnover of the glucose carrier of the human erythrocyte. Biochim Biophys Acta 903:547–550
Lucarelli G, Porcellini A, Carnevali C, Carmena A, Stohlman FJ (1968) Fetal and neonatal erythropoiesis. Ann N Y Acad Sci 149:544–559
Lutz PL (1980) On the oxygen affinity of bird blood. Am Zool 20:187–198
Lykkeboe G, Weber RE (1978) Changes in the respiratory properties of the blood in the carp, *Cyprinus carpio*, induced by diurnal variation in ambient oxygen tension. J Comp Physiol 128:117–125
Lyman CP, Weiss LP, O'Brien RC, Barbeau AA (1957) The effect of hibernation on the replacement of blood in the golden hamster. J Exp Zool 136:471–485
Macey DJ, Potter IC (1982) The effect of temperature on the oxygen dissociation curves of whole blood of larval and adult lampreys (*Geotria australis*). J Exp Biol 97:253–261
Macey RI (1980) Mathematical models of membrane transport processes. In: Andreoli TE, Hoffman JF, Fanestil DD (eds) Membrane physiology. Plenum, New York, pp 125–146
Macknight ADC, Leaf A (1977) Regulation of cellular volume. Physiol Rev 57:510–573
Maclean GS (1981) Blood viscosity of the two mammalian hibernators: *Spermorphilus tridecemlineatus* and *Tamias striatus*. Physiol Zool 54:122–131
Maginniss LA (1985) Red cell organic phosphates and Bohr effects in house sparrow blood. Respir Physiol 59:93–103
Maginniss LA, Hitzig BM (1987) Acid-base status and electrolytes in red blood cells and plasma of turtles submerged at 3°C. Am J Physiol 253:R64-R70
Maginniss LA, Song YK, Reeves RB (1980) Oxygen equilibria of ectotherm blood containing multiple hemoglobins. Respir Physiol 42:329–343
Maginniss LA, Tapper SS, Miller LS (1983) Effect of chronic cold and submergence on blood oxygen transport in the turtle, *Chrysemys picta*. Respir Physiol 53:15–29
Mahe Y, Garcia-Romeu F, Motais R (1985) Inhibition by amiloride of both adenylate cyclase activity and the Na^+/H^+ antiporter in fish erythrocytes. Eur J Pharmacol 116:199–206
Maines MD, Kappas A (1977) Metals as regulators of heme metabolism. Science 198:1215–1221
Mairbäurl H, Humpeler E (1981) In vitro influences of adrenaline on erythrocyte metabolism and oxygen affinity of hemoglobin. In: Brewer GJ (ed) The red cell: fifth Ann Arbor conference. Alan R Liss, New York, pp 311–319
Mairbäurl H, Humpeler E, Schwaberger G, Pessenhofer H (1983) Training-dependent changes of red cell density and erythrocytic oxygen transport. J Appl Physiol 55:1403–1407
Majerus PW, Wilson DB, Connolly TM, Bross TE, Neufeld EJ (1985) Phosphoinositide turnover provides a link in stimulus-response coupling. Trends Biochem Sci 10:168–171
Makonnen E, Est M (1987) Inhibition of red blood cell Na^+, K^+-ATPAse from different species by some cardiac glycosides. Iugosl Physiol Pharmacol Acta 23:263–268
Malan A (1985) Intracellular pH in response to ambient changes: homeostatic or adaptive responses. In: Gilles R (ed) Circulation, respiration, and metabolism. Current comparative approaches. Springer, Berlin Heidelberg New York, pp 464–473
Maniatis T, Fritsch EF, Lauer J, Lawn RM (1980) The molecular genetics of human hemoglobins. Annu Rev Genet 14:145–178
Manwell C (1958a) A "fetal-maternal" shift in the ovoviviparous spiny dogfish *Squalus suckleyi* (Girard). Physiol Zool 31:93–100

Manwell C (1958b) Ontogeny of hemoglobin in the skate *Raja binoculata*. Science 128:419–420
Maples PB, Dorn AR, Broyles RH (1983) Embryonic and larval hemoglobins during the early development of the bullfrog, *Rana catesbeiana*. Dev Biol 96:515–519
Maples PB, Palmer JC, Broyles RH (1986) In vivo regulation of hemoglobin phenotypes of developing *Rana catesbeiana*. Dev Biol 117:337–341
Marchesi VT (1984) Structure and function of the erythrocyte membrane skeleton. In: Kruckeberg WC, Eaton JW, Aster J, Brewer GJ (eds) Erythrocyte membranes: recent clinical and experimental advances, vol 3. Alan R Liss, New York, pp 1–12
Marchesi VT (1985) Stabilizing infrastructure of cell membranes. Annu Rev Cell Biol 1:531–561
Maren TH (1967) Carbonic anhydrase: chemistry, physiology, and inhibition. Physiol Rev 47:595–781
Maren TH, Friedland BR, Rittmaster RS (1980) Kinetic properties of primitive vertebrate carbonic anhydrases. Comp Biochem Physiol 67B:69–74
Maren TH, Sanyal G (1983) The activity of sulfonamides and anions against the carbonic anhydrases of animals, plants and bacteria. Annu Rev Pharmacol Toxicol 23:439–459
Marino D, Sarkadi B, Gardos G, Bolis L (1981) Calcium-induced alkali cation transport in nucleated red cells. Mol Physiol 1:295–300
Marks PA (1964) Glucose-6-phosphate dehydrogenase: its properties and role in mature erythrocytes. In: Bishop C, Surgenor DM (eds) The red blood cell. Academic Press, London, pp 211–242
Marks PA, Fantoni A, Chapelle De La A (1968) Hemoglobin synthesis and differentiation of erythroid cells. Vitam Horm 26:331–355
Martin JP, Bonaventura J, Brunori M, Fyhn HJ, Fyhn UEH, Garlick RL, Powers DA, Wilson MT (1979) The isolation and characterization of the hemoglobin components of Mylossoma sp., an Amazonian teleost. Comp Biochem Physiol 62A:155–162
Matsuura MSA, Ogo SH, Focesi A, Jr (1987) Dimer-tetramer transition in hemoglobins from *Liophis miliaris* — I. Effect of organic polyphosphates. Comp Biochem Physiol 86A:683–687
Mattison AGM, Fänge R (1977) Light- and electronmicroscopic observations on the blood cells of the Atlantic hagfish *Myxine glutinosa* (L.). Acta Zool 58:205–221
Mattsoff L, Nikinmaa M (1987) Effects of plasma proteins on the dehydroabietic-acid-induced red cell breakdown. Ecotoxicol Environ Saf 14:157–163
Mattsoff L, Nikinmaa M (1988) Effects of external acidification on the blood acid-base status and ion concentrations of lamprey. J Exp Biol 136:351–361
Mavilio F, Giampaolo A, Care A, Migliaccio G, Calandrini M, Russo G, Pagliardi GL, Mastroberardino G, Marinúcci M, Peschle C (1983) Molecular mechanisms of human hemoglobin switching: selective undermethylation and expression of globin genes in embryonic, fetal, and adult erythroblasts. Proc Natl Acad Sci USA 80:6907–6911
May WS Jr, Cuatrecasas P (1985) Transferrin receptor: its biological significance. J Membr Biol 88:205–215
Maynard JR, Coleman JE (1971) Elasmobranch carbonic anhydrase. J Biol Chem 246:4455–4464
Mayo KH, Chien JCW (1980) Effect of temperature on functional properties of carp hemoglobin. J Mol Biol 142:63–73
McConnell FM, Goldstein L (1988) Intracellular signals and volume regulatory response in skate erythrocytes. Am J Physiol 255:R982–R987
McLeay DJ (1973) Effects of a 12-hr and 25-day exposure to kraft pulp mill effluent on the blood and tissues of juvenile coho salmon (*Oncorhynchus kisutch*). J Fish Res Board Can 30:395–400
McLeod TF, Sigel MM, Yunis AA (1978) Regulation of erythropoiesis in the Florida gar, *Lepisosteus platyrhinchus*. Comp Biochem Physiol 60A:145–150
McManus TJ, Schmidt WF III (1978) Ion and co-ion transport in avian red cells. In: Hoffman JF (ed) Membrane transport processes. Raven, New York, pp 79–106
Meints RH, Carver FJ (1973) Triiodothyronine and hydrocortisone effects on *Rana pipiens* erythropoiesis. Gen Comp Endocrinol 21:9–13
Meints RH, Carver FJ, Gerst JW, McLaughlin DW (1975) Erythropoietic activity in the turtle: the influence of hemolytic anemia, hypoxia and hemorrhage on hemopoietic function. Comp Biochem Physiol 50A:419–422

Meiselman HJ, Evans EA, Hochmuth RM (1978) Membrane mechanical properties of ATP-depleted human erythrocytes. Blood 52:499–504
Mellman I, Fuchs R, Helenius A (1986) Acidification of the endocytic and exocytic pathways. Annu Rev Biochem 55:663–700
Meraldi J-P, Slichter J (1981) A statistical mechanical treatment of fatty acyl chain order in phospholipid bilayers and correlation with experimental data. Biochim Biophys Acta 645:183–192
Micheli V, Pieragalli D, Toti P, Forconi S, Guerrini M, Acciavatti A, Galigani C, Weber G, Ricci C, Di Perri T (1987) Haemorheological changes during human erythrocyte life span. Clin Hemorheol 7:267–272
Middelkoop E, Lubin BH, Bevers EM, Op Den Kamp JAF, Comfurius P, Chiu DT-Y, Zwaal RFA, Van Deenen LLM, Roelofsen B (1988) Studies on sickled erythrocytes provide evidence that the asymmetric distribution of phosphatidylserine in the red cell membrane is maintained by both ATP-dependent translocation and interaction with membrane skeletal proteins. Biochim Biophys Acta 937:281–288
Miles PR, Lee P (1972) Sodium and potassium content and membrane transport properties in red blood cells from newborn puppies. J Cell Physiol 79:367–376
Milligan CL, Wood CM (1986) Intracellular and extracellular acid-base status and H^+ exchange with the environment after exhaustive exercise in the rainbow trout. J Exp Biol 123:93–121
Milligan CL, Wood CM (1987) Regulation of blood oxygen transport and red cell pHi after exhaustive activity in rainbow trout (*Salmo gairdneri*) and starry flounder (*Platichthys stellatus*). J Exp Biol 133:263–282
Milsom WK, Johansen K, Millard RW (1973) Blood respiratory properties in some antarctic birds. Condor 75:427:474
Mische SM, Morrow JS (1988) Post-translational regulation of the erythrocyte cortical cytoskeleton. Protoplasma 145:167–175
Mohandas N, Clark MR, Feo C, Jacobs MS, Shohet SB (1981) Factors that limit whole cell deformability in erythrocytes after calcium loading and ATP depletion. In: Brewer GJ (ed) The red cell: fifth Ann Arbor conference. Alan R Liss, New York, pp 423–434
Monette FC (1983) Cell amplification in erythropoiesis: in vitro perspectives. In: Dunn CDR (ed) Current concepts in erythropoiesis. Wiley, New York, pp 21–57
Monod J, Wyman J, Changeux J-P (1965) On the nature of allosteric transitions: a plausible model. J Mol Biol 12:88–118
Montrose MH, Kimmich GA (1986) Quantitative use of weak bases for estimation of cellular pH gradients. Am J Physiol 250:C418–C422
Morris RJ, Gibson QH (1982) Cooperative ligand binding to hemoglobin — Effects of temperature and pH on a hemoglobin with spectrophotometrically distinct chains (*Thunnus thynnus*). J Biol Chem 257:4869–4874
Morris RJ, Neckameyer WS, Gibson QH (1981) Multiple T state conformations in a fish hemoglobin. Carbon monoxide binding to hemoglobin in *Thunnus thynnus*. J Biol Chem 256:4598–4603
Morrow JS, Marchesi VT (1981) Self-assembly of spectrin oligomers in vitro: a basis for dynamic cytoskeleton. J Cell Biol 88:463–468
Morrow JS, Speicher DW, Knowles WJ, Hsu CJ, Marchesi VT (1980) Identification of functional domains of human erythrocyte spectrin. Proc Natl Acad Sci USA 77:6592–6596
Moss B, Ingram VM (1968a) Hemoglobin synthesis during amphibian metamorphosis I. Chemical studies on the hemoglobins from the larval and adult stages of *Rana catesbeiana*. J Mol Biol 32:481–492
Moss B, Ingram VM (1968b) Hemoglobin synthesis during amphibian metamorphosis II. Synthesis of adult hemoglobin following thyroxine administration. J Mol Biol 32:493–504
Motais R, Garcia-Romeu F (1987) Cell volume control by catecholamines in erythrocytes. In: Kirsch R, Lahlou B (eds) Comparative physiology of environmental adaptations, vol 1. Adaptations to salinity and dehydration. Karger, Basel, pp 13–25
Motais R, Garcia-Romeu F, Borgese F (1987) The control of Na^+/H^+ exchange by molecular oxygen in trout erythrocytes. J Gen Physiol 90:197–207
Motais R, Fievet B, Garcia-Romeu F, Thomas S (1989) Na^+-H^+ exchange and pH regulation in red blood cells: role of uncatalyzed H_2CO_3 dehydration. Am J Physiol 256:C728–C735

Mrabet NT, McDonald MJ, Turci S, Sarkar R, Szabo A, Bunn HF (1986) Electrostatic attraction governs the dimer assembly of human hemoglobin. J Biol Chem 261:5222–5228
Mullins LJ, Noda K (1963) The influence of sodium-free solutions on the membrane potential of frog muscle fibers. J Gen Physiol 47:117–132
Mumm DP, Atha DH, Riggs A (1978) The hemoglobin of the common sting-ray, *Dasyatis sabina*: structural and functional properties. Comp Biochem Physiol 60B:189–193
Murphy DB, Grasser WA, Wallis KT (1986) Immunofluorescence examination of beta tubulin expression and marginal band formation in developing chicken erythroblasts. J Cell Biol 102:628–635
Musacchia XJ, Volkert WA (1971) Blood gases in hibernating and active ground squirrels: HbO_2 affinity at 6 and 38 C. Am J Physiol 221:128–130
Naftalin RJ, Holman GD (1977) Transport of sugars in human red cells. In: Ellory JC, Lew VL (eds) Membrane transport in red cells. Academic Press, London, pp 257–300
Nakao M, Nakao T, Yamazoe S (1960) Adenosine triphosphate and maintenance of shape of human red cells. Nature 187:945–946
Nakashima M, Noda H, Hasegaea M, Ikai A (1985) The oxygen affinity of mammalian hemoglobins in the absence of 2,3-diphosphoglycerate in relation to body weight. Comp Biochem Physiol 82A:583–589
Nash AR, Fisher WK, Thompson EOP (1976) Haemoglobins of the shark *Heterodontus portusjacksoni* II. Amino acid sequence of the α-chain. Aust J Biol Sci 29:73–97
Nash GB, Meiselman HJ (1983) Red cell and ghost viscoelasticity. Effects of hemoglobin concentration and in vivo aging. Biophys J 43:63–73
Nash GB, Wyard SJ (1981) Erythrocyte membrane elasticity during in vivo aging. Biochim Biophys Acta 643-269-275
Nathans J, Hogness DS (1983) Isolation, sequence analysis, and intron-exon arrangement of the gene encoding bovine rhodopsin. Cell 34:807–814
Neumann F-J, Schmid-Schönbein H, Ohlenbusch H (1987) Temperature-dependence of red cell aggregation. Pflügers Arch 408:524–530
Newman R, Schneider C, Sutherland R, Vodinelich L, Greaves M (1982) The transferrin receptor. Trends Biochem Sci 7:397–400
Ng WG (1971) Galactose metabolism of the red cell. Exp Eye Res 11:402–414
Nicol SC, Melrose W, Stahel CD (1988) Haematology and metabolism of the blood of the little penguin (*Eudyptula minor*). Comp Biochem Physiol 89A:383–386
Nicola NA, Johnson GR (1982) The production of committed hemopoietic colony-forming cells from multipotential precursors *in vitro*. Blood 60:1019–1029
Nijhof W, Wierenga PK, Sahr K, Beru N, Goldwasser E (1987) Induction of globin mRNA transcription by erythropoietin in differentiating erythroid precursor cells. Exp Hematol 15:779–784
Nikinmaa M (1981) Respiratory adjustments of rainbow trout (*Salmo gairdneri* Richardson) to changes in environmental temperature and oxygen availability. PhD Thesis, University of Helsinki, 56 pp
Nikinmaa M (1982a) The effects of adrenaline on the oxygen transport properties of *Salmo gairdneri* blood. Comp Biochem Physiol 71A:353–356
Nikinmaa M (1982b) Effects of adrenaline on red cell volume and concentration gradient of protons across the red cell membrane in the rainbow trout, *Salmo gairdneri*. Mol Physiol 2:287–297
Nikinmaa M (1983a) Kalojen hengityksen sopeutuminen veden alhaiseen ja vaihtelevaan happipitoisuuteen. Luonnon Tutkija 87:53–57
Nikinmaa M (1983b) Adrenergic regulation of haemoglobin oxygen affinity in rainbow trout red cells. J Comp Physiol 152B:67–72
Nikinmaa M (1986a) Red cell pH of lamprey (*Lampetra fluviatilis*) is actively regulated. J Comp Physiol 156B:747–750
Nikinmaa M (1986b) Control of red cell pH in teleost fishes. Ann Zool Fenn 23:223–235
Nikinmaa M, Huestis WH (1984a) Shape changes in goose erythrocytes. Biochim Biophys Acta 773:317–320
Nikinmaa M, Huestis WH (1984b) Adrenergic swelling in nucleated erythrocytes: cellular mechanisms in a bird, domestic goose, and two teleosts, striped bass and rainbow trout. J Exp Biol 113:215–224

Nikinmaa M, Railo E (1987) Anion movements across lamprey (*Lampetra fluviatilis*) red cell membrane. Biochim Biophys Acta 899:134–136
Nikinmaa M, Soivio A (1979) Oxygen dissociation curves and oxygen capacities of blood of a freshwater fish, *Salmo gairdneri*. Ann Zool Fenn 16:217–221
Nikinmaa M, Soivio A (1982) Blood oxygen transport of hypoxic *Salmo gairdneri*. J Exp Zool 219:173–178
Nikinmaa M, Weber RE (1984) Hypoxic acclimation in the lamprey, *Lampetra fluviatilis*: Organismic and erythrocytic responses. J Exp Biol 109:109–119
Nikinmaa M, Tuurala H, Soivio A (1980) Thermoacclimatory changes in blood oxygen binding properties and gill secondary lamellar structure of *Salmo gairdneri*. J Comp Physiol 140:255–260
Nikinmaa M, Cech JJ Jr, McEnroe M (1984) Blood oxygen transport in stressed striped bass (*Morone saxatilis*): role of beta-adrenergic responses. J Comp Physiol 154B:365–369
Nikinmaa M, Kunnamo-Ojala T, Railo E (1986) Mechanisms of pH regulation in lamprey (*Lampetra fluviatilis*) red blood cells. J Exp Biol 122:355–367
Nikinmaa M, Cech JJ Jr, Ryhänen E-L, Salama A (1987a) Red cell function of carp (*Cyprinus carpio*) in acute hypoxia. Exp Biol 47:53–58
Nikinmaa M, Steffensen JF, Tufts BL, Randall DJ (1987b) Control of red cell volume and pH in trout: effects of isoproterenol, transport inhibitors, and extracellular pH in bicarbonate/carbon dioxide-buffered media. J Exp Zool 242:273–281
Nilsson S (1978) Symphathetic innervation of the spleen of the cane toad, *Bufo marinus*. Comp Biochem Physiol 61C:133–149
Nilsson S (1983) Autonomic nerve function in the vertebrates. Springer, Berlin Heidelberg New York, 253 pp
Nilsson S, Grove DJ (1974) Adrenergic and cholinergic innervation of the spleen of the cod, *Cadus morhua*. Eur J Pharmacol 28:135–143
Nilsson S, Holmgren S, Grove DJ (1975) Effects of drugs and nerve stimulation on the spleen and arteries of two species of dogfish, *Scyliorhinus canicula* and *Squalus acanthias*. Acta Physiol Scand 95:219–230
Nishizuka Y (1984) Turnover of inositol phospholipids and signal transduction. Science 225:1365–1370
Noble RW, Kwiatkowski LD, De Young A, Davis BJ, Haedrich RL, Tam L-T, Riggs AF (1986) Functional properties of hemoglobins from deep-sea fish: correlations with depth distribution and presence of a swimbladder. Biochim Biophys Acta 870:552–563
Noble RW, Parkhurst LJ, Gibson QH (1970) The effect of pH on the reactions of O_2 and CO with the haemoglobin of the carp, *Cyprinus carpio*. J Biol Chem 245:6628–6633
Oberthür W, Braunitzer G (1984) Hämoglobin vom gemeinen star (*Sturnus vulgaris, passeriformes*): die primarstructur der α^A- und β-ketten der hauptkomponents. Hoppe-Seyler's Z Physiol Chem 362:1101–1112
Oberthür W, Braunitzer G, Grimm F, Kösters J (1983) Hämoglobine des steinadlers (*Aquila chrysaetos, accipitriformes*): die aminosaure-sequenz der α^A und β^B-ketten der hauptkomponente. Hoppe-Seyler's Z Physiol Chem 364:851–858
O'Donnell S, Mandaro R, Schuster TM, Arnone A (1979) X-ray diffraction and solution studies of specifically carbamylated human hemoglobin A. J Biol Chem 254:12204–12208
Ogawa M, Porter PN, Nakahata T (1983) Renewal and commitment to differentiation of hemopoietic stem cells: an interpretive review. Blood 61:823–829
Ogawa M, Pharr PN, Suda T (1985) Stochastic nature of stem cell functions in culture. In: Cronkite EP, Dainiak N, McCaffrey RP, Palek J, Quesenberry PJ (eds) Hematopoietic stem cell physiology. Alan R Liss, New York, pp 11–19
Ohno S, Atkin NB (1966) Comparative DNA values and chromosome complements of eight species of fishes. Chromosoma (Berlin) 18:455–466
Ohtsuka Y, Kondo T, Kawakami Y (1988) Oxidative stresses induced the cystine transport activity in human erythrocytes. Biochem Biophys Res Commun 155:160–166
Oliver JM, Paterson ARP (1971) Nucleoside transport. I. A mediated process in human erythrocytes. Can J Biochem 49:262–270
Olson KR (1984) Distribution of flow and plasma skimming in isolated perfused gills of three teleosts. J Exp Biol 109:97–108

Op Den Kamp JAF, Roelofsen B, Deenen Van LLM (1985) Structural and dynamic aspects of phosphatidylcholine in the human erythrocyte membrane. Trends Biochem Sci 10:320–323
Orskov S (1956) Experiments on the influence of adrenaline on the potassium absorption of red blood cells from pigeons and frogs. Acta Physiol Scand 37:299–306
Ott P, Hope MJ, Verkleij AJ, Roelofsen B, Brodbeck U, Van Deenen LLM (1981) Effect of dimyristoylphosphatidylcholine on intact erythrocytes: Release of spectrin-free vesicles without ATP depletion. Biochim Biophys Acta 641:79–87
Palek J, Liu SC (1981) Altered red cell cytoskeletal protein associations leading to membrane instability. In: Brewer GJ (ed) The red cell: fifth Ann Arbor conference. Alan R Liss, New York, pp 385–402
Palek J, Stewart G, Lionetti FJ (1974) The dependence of shape of human erythrocyte ghosts on calcium, magnesium and ATP. Blood 44:583–597
Paleus S, Vesterberg O, Liljeqvist G (1971) The hemoglobins of *Myxine glutinosa* L. -I. Preparation and crystallization. Comp Biochem Physiol 39B:551–557
Palfrey HC, Greengard P (1981) Hormone-sensitive ion transport systems in erythrocytes as models for epithelial ion pathways. Ann N Y Acad Sci 372:291–308
Palfrey HC, Greengard P, Feit PW (1980) Specific inhibition by "loop" diuretics of an anion-dependent $Na^+ + K^+$ cotransport system in avian erythrocytes. Ann N Y Acad Sci 341:134–138
Papayannopoulou T, Kalmantis T, Stamatoyannopoulos G (1979) Cellular regulation of hemoglobin switching; Evidence for inverse relationship between fetal hemoglobin synthesis and degree of maturity of human erythroid cells. Proc Natl Acad Sci USA 76:6420–6424
Parker JC (1977) Solute and water transport in dog and cat red blood cells. In: Ellory JC, Lew VL (eds) Membrane transport in red cells. Academic Press, London pp 427–465
Parker JC (1978) Sodium and calcium movements in dog red blood cells. J Gen Physiol 71:1–17
Parker JC (1979) Active and passive Ca movements in dog red blood cells and resealed ghosts. Am J Physiol 237:C10–C16
Parker JC (1983a) Volume-responsive sodium movements in dog red blood cells. Am J Physiol 244:C324–C330
Parker JC (1983b) Passive calcium movements in dog red blood cells: anion effects. Am J Physiol 244:C318–C323
Parker JC (1986) Interactions of lithium and protons with the sodium-proton exchanger of dog red blood cells. J Gen Physiol 87:189–200
Parker JC, Glosson PS (1987) Interactions of sodium-proton exchange mechanism in dog red blood cells with N-phenylmaleimide. Am J Physiol 253:C60–C65
Parkhurst LJ, Goss DJ, Perutz MF (1983) Kinetic and equilibrium studies on the role of the β-147 histidine in the Root effect and cooperativity in carp hemoglobin. Biochemistry 22:5401–5409
Parks RE Jr, Brown PR, Cheng Y-C, Agarwal KC, Kong CM, Agarwal RP, Parks CC (1973) Purine metabolism in primitive erythrocytes. Comp Biochem Physiol 45B:355–364
Passow H (1986) Molecular aspects of band 3 protein-mediated anion transport across the red blood cell membrane. Rev Physiol Biochem Pharmacol 103:61–203
Pasternack GR, Anderson RA, Leto TL, Marchesi VT (1985) Interaction between protein 4.1 and band 3. An alternative binding site for an element of the membrane-skeleton. J Biol Chem 260:3676–3683
Pasternack M, Nikinmaa M (1988) Anion exchange in fish red cells. Acta Physiol Scand 134 (suppl. 575):P74
Paterson S, Armstrong NJ, Iacopetta BJ, McArdle HJ, Morgan EH (1984) Intravesicular pH and iron uptake by immature erythroid cells. J Cell Physiol 120:225–232
Pati AK, Thapliyal JP (1984) Erythropoietin, testosterone, and thyroxine in the erythropoietic response of the snake, *Xenochrophis piscator*. Gen Comp Endocrinol 53:370–375
Pelicci PG, Tabilio A, Thomopoulos P, Titeux M, Vainchenker W, Rochant H, Testa U (1982) Hemin regulates the expression of transferrin receptors in human hematopoietic cell lines. FEBS Lett 145:350–354
Pennelly RR, Riggs A, Noble RW (1978) The kinetics and equlibria of squirrel-fish hemoglobin. A Root effect hemoglobin complicated by large subunit heterogeneity. Biochim Biophys Acta 533:120–129
Percy Lord R, Potter IC (1977) Changes in haemopoietic sites during the metamorphosis of the lampreys *Lampetra fluviatilis* and *Lampetra planeri*. J Zool (Lond) 183:111–123

Perrella M, Bresciani D, Rossi-Bernardi L (1975) The binding of CO_2 to human hemoglobin. J Biol Chem 250:5413–5418
Perrone JR, Hackney JF, Dixon JF, Hokin LE (1975) Molecular properties of purified sodium + potassium-activated adenosine triphosphatases and their subunits from the rectal gland of *Squalus acanthias* and the electric organ of *Electrophorus electricus*. J Biol Chem 250:4178–4184
Perry SF (1986) Carbon dioxide excretion in fishes. Can J Zool 64:565–572
Perry SF, Davie PS, Daxboeck C, Randall DJ (1982) A comparison of CO_2 excretion in spontaneously ventilating blood-perfused trout preparation and saline-perfused gill preparations: contribution of the branchial epithelium and red blood cell. J Exp Biol 101:47–60
Perutz MF (1970) Stereochemistry of cooperative effects in haemoglobin. Nature 228:726–739
Perutz MF (1972) Nature of haem-haem interaction. Nature 237:495–499
Perutz MF (1978) Hemoglobin structure and respiratory transport. Scient Am 239:68–86
Perutz MF (1979) Regulation of oxygen affinity of hemoglobin: influence of structure of the globin on the heme iron. Annu Rev Biochem 48:327–86
Perutz MF, Brunori M (1982) Stereochemistry of cooperative effects in fish and amphibian haemoglobins. Nature 299:421–426
Perutz MF, Muirhead H, Mazzarella L, Crowther RA, Greer J, Kilmartin JV (1969) Identification of residues responsible for the alkaline Bohr effect in haemoglobin. Nature 222:1240–1243
Perutz MF, Kilmartin JV, Nishikura K, Fogg JH, Butler PJG, Rollema HS (1980) Identification of the residues contributing to the Bohr effect of human haemoglobin. J Mol Biol 138:649–670
Perutz MF, Bauer C, Gros G, Leclercq F, Vandecasserie C, Schnek AG, Braunitzer G, Friday AE, Joysey KA (1981) Allosteric regulation of crocodilian haemoglobin. Nature 291:682–684
Perutz MF, Fermi G, Shih T-B (1984) Structure of deoxyhemoglobin Cowtown [His HC3(146)β-Leu]: origin of the alkaline Bohr effect and electrostatic interactions in hemoglobin. Proc Natl Acad Sci USA 81:4781–4784
Perutz MF, Gronenborn AM, Clore GM, Fogg JH, Shih DT (1985) The pKa values of two histidine residues in human haemoglobin, the Bohr effect, and the dipole moments of α-helices. J Mol Biol 183:491–498
Peschle C, Migliaccio G, Migliaccio AR, Covelli A, Giuliani A, Mavilio F, Mastroberardino G (1983) Hemoglobin switching in humans. In: Dunn CDR (ed) Current concepts in erythropoiesis. Wiley, New York, pp 339–387
Peschle C, Mavilio F, Care A, Migliaccio G, Migliaccio AR, Salvo G, Samoggia P, Petti S, Guerriero R, Marinucci M, Lazzaro D, Russo G, Mastroberardino G (1985) Hemoglobin switching in human embryos: asynchrony of $\xi \rightarrow \alpha$ and $\varepsilon \rightarrow \vartheta$ globin switches in primitive and definitive erythropoietic lineage. Nature 313:235–238
Peters WHM, Swarts HGP, De Pont JJHHM, Schuurmans-Stekhoven FMAH, Bonting SL (1981) ($Na^+ + K^+$)ATPase has one functioning phosphorylation site per subunit. Nature 290:338–339
Petschow D, Würdinger I, Baumann R, Duhm J, Braunitzer G, Bauer C (1977) Causes of high blood O2 affinity of animals living at high altitude. J Appl Physiol 42:139–143
Pfafferott C, Maiselman HJ, Hochstein P (1982) The effect of malonyldialdehyde on erythrocyte deformability. Blood 59:12–15
Pfenniger KH, Johnson MP (1983) Membrane biogenesis in the sprouting neuron. I. Selective transfer of newly synthesized phospholipid into the growing neurite. J Cell Biol 97:1038–1042
Pfeuffer E, Dreher RF, Metzger M, Pfeuffer T (1985) Catalytic unit of adenylate cyclase: purification and identification by affinity crosslinking. Proc Natl Acad Sci USA 82:3086–3090
Phelps C, Farmer M, Fyhn HJ, Fyhn UEH, Garlick RL, Noble RW, Powers DA (1979) Equilibria and kinetics of oxygen and carbon monoxide ligand binding to the hemoglobin of South American lungfish, *Lepidosiren paradoxa*. Comp Biochem Physiol 62A:139–143
Pickering AD (1986) Changes in blood cell composition of the brown trout, *Salmo trutta* L., during the spawning season. J Fish Biol 29:335–347
Pinder JC, Clark SE, Baines AJ, Morris E, Gratzer WB (1981) The construction of the red cell cytoskeleton. In: Brewer GJ (ed) The red cell: fifth Ann Arbor conference. Alan R Liss, New York, pp 343–354
Pionetti JM, Bouverot P (1977) Effects of acclimation to altitude on oxygen affinity and organic phosphate concentrations in pigeon blood. Life Sci 20:1207–1212

Plagemann PGW, Wohlhueter RM (1980) Permeation of nucleosides, nucleic acid bases, and nucleotides in animal cells. Curr Top Membr Transp 14:225–330
Plagemann PGW, Woffendin C, Puziss MB, Wohlhueter RM (1987) Purine and pyrimidine transport and permeation in human erythrocytes. Biochim Biophys Acta 905:17–29
Plagemann PGW, Wohlhueter RM, Woffendin C (1988) Nucleoside and nucleobase transport in animal cells. Biochim Biophys Acta 947:405–443
Plishker GA, White PH, Cadman ED (1986) Involvement of a cytoplasmic protein in calcium-dependent potassium efflux in red blood cells. Am J Physiol 251:C535–C540
Polvani C, Blostein R (1988) Protons as substitutes for sodium and potassium in the sodium pump reaction. J Biol Chem 263:16757–16763
Ponka P, Schulman HM (1985) Acquisition of iron from transferrin regulates reticulocyte heme synthesis. J Biol Chem 260:14717–14721
Potter IC, Brown ID (1975) Changes in haemoglobin electropherograms during the life cycle of two closely related lampreys. Comp Biochem Physiol 51B:517–519
Potter IC, Robinson ES, Brown ID (1974) Studies on the erythrocytes of larval and adult lampreys (*Lampetra fluviatilis*). Acta Zool 55:173–177
Potter IC, Percy Lord R, Barber DL, Macey DJ (1982) The morphology, development and physiology of blood cells. In: Hardisty MW, Potter IC (eds) Biology of lampreys, vol 4A. Academic Press, London, pp 233–292
Pottinger TG, Pickering AD (1987) Androgen levels and erythrocytosis in maturing brown trout, *Salmo trutta*. Fish Physiol Biochem 3:121–126
Pough FH (1969) Environmental adaptations in the blood of lizards. Comp Biochem Physiol 31:885–901
Pough FH (1980) Blood oxygen transport and delivery in reptiles. Am Zool 20:173–185
Pough FH, Lillywhite HB (1984) Blood volume and blood oxygen capacity of sea snakes. Physiol Zool 57:32–39
Powers DA (1972) Hemoglobin adaptation for fast and slow water habitats in sympatric catostomid fishes. Science 177:360–362
Powers DA (1977) Structure, function, and molecular ecology of fish hemoglobins. Ann N Y Acad Sci 241:472–490
Powers DA (1980) Molecular ecology of teleost fish hemoglobins: strategies for adapting to changing environment. Am Zool 20:139–162
Powers DA, Edmundson AB (1972) Multiple hemoglobins of Catostomid fish — II. The amino acid sequence of the major α chain from *Catostomus clarkii* hemoglobins. J Biol Chem 247: 6694–6707
Powers DA, Fyhn HJ, Fyhn UEH, Martin JP, Garlick RL, Wood SC (1979a) A comparative study of the oxygen equilibria of blood from 40 genera of Amazonian fishes. Comp Biochem Physiol 62A:67–85
Powers DA, Martin JP, Garlick RL, Fyhn HJ, Fyhn UEH (1979b) The effect of temperature on the oxygen equilibria of fish hemoglobins in relation to environmental thermal variability. Comp Biochem Physiol 62A:87–94
Poyart CF, Guesnon P, Bohn BM (1981) The measurement of the intrinsic alkaline Bohr effect of various human haemoglobins by isoelectric focusing. Biochem J 195:493–501
Primmett DRN, Randall DJ, Mazeaud M, Boutilier RG (1986) The role of catecholamines in erythrocyte pH regulation and oxygen transport in rainbow trout (*Salmo gairdneri*) during exercise. J Exp Biol 122:139–148
Proudfoot N (1984) The end of message and beyond. Nature 307:412–413
Pulsford A, Fänge R, Morrow WJW (1982) Cell types and interactions in the spleen of the dogfish *Scyliorhinus canicula* L.: an electron microscopic study. J Fish Biol 21:649–662
Qvist J, Weber RE, DeVries AL, Zapol WM (1977) pH and haemoglobin oxygen affinity in blood from the antarctic cod *Dissostichus mawsoni*. J Exp Biol 67:77–88
Railo E, Nikinmaa M, Soivio A (1985) Effects of sampling on blood parameters in the rainbow trout, *Salmo gairdneri* Richardson. J Fish Biol 26:725–732
Ralston GB (1975) Proteins of the camel erythrocyte membrane. Biochim Biophys Acta 401: 83–94
Rampling MW, Whittingstall P (1987) The effect of temperature on the viscosity characteristics of erythrocyte suspensions. Clin Hemorheol 7:745–755

Randall DJ, Daxboeck C (1984) Oxygen and carbon dioxide transfer across fish gills. In: Hoar WS, Randall DJ (eds) Fish physiology, vol 10A. Academic Press, London, pp 263–314
Rapoport SM (1985) Mechanisms of the maturation of the reticulocyte. In: Gilles R (ed) Circulation, respiration and metabolism. Current comparative approaches. Springer, Berlin Heidelberg New York, pp 333–342
Rapoport SM (1986) The reticulocyte. CRC, Boca Raton, Florida, 238 pp
Rapoport SM, Rost J, Schultze M (1971) Glutamine and glutamate as respiratory substrates of rabbit reticulocytes. Eur J Biochem 23:166–170
Rapoport SM, Schmidt J, Prehn S (1985) Maturation of rabbit reticulocytes: suspectibility of mitochondria to ATP-dependent proteolysis is determined by the maturational state of reticulocyte. FEBS Lett 183:370–374
Ray TK, Dutta-Roy AK, Sinha AK (1986) Regulation of insulin receptor activity of human erythrocyte membrane by prostaglandin E_1. Biochim Biophys Acta 856:421–427
Raynard RS (1987) Thermal compensation of the Na^+ pump of rainbow trout (*Salmo gairdneri*) erythrocytes. In: Bowler K, Fuller BJ (eds) Temperature and animal cells. Company of Biologists, Cambridge, pp 455–456
Reeves RB (1976) Temperature induced changes in blood acid-base status, Donnan r_{CL} and red cell volume. J Appl Physiol 40:762–767
Reeves RB (1980) A rapid micro method for obtaining oxygen equilibrium curves on whole blood. Respir Physiol 42:299–315
Reeves RB (1984) Blood oxygen affinity in relation to yolk-sac and chorioallantoic gas exchange in developing chick embryo. In: Seymour RS (ed) Respiration and metabolism of embryonic vertebrates. Junk, Dordrecht, pp 231–244
Reeves RB, Rahn H (1979) Patterns in vertebrate acid-base regulation. In: Wood SC, Lenfant C (eds) Evolution of respiratory processes. Dekker, New York, pp 225–252
Reid HL, Barnes AJ, Lock PJ, Dormandy JA, Dormandy TL (1976) A simple method for measuring erythrocyte deformability. J Clin Pathol (Lond) 29:855–858
Reid ME, Chasis JA, Mohandas N (1987) Identification of a functional role for human erythrocyte sialoglycoproteins β and γ. Blood 69:1068–1072
Reinhart WH, Chien S (1985) Roles of cell geometry and cellular viscosity in red cell passage through narrow pores. Am J Physiol 248:C473–C479
Reinhart WH, Chien S (1986) Red cell rheology in stomatocyte-echinocyte transformation: roles of cell geometry and cell shape. Blood 67:1110–1118
Reinhart WH, Usami S, Schmalzer EA, Lee MML, Chien S (1984) Evaluation of red blood cell filterability test: influences of pore size, haematocrit level, and flow rate. J Lab Clin Med 104:501–516
Reinke W, Johnson PC, Gaehtgens P (1986) Effect of shear rate variation on apparent viscosity of human blood in tubes of 29 to 94 μm diameter. Circ Res 59:124–132
Repasky EA, Eckert BS (1981) A reevaluation of the process of enucleation in mammalian erythroid cells. In: Brewer GJ (ed) The red cell: fifth Ann Arbor conference. Alan R Liss, New York, pp 679–690
Reuss L, Cassel D, Rothenberg P, Whiteley B, Mancuso D, Glaser L (1986) Mitogens and ion fluxes. Curr Top Membr Transp 27:3–54
Riddick DH, Kregenow FM, Orloff J (1971) The effect of norepinephrine and dibutyryl cyclic adenosine monophosphate on cation transport in duck erythrocytes. J Gen Physiol 57:752–766
Riggs A (1970) Properties of fish hemoglobins. In: Hoar WS, Randall DJ (eds) Fish physiology, vol 4. Academic Press, London, pp 209–252
Riggs A (1971) Mechanism of the enhancement of the Bohr effect in mammalian hemoglobins by diphosphoglycerate. Proc Natl Acad Sci USA 68:2062–2065
Riggs AF (1972) The haemoglobins. In: Hardisty MW, Potter IC (eds) The biology of lampreys, vol 2. Academic Press, London, pp 261–286
Riggs AF (1988) The Bohr effect. Annu Rev Physiol 50:181–204
Ristori MT, Laurent P (1985) Plasma catecholamines and glucose during moderate exercise in the trout: comparison with bursts of violent activity. Exp Biol 44:247–253
Röcker L, Laniado M, Kirsch K (1983) The effect of physical exercise on plasma volume and red blood cell mass. In: Dunn CDR (ed) Current concepts in erythropoiesis. Wiley, New York, pp 245–277

Rodewald K, Stangl A, Braunitzer G (1984) Primary structure, biochemical and physiological aspects of hemoglobin from South American lungfish (*Lepidosiren paradoxus*, Dipnoi). Hoppe-Seyler's Z Physiol Chem 365:639–649
Roelofsen B (1981) The (non)specificity in the lipid-requirement of calcium and (sodium plus potassium)-transporting adenosine triphosphatases. Life Sci 29:2235–2247
Roelofsen B, Van Deenen LLM (1973) Lipid requirement of membrane bound ATPase. Studies on human erythrocyte ghosts. Eur J Biochem 40:245–257
Romano L, Passow H (1984) Characterization of anion transport system in trout red blood cell. Am J Physiol 246:C330–C338
Root R (1931) The respiratory function of the blood of marine fishes. Biol Bull mar biol Lab, Woods Hole 61:427–457
Rosa R, Rosa CD, Ocampos D, Bacila M (1983) The profile of the glycolytic system and the metabolic activity of chicken erythrocytes. Comp Biochem Physiol 75B:141–145
Rose IA (1971) Regulation of human red cell glycolysis: a review. Exp Eye Res 11:264–272
Rosenberg R, Young JD, Ellory JC (1980) L-tryptophan transport in human red blood cells. Biochim Biophys Acta 598:375–384
Rosokivi V, Suomalainen P (1973) Studies on the physiology of the hibernating hedgehog 17. The blood cell count of hedgehog at different times of the year and in different phases of the hibernating cycle. Ann Acad Sci Fenn 198:1–8
Ross J, Sautner D (1976) Induction of globin mRNA accumulation by hemin in cultured erythroleukemic cells. Cell 8:513–520
Rosse WF, Waldmann T (1966) Factors controlling erythropoiesis in birds. Blood 27:654–661
Rosse WF, Waldmann T, Hull E (1963) Factors stimulating erythropoiesis in frogs. Blood 22:66–72
Rothstein S, Jürgens KD, Bartels H, Baumann R (1984) Oxygen and carbon dioxide transport in the blood of the muskrat (*Ondatra zibethica*). Respir Physiol 57:15–22
Rouault T, Rao K, Harford J, Mattia E, Klausner RD (1985) Hemin, chelatable iron, and the regulation of transferrin receptor biosynthesis. J Biol Chem 260:14862–14866
Roughton FJW (1964) Transport of oxygen and carbon dioxide. In: Fenn WO, Rahn H (eds) Handbook of physiology, respiration, vol 1. American Physiological Society, Washington DC, pp 767–825
Royer WE Jr, Love WE, Fenderson FF (1985) Cooperative dimeric and tetrameric clam haemoglobins are novel assemblages of myoglobin folds. Nature 316:277–280
Rudolph SA, Beam KG, Greengard P (1978) Studies of protein phosphorylation in relation to hormonal control of ion transport in intact cells. In: Hoffman JF (ed) Membrane transport processes. Raven, New York, pp 107–123
Rudolph SA, Greengard P (1980) Effects of catecholamines and prostaglandin E_1 on cyclic AMP, cation fluxes and protein phosphorylation in the frog erythrocyte. J Biol Chem 255:8534–8540
Ruiz-Ruano A, Martin M, Luque J (1984) Synthesis and levels of organic phosphates in erythrocytes during avian development: specific formation of BPG and IP_5 in two distinct populations from young chicks. Cell Biochem Funct 2:257–262
Russu IM, Ho NT, Ho C (1980) Role of the β146-histidyl residue in the alkaline Bohr effect of hemoglobin. Biochemistry 19:1043–1052
Russu IM, Ho NT, Ho C (1982) A proton nuclear magnetic resonance investigation of histidyl residues in human normal adult hemoglobin. Biochemistry 21:5031–5043
Rutherford T, Thompson GG, Moore MR (1979) Heme biosynthesis in Friend erythroleukemia cells: Control by ferrochelatase. Proc Natl Acad Sci USA 76:833–836
Saffran WA, Gibson QH (1978) The effect of pH on CO binding to menhaden haemoglobin. J Biol Chem 253:3171–3179
Sahr K, Goldwasser E (1983) The effects of erythropoietin on the biosynthesis of translatable globin mRNA. In: Goldwasser E (ed) Regulation of hemoglobin biosynthesis. Elsevier, Amsterdam, pp 153–161
Salama A (1986) Adrenaliinin vaikutukset kirjolohen (*Salmo gairdneri* Richardson) punasolujen aineenvaihduntaan. (Effects of adrenaline on red cell metabolism in rainbow trout, in Finnish). MSc Thesis, University of Helsinki, January 1986 74 pp
Salama A, Nikinmaa M (1988) The adrenergic responses of carp (*Cyprinus carpio*) red cells: effects of Po_2 and pH. J Exp Biol 136:405–416

Salama A, Nikinmaa M (1989) Species differences in the adrenergic responses of fish red cells: studies on whitefish, pikeperch, trout and carp. Fish Physiol Biochem 6:167–173
Salerno C, Werner A, Siems W, Gerber G (1987) Incorporation of purine bases in human erythrocytes. Biomed Biochim Acta 46:278–279
Salter DW, Baldwin SA, Lienhard GE, Weber MJ (1982) Proteins antigenically related to the human erythrocyte glucose transporter in normal and Rous sarcoma virus transformed chicken embryo fibroblasts. Proc Natl Acad Sci USA 79:1540–1544
Samaja M, Rovida E (1983) A new method to measure the haemoglobin oxygen saturation by the oxygen electrode. J Biochem Biophys Methods 7:143–152
Sanyal G (1984) Comparative carbon dioxide hydration kinetics and inhibition of carbonic anhydrase isozymes in vertebrates. Ann N Y Acad Sci 429:165–178
Sanyal G, Maren TH (1981) Thermodynamics of carbonic anhydrase catalysis: a comparison between human isozymes B and C. J Biol Chem 256:608–612
Sanyal G, Pessah NI, Swenson ER, Maren TH (1982) The carbon dioxide hydration activity of purified teleost red cell carbonic anhydrase. Inhibition by sulfonamides and anions. Comp Biochem Physiol 73B:937–944
Sariban-Sohraby S, Benos DJ (1986) The amiloride-sensitive sodium channel. Am J Physiol 250:C175-C190
Sarkadi B, Gardos G (1985) Calcium-induced potassium transport in cell membranes. In: Martonosi AN (ed) The enzymes of biological membranes, 2nd edn. Plenum, New York, pp 193–234
Sasaki R, Ikura K, Narita H, Yanagawa S, Chiba H (1985) 2,3-Bisphosphoglyserate in erythroid cells. In: Ochs RS, Hanson RW, Hall J (eds) Metabolic regulation. Elsevier, Amsterdam, pp 229–234
Sassa S, Granick S (1971) Delta-aminolevulinic acid synthetase and its control in liver and red cells. In: Travnicek T, Neuwirt J (eds) The regulation of erythropoiesis and haemoglobin synthesis. Universita Karlova, Praha, pp 299–303
Sato SB, Ohnishi S-I (1983) Interaction of a peripheral protein of the erythrocyte membrane, band 4.1, with phosphatidylserine-containing liposomes and erythrocyte inside-out vesicles. Eur J Biochem 130:19–25
Savitz D, Sidel VW, Solomon AK (1964) Osmotic properties of human red cells. J Gen Physiol 48:79–94
Schalekamp M, Van Goor D (1984) Haemoglobin and globin synthesis in the isolated primitive and definitive erythroid cells of embryos. Evidence for a non-clonal mechanism at the haemoglobin switch. J Embryol Exp Morphol 84:125–148
Schalekamp M, De Jonge P, Van Goor D (1982) Is erythroid cell differentiation a matter of all-or-none transcription only? In: Akoyunoglou G, Evangelopoulos AE, Georgatsos J, Palaiologos G, Trakatellis A, Tsiganos CP (eds) Cell Function and Differentiation, Part A. Alan R Liss, New York, pp 25–33
Schatzmann HJ (1982) The plasma membrane calcium pump of erythrocytes and other animal cells. In: Carafoli E (ed) Membrane transport of calcium. Academic Press, London, pp 41–108
Schatzmann HJ (1986) The human red blood cell calcium pump. Fortschr Zool 33:435–442
Schlegel RA, Prendergast TW, Williamson P (1985) Membrane phospholipid asymmetry as a factor in erythrocyte-endothelial cell interactions. J Cell Physiol 123:215–218
Schmid-Schönbein H, Gaehtgens P (1981) What is red cell deformability. Scand J Clin Lab Invest 41 Suppl 156:13–26
Schmidt JA, Marshall J, Hayman MJ, Beug H (1986) Primitive series embryonic chick erythrocytes express the transferrin receptor. Exp Cell Res 164:71–78
Schmidt WF III, McManus TJ (1977a) Ouabain-insensitive salt and water movements in duck red cells. I. Kinetics of cation transport under hypertonic conditions. J Gen Physiol 70:59–79
Schmidt WF III, McManus TJ (1977b) Ouabain-insensitive salt and water movements in duck red cells. II. Norepinephrine stimulation of sodium plus potassium cotransport. J Gen Physiol 70:81–97
Schmidt-Nielsen K, Larimer JL (1958) Oxygen dissociation curves of mammalian blood in relation to body size. Am J Physiol 195:424–428
Schnek AG, Paul C, Leonis J (1985) Evolution and adaptation of avian and crocodilian hemoglobins. In: Lamy J, Truchot J-P, Gilles R (eds) Respiratory pigments in animals. Relation structure-function. Springer, Berlin Heidelberg New York, pp 141–158

Scholander PV, Dam Van L (1954) Secretion of gases against high pressure in the swimbladders of deep sea fishes I. Oxygen dissociation in blood. Biol Bull mar biol Lab Woods Hole 107:247–259
Schroit AJ, Madsen JW, Tanaka Y (1985) In vivo recognition and clearance of red blood cells containing phosphatidylserine in their plasma membrane. J Biol Chem 260:5131–5138
Schuurmans-Stekhoven FMAH, Bonting SL (1981) Sodium-potassium-activated adenosine triphosphatase. In: Bonting SL, de Pont JJHHM (eds) Membrane transport. Elsevier, Amsterdam, pp 159–182
Schwartz RS, Chiu DT-Y, Lubin B (1984) Studies on the organization of plasma membrane phospholipids in human erythrocytes. In: Kruckeberg WC, Eaton JW, Aster J, Brewer GJ (eds) Erythrocyte membranes: recent clinical and experimental advances, vol 3. Alan R Liss, New York, pp 89–122
Schwartz RS, Chiu DT-Y, Lubin B (1985) Plasma membrane phospholipid organization in human erythrocytes. Curr Top Hematol 5:63–112
Schweiger HA (1962) Pathways of metabolism in nucleate and anucleate erythrocytes. Int Rev Cytol 13:135–201
Seider MJ, Kim HD (1979) Cow red blood cells. I. Effect of purines, pyrimidines, and nucleosides in bovine red cell glycolysis. Am J Physiol 236:C255–C261
Seigneuret M, Devaux PF (1984) ATP-dependent asymmetric distribution of spin-labeled phospholipids in the erythrocyte membrane: relation to shape changes. Proc Natl Acad Sci USA 81:3751–3755
Seligman PA (1983) Structure and function of the transferrin receptor. Progress Hematol 13:131–147
Sha'afi RI (1977) Water and small nonelectrolyte permeation in red cells. In: Ellory JC, Lew VL (eds) Membrane transport in red cells. Academic Press, London, pp 221–256
Shaanan B (1982) The iron-oxygen bond in human oxyhaemoglobin. Nature 296:683–684
Shaanan B (1983) The structure of human oxyhaemoglobin at 2.1 Å resolution. J Mol Biol 171: 31–59
Shaklai N, Benitez L, Ranney HM (1978) Binding of 2,3-diphosphoglycerate by spectrin and its effect on oxygen affinity of hemoglobin. Am J Physiol 234:C36–C40
Sharp GD (1975) A comparison of the O_2 dissociation properties of some scombrid hemoglobins. Comp Biochem Physiol 51A:683–691
Sheetz MP, Singer SJ (1974) Biological membranes as bilayer couples. A molecular mechanism of drug-echinocyte interactions. Proc Natl Acad Sci USA 81:3751–3755
Sheetz MP, Schindler M, Koppel DE (1980) Lateral mobility of integral membrane proteins is increased in spherocytic erythrocytes. Nature 285:510–512
Sherwood JB (1984) The chemistry and physiology of erythropoietin. Vitam Horm 41:161–211
Shiga T, Sekiya M, Maeda N, Kon K, Okazaki M (1985) Cell age-dependent changes in deformability and calcium accumulation in human erythrocytes. Biochim Biophys Acta 814:289–299
Shoemaker DG, Hoffman JF (1985) Membrane bound ATP and the operation of the red cell Na,K-pump. In: Glynn IM, Ellory JC (eds) The sodium pump, The Company of Biologists, Cambridge, pp 723–725
Shoemaker DG, Bender CA, Gunn RB (1988) Sodium-phosphate cotransport in human red blood cells. Kinetics and role in membrane metabolism. J Gen Physiol 92:449–474
Sibley DR, Strasser RH, Caron MG, Lefkowitz RJ (1985) Homologous desensitization of adenylate cyclase is associated with phosphorylation of β-adrenergic receptor. J Biol Chem 260:3883–3886
Sibley DR, Benovic JL, Caron MG, Lefkowitz RJ (1987) Regulation of transmembrane signaling by receptor phosphorylation. Cell 48:913–922
Siebens AW, Kregenow FM (1985) Volume-regulatory responses of *Amphiuma* red cells in anisotonic media. J Gen Physiol 86:527–564
Sieff C, Bicknell D, Caine G, Robinson J, Lam G, Greaves M (1982) Changes in cell surface antigen expression during hematopoietic differentiation. Blood 60:703–713
Siegel DL, Branton D (1985) Partial purification and characterization of an actin-bundling protein, band 4.9, from human erythrocytes. J Cell Biol 100:775–785
Siekierka J, Ochoa S (1983) Mechanism of translational inhibition associated with phosphorylation of the α subunit of the eykaryotic initiation factor-2. In: Regulation of hemoglobin biosynthesis, Goldwasser E (ed), Elsevier, New York, pp 253–266

Siems W, Müller M, Dumdey R, Holzhütter H-G, Rathmann J, Rapoport SM (1982) Quantification of pathways of glucose utilization and balance of energy metabolism of rabbit reticulocytes. Eur J Biochem 124:567–576
Siggaard-Andersen O (1974) The acid-base status of the blood, 4th edn, Munksgaard, Copenhagen, 229 pp
Silverman DN, Vincent SH (1983) Proton transfer in the catalytic mechanism of carbonic anhydrase. CRC Crit Rev Biochem 14:207–255
Simchon S, Jan K-M, Chien S (1987) Influence of reduced red cell deformability on regional blood flow. Am J Physiol 253:H898–H903
Simpson CF, Kling JM (1967) The mechanism of denucleation in circulating erythroblasts. J Cell Biol 35:237–245
Singer JA, Jennings LK, Jackson CW, Dockter ME, Morrison M, Walker WS (1986) Erythrocyte homeostasis: antibody-mediated recognition of the senescent state by magrophages. Proc Natl Acad Sci USA 83:5498–5501
Skalak R, Chien S (1981) Capillary flow: history, experiments and theory. Biorheology 18:307–330
Sleet RB, Weber LJ (1983) Blood volume of a marine teleost before and after arterial cannulation. Comp Biochem Physiol 76A:791–794
Slicher AM, Pickford GE (1968) Temperature-controlled stimulation of hemopoiesis in a hypophysectomized cypridont fish *Fundulus heteroclitus*. Physiol Zool 41:293–297
Smallwood JI, Waisman DM, Lafreniere D, Rasmussen H (1983) Evidence that the erythrocyte calcium pump catalyzes a Ca^{2+}:nH^{+} exchange. J Biol Chem 258:11092–11097
Smith L, Hochmuth GM (1982) Effect of wheat germ agglutinin on the viscoelastic properties of erythrocyte membrane. J Cell Biol 94:7–11
Snapp BD, Heller HC (1981) Suppression of metabolism during hibernation in ground squirrels (*Citellus lateralis*). Physiol Zool 54:297–307
Snyder GK (1983) Respiratory adaptations in diving mammals. Respir Physiol 54:269–294
Snyder GK, Black CP, Birchard GF, Lucich R (1982) Respiratory properties of blood from embryos of highland vs. lowland geese. J Appl Physiol 53:1432–1438
Sobue K, Muramoto Y, Fujita M (1981) Calmodulin-binding protein of erythrocyte cytoskeleton. Biochem Biophys Res Comm 100:1063–1070
Soivio A (1967) Hibernation in the hedgehog (*Erinaceus europaeus* L.). Ann Acad Sci Fenn 110:1–71
Soivio A, Nikinmaa M (1981) The swelling of erythrocytes in relation to the oxygen affinity of the blood of the rainbow trout, *Salmo gairdneri* Richardson. In: Pickering AD (ed) Stress and fish. Academic Press, London, pp 103–119
Soivio A, Nikinmaa M, Westman K (1980) The blood oxygen binding properties of hypoxic *Salmo gairdneri*. J Comp Physiol B 136:83–87
Soivio A, Nikinmaa M, Nyholm K, Westman K (1981) The role of gills in the responses of *Salmo gairdneri* during moderate hypoxia. Comp Biochem Physiol 70A:133–139
Solomon AK, Chasan B, Dix JA, Lukacovic MF, Toon MR, Verkman AS (1983) The aqueous pore in the red cell membrane: band 3 as a channel for anions, cations, nonelectrolytes, and water. Ann N Y Acad Sci 414:97–124
Solomon AK, Toon MR, Dix JA (1986) Osmotic properties of human red cells. J Membr Biol 91:259–273
Southard JN, Berry CR Jr, Farley TM (1986) Multiple hemoglobins of the cutthroat trout, *Salmo clarki*. J Exp Zool 239:7–16
Srivastava SK (1971) Metabolism of red cell glutathione. Exp Eye Res 11:294–305
Stadel JM, Nambi P, Shorr RGL, Sawyer DF, Caron MG, Lefkowitz RJ (1983) Catecholamine-induced desensitization of turkey erythrocyte adenylate cyclase is associated with phosphorylation of the β-adrenergic receptor. Proc Natl Acad Sci USA 80:3173–3177
Stalder J, Groudine M, Dodgson JB, Engel JD, Weintraub H (1980) Hb switching in chickens. Cell 19:973–980
Stamatoyannopoulos G, Nakamoto B, Kurachi S, Papayannopoulou T (1983) Direct evidence for interaction between human erythroid progenitor cells and a hemoglobin switching activity present in fetal sheep serum. Proc Natl Acad Sci USA 80:5650–5654
Stamatoyannopoulos G, Constantoulakis P, Brice M, Kurachi S, Papayannopoulou T (1987) Coexpression of embryonic, fetal, and adult globins in erythroid cells of human embryos: relevance to the cell-lineage models of globin switching. Dev Biol 123:191–197

Steffensen JF, Lomholt JP, Vogel WOP (1986) *In vivo* observations on a specialized microvasculature, the primary and secondary vessels in fishes. Acta Zool 67:193–200

Steinke JM, Shepherd AP (1987) Reflectance measurements of hematocrit and oxyhemoglobin saturation. Am J Physiol 253:H147-H153

Sternweis PC, Northup JK, Smigel MD, Gilman AG (1981) The regulatory component of adenylate cyclase. J Biol Chem 256:11517–11526

Stevens ED (1968) The effect of exercise on the distribution of blood to various organs in rainbow trout. Comp Biochem Physiol 25:615–625

Stohlman FJ (1970) Fetal erythropoiesis. In: Gordon AS (ed) Regulation of hematopoiesis, vol 1. Red cell production. Appleton-Century-Crofts, New York, pp 471–485

Stoltz JF, Donner M (1987) Hemorheology: importance of erythrocyte aggregation. Clin Hemorheol 7:15–23

Stoltz JF, Ravey JC, Larcan A, Mazeron P, Lucius M, Guillot M (1981) Deformation and orientation of red blood cells in a simple shear flow. Scand J clin Lab Invest 41, Suppl. 156:67–75

Stoner LC, Kregenow FM (1980) A single-cell technique for the measurement of membrane potential, membrane conductance, and the efflux of rapidly penetrating solutes in *Amphiuma* erythrocytes. J Gen Physiol 76:455–478

Storch J, Kleinfeld AM (1985) The lipid structure of biological membranes. Trends Biochem Sci 10:418–421

Strader CD, Sigal IS, Register RB, Candelore MR, Rands E, Dixon RAF (1987) Identification of residues required for ligand binding to the β-adrenergic receptor. Proc Natl Acad Sci USA 84:4384–4388

Stryer L, Bourne HR (1986) G-proteins: a family of signal transducers. Annu Rev Cell Biol 2:391–419

Sullivan B (1974a) Reptilian hemoglobins. In: Florkin M, Scheer BT (eds) Chemical zoology, vol 9. Academic Press, London, pp 377–398

Sullivan B (1974b) Amphibian hemoglobins. In: Florkin M, Scheer BT (eds) Chemical zoology, vol 9. Academic Press, London, pp 77–122

Sullivan B, Riggs A (1964) Haemoglobin: Reversal of oxidation and polymerization in turtle red cells. Nature 204:1098–1099

Sullivan B, Riggs A (1967a) Structure, function and evolution of turtle hemoglobins — II. Electrophoretic studies. Comp Biochem Physiol 23:449–458

Sullivan B, Riggs A (1967b) Structure, function and evolution of turtle hemoglobins — I. Distribution of heavy hemoglobins. Comp Biochem Physiol 23:437–447

Sutera SP, Gardner RA, Boylan CW, Carroll GL, Chang KC, Marvel JS, Kilo C, Gonen B, Williamson JR (1985) Age-related changes in deformability of human erythrocytes. Blood 65:275–282

Suzuki T, Agar NS, Suzuki M (1984) Red cell metabolism: a comparative study of some mammalian species. Comp Biochem Physiol 79B:515–520

Swan JA, Solomon F (1984) Reformation of the marginal band of avian erythrocytes in vitro using calf-brain tubulin: Peripheral determinants of microtubule form. J Cell Biol 99:2108–2113

Swenson ER, Maren TH (1987) Roles of gill and red cell carbonic anhydrase in elasmobranch HCO_3^- and CO_2 excretion. Am J Physiol 253:R450–R458

Taber E, Davis DE, Domm LV (1943) Effect of sex hormones on the erythrocyte numbers in the blood of the domestic fowl. Am J Physiol 138:479–487

Tagle DA, Miyamoto MM, Goodman M, Hofmann O, Braunitzer G, Göltenboch R, Jalanka H (1986) Hemoglobin of pandas: phylogenetic relationships of carnivores as ascertained with protein sequence data. Naturwissenschaften 73:512–514

Tähti H (1978) Periodicity of hibernation in the hedgehog (*Erinaceus europaeus L.*). Ph D Thesis University of Helsinki, 42 pp

Tähti H, Nikinmaa M, Soivio A (1981) Cheyne-Stokes breathing pattern as respiratory adaptation to deep hibernation hypothermia. Acta Univ Carol -Biol 1979:229–231

Takatani S, Noda H, Kohno H, Takano H, Akutsu T (1988a) Continuous measurement of oxygen delivery and oxygen consumption in awake live animals. In: Mochizuki M, Honig CR, Koyama T, Goldstick TK, Bruley DF (eds) Oxygen transport to tissue X. Plenum, New York, pp 245–256

Takatani S, Noda H, Takano H, Akutsu T (1988b) A miniature hybrid reflection type optical sensor for measurement of hemoglobin content and oxygen saturation of whole blood. IEEE Trans Biomed Eng 35:187–198

Taketa F (1974) Organic phosphates and hemoglobin structure-function relationships in the feline. Ann N Y Acad Sci 241:524–537
Tam L-T, Riggs AF (1984) Oxygen binding and aggregation of bullfrog hemoglobin. J Biol Chem 259:2610–2616
Tam L-T, Gray GP, Riggs AF (1986) The hemoglobins of the bullfrog *Rana catesbeiana*. The structure of the β chain of component C and the role of the α chain in the formation of intermolecular disulfide bonds. J Biol Chem 261:8290–8294
Tang Y, Nolan S, Boutilier RG (1988) Acid-base regulation following acute acidosis in seawater-adapted rainbow trout *Salmo gairdneri*: a possible role for catecholamines. J Exp Biol 134:297–312
Tardieu A, Luzzati V, Reman FC (1973) Structure and polymorphism of the hydrocarbon chains of lipids: a study of lecithin-water phases. J Mol Biol 75:711–733
Tempel GE, Musacchia XJ (1975) Erythrocyte 2,3-diphosphoglycerate concentrations in hibernating, hypothermic, and rewarming hamsters. Proc Soc Exp Biol Med 148:588–592
Templeton BA, Chilson OP (1981) Adenine transport by mature rabbit erythrocytes. J Biol Chem 256:285–290
Ten Eyck LF (1972) Stereochemistry of hemoglobin. In: Rorth M, Astrup P (eds) Oxygen affinity of hemoglobin and red cell acid-base status, Alfred Benzon Symposium IV. Munksgaard, Copenhagen, pp 19–31
Testa U (1985) Transferrin receptors: Structure and function. Curr Top Hematol 5:127–161
Tetens V (1987) Regulation of blood O_2 affinity during acute hypoxic exposure of rainbow trout, *Salmo gairdneri*: organismal and cellular processes. PhD Thesis, Aarhus University 12 pp
Tetens V, Christensen NJ (1987) Beta-adrenergic control of blood oxygen affinity in acutely hypoxia exposed rainbow trout. J Comp Physiol B 157:667–675
Tetens V, Lykkeboe G (1981) Blood respiratory properties of rainbow trout, *Salmo gairdneri*: responses to hypoxia acclimation and anoxic incubation of blood in vitro. J Comp Physiol B 145:117–125
Tetens V, Lykkeboe G (1985) Acute exposure of rainbow trout to mild and deep hypoxia: O_2 affinity and O_2 capacitance of arterial blood. Respir Physiol 61:221–235
Tetens V, Wells RMG (1984) Oxygen binding properties of blood and hemoglobin solutions in the carpet shark (*Cephaloscyllium isabella*): roles of ATP and urea. Comp Biochem Physiol 79A:165–168
Tetens V, Wells RMG, DeVries AL (1984) Antarctic fish blood: respiratory properties and the effect of thermal acclimation. J Exp Biol 109:265–279
Tetens V, Lykkeboe G, Christensen NJ (1988) Potency of adrenaline and noradrenaline for β-adrenergic proton extrusion from red cells of rainbow trout, *Salmo gairdneri*. J Exp Biol 134:267–280
Thapliyal JP, Pati AK, Gupta BBP (1982) The role of erythropoietin, testosterone, and 1-thyroxine in the tissue oxygen comsumption and erythropoiesis of spotted munia, *Lonchura punctulata*. Gen Comp Endocrinol 48:84–88
Thomas P, Limbrick AR, Allan D (1983) Limited breakdown of cytoskeletal proteins by an endogenous protease controls Ca^{2+}-induced membrane fusion events in chicken erythrocytes. Biochim Biophys Acta 730:351–358
Thomas RC (1984) Experimental displacement of intracellular pH and the mechanism of its subsequent recovery. J Physiol (Lond) 354:3P–22P
Thompson NL, Axelrod D (1980) Reduced lateral mobility of a fluorescent lipid probe in cholesterol-depleted erythrocyte membrane. Biochim Biophys Acta 597:155–165
Thorburn DR, Kuchel PW (1985) Regulation of the human-erythrocyte hexose-monophosphate shunt under conditions of oxidative stress. A study using NMR spectroscopy, a kinetic isotope effect, a reconstituted system and computer simulation. Eur J Biochem 150:371–386
Till JE, McCulloch EA, Siminovitch L (1964) A stochastic model of stem cell proliferation, based on the growth of spleen colony forming cells. Proc Natl Acad Sci USA 51:29–36
Ting A, Lee JW, Vidaver GA (1979) Calcium transport by pigeon erythrocyte membrane vesicles. Biochim Biophys Acta 555:239–248
Tolkowsky AM, Levitzki A (1978) Collision coupling of the β-adrenergic receptor with adenylate cyclase. Biochemistry 17:3795–3810
Tooze J, Davies HG (1965) Cytolysomes in amphibian erythrocytes. J Cell Biol 24:146–150

Torracca AMV, Raschetti R, Salvioli R, Ricciardi G, Winterhalter KH (1977) Modulation of the Root effect in goldfish by ATP and GTP. Biochim Biophys Acta 496:367–373

Torrance JD, Lenfant C (1970) Methods for the determination of O_2 dissociation curves, including Bohr effect. Respir Physiol 8:127–136

Tosteson DC (1972) Functions of ion transport across the red cell membrane. In: Rorth M. Astrup P (eds) Oxygen affinity of hemoglobin and red cell acid-base status, Alfred Benzon Symposium IV. Munksgaard, Copenhagen, pp 252–264

Tosteson DC, Hoffman JF (1960) Regulation of cell volume by active cation transport in high and low potassium sheep red cells. J Gen Physiol 44:169–194

Traykov TT, Jain RK (1987) Effect of glucose and galactose on red blood cell membrane deformability. Int J Microcirc Clin Exp 6:35–44

Trentin JJ (1970) Influence of hematopoietic organ stroma (hematopoeitic inductive microenvironments) on stem cell differentiation. In: Gordon AS (ed) Regulation of hematopoiesis. Appleton-Century-Crofts, New York pp 161–186

Trivedi B, Danforth WH (1966) Effect of pH on the kinetics of frog muscle phosphofructokinase. J Biol Chem 241:4110–4114

Trowbridge IS, Newman RA, Domingo DL, Sauvage C (1984) Transferrin receptors: structure and function. Biochem Pharmacol 33:925–932

Truchot JP (1987) Comparative aspects of extracellular acid-base balance. Springer, Berlin Heidelberg New York, 248 pp

Tsuji A, Ohnishi S-I (1986) Restriction of the lateral motion of band 3 in the erythrocyte membrane by the cytoskeletal network: dependence on spectrin association state. Biochemistry 25:6133–6139

Tsuyuki H, Ronald AP (1971) Molecular basis for multiplicity of Pacific salmon hemoglobins: evidence for in vivo existence of molecular species with up to four different polypeptides. Comp Biochem Physiol 39B:503–522

Tucker VA (1967) Method for oxygen content and dissociation curves on microliter blood samples. J Appl Physiol 23:410–414

Tufts BL, Boutilier RG (1989) The absence of anion exchange in agnathan erythrocytes: implications for CO_2 transport and ion distributions in the blood of the lamprey *Petromyzon marinus.* J Exp Biol 144:565–576

Tufts BL, Randall DJ (1989) The functional significance of adrenergic pH regulation in fish erythrocytes. Can J Zool 67:235–238

Tufts BL, Mense DC, Randall DJ (1987a) The effects of forced activity on circulating catecholamines and pH and water content of erythrocytes in the toad. J Exp Biol 128:411–418

Tufts BL, Nikinmaa M, Steffensen JF, Randall DJ (1987b) Ion exchange mechanisms on the erythrocyte membrane of the aquatic salamander, *Amphiuma tridactylum.* J Exp Biol 133:329–338

Tun N, Houston AH (1986) Temperature, oxygen, photoperiod, and the hemoglobin system of the rainbow trout, *Salmo gairdneri.* Can J Zool 64:1883–1888

Tycho B, Maxfield F (1982) Rapid acidification of endocytic vesicles containing a_2-macroglobulin. Cell 28:643–651

Tyler J, Hargreaves W, Branton D (1979) Purification of two spectrin-binding proteins. Biochemical and electron microscopic evidence for site-specific reassociation between spectrin and bands 2.1 and 4.1. Proc Natl Acad Sci USA 76:5192–5196

Tyler J, Reinhardt B, Branton D (1980) Associations of erythrocyte membrane proteins. Binding of purified bands 2.1 and 4.1 to spectrin. J Biol Chem 255:7034–7039

Ueberschär S, Bakkar-Grunwald T (1983) Bumetanide-sensitive potassium transport and volume regulation in turkey erythrocytes. Biochim Biophys Acta 731:243–250

Usami S, Chien S, Bertles JF (1975) Deformability of sickle cells as studied by microsieving. J Lab Clin Med 86:274–279

Vadgama JV, Christensen HN (1985) Discrimination of Na^+-independent transport systems, L, T, and asc in erythrocytes. Na^+ independence of the latter a consequence of cell maturation? J Biol Chem 260:2912–2921

Vadgama JV, Castro M, Christensen HN (1987) Characterization of amino acid transport during erythroid cell differentiation. J Biol Chem 262:13273–13284

Van der Ploeg LHT, Flavell RA (1980) DNA methylation in the human $\delta\beta$-globin locus in erythroid and nonerythroid tissues. Cell 19:947–958
Van Slyke D, Wu H, McLean F (1923) Studies of gas and electrolyte equilibria in the blood. V. Factors controlling the electrolyte and water distribution in the blood. J Biol Chem 56:765–849
Vermette MG, Perry SF (1988) Effects of prolonged epinephrine infusion on blood respiratory and a acid-base states in the rainbow trout: alpha and beta effects. Fish Physiol Biochem 4:189–202
Vidaver GA, Shepherd SL (1968) Transport of glycine by hemolyzed and restored pigeon red blood cells. J Biol Chem 243:6140–6150
Villalobo A, Roufogalis BD (1986) Proton countertransport by the reconstituted erythrocyte Ca^{2+}-translocating ATPase: evidence using ionophoretic compounds. J Membr Biol 93:249–258
Virtanen E, Salama A, Lönn B-E (1988) Adaptations in the capacity of ionic and osmotic regulation in young Baltic salmon (*Salmo salar* L.) in brackish water. Comp Biochem Physiol 91A:79–86
Wagner PD (1977) Diffusion and chemical reaction in pulmonary gas exchange. Physiol Rev 57:257–312
Waisman DM, Gimble JM, Goodman DBP, Rasmussen H (1981) Studies of the Ca^{2+} transport mechanism of human erythrocyte inside out plasma membrane vesicles. I. Regulation of the Ca^{2+} pump by calmodulin. J Biol Chem 256:409–414
Wallick ET, Lane LK, Schwartz A (1979) Biochemical mechanism of the sodium pump. Annu Rev Physiol 41:397–411
Walmsley AR (1988) The dynamics of the glucose transporter. Trends Biochem Sci 13:226–231
Watt KWK, Maruyama T, Riggs A (1980) Hemoglobins of the tadpole of the bullfrog, *Rana catesbeiana*. J Biol Chem 255:3294–3301
Watts C, Wheeler KP (1978) Protein and lipid components of the pigeon erythrocyte membrane. Biochem J 173:899–907
Waugh RE, Evans EA (1976) Viscoelastic properties of erythrocyte membranes of different vertebrate animals. Microvasc Res 12:291–304
Waugh RE, Evans EA (1979) Thermoelasticity of red blood cell membrane. Biophys J 26:115–131
Waugh SM, Low PS (1985) Hemichrome binding to band 3: Nucleation of Heinz bodies on the erythrocyte membrane. Biochemistry 24:34–39
Waugh SM, Walder JA, Low PS (1987) Partial characterization of the copolymerization reaction of erythrocyte membrane band 3 with hemichromes. Biochemistry 26:1777–1783
Webb PW, Brett JR (1972) Oxygen consumption of embryos and parents, and oxygen transfer characteristics within the ovary of two species of viviparous seaperch, *Rhacochilus vacca* and *Embiotoca lateralis*. J Fish Res Board Can 29:1543–1553
Weber RE (1978) Functional interaction between fish hemoglobin, erythrocyte nucleoside triphosphates and magnesium. Scand J Physiol 102:20A–21A
Weber RE (1982) Intraspecific adaptation of hemoglobin function in fish to oxygen availability. In: Addink ADF, Spronk N (eds) Exogenous & endogenous influences on metabolic and neural control. Pergamon, Oxford, pp 87–102
Weber RE (1983) TMAO (trimethylamine oxide)-independence of oxygen affinity and its urea and ATP sensitivities in an elasmobranch hemoglobin. J Exp Zool 228:551–554
Weber RE, Hartvig M (1984) Specific fetal hemoglobin underlies the fetal-maternal shift in blood oxygen affinity in a viviparous teleost. Mol Physiol 6:27–32
Weber RE, Lykkeboe G (1978) Respiratory adaptations in carp blood. Influences of hypoxia, red cell organic phosphates, divalent cations and CO_2 on hemoglobin-oxygen affinity. J Comp Physiol 128:127–137
Weber RE, White FN (1986) Oxygen binding in alligator blood related to temperature, diving, and "alkaline tide". Am J Physiol 251:R901–R908
Weber RE, Wilde De JAM (1976) Multiple hemoglobins in plaice and flounder and their functional properties. Comp Biochem Physiol 54B:443–437
Weber RE, Lykkeboe G, Johansen K (1976a) Physiological properties of eel haemoglobin: hypoxic acclimation, phosphate effects and multiplicity. J Exp Biol 64:75–88
Weber RE, Wood SC, Lomholt JP (1976b) Temperature acclimation and oxygen binding properties of blood and multiple haemoglobins of rainbow trout. J Exp Biol 65:333–345
Weber RE, Johansen K, Lykkeboe G, Maloiy GMO (1977) Oxygen-binding properties of hemoglobins from estivating and active African lungfish. J Exp Zool 199:85–96

Weber RE, Wood SC, Davis BJ (1979) Acclimation to hypoxic water in facultative air-breathing fish: Blood oxygen affinity and allosteric effectors. Comp Biochem Physiol 62A:125–129
Weber RE, Wells RMG, Rossetti JE (1983a) Allosteric interactions governing oxygen equilibria in the haemoglobin system of the spiny dogfish, *Squalus acanthia*. J Exp Biol 103: 109–120
Weber RE, Wells RMG, Tougaard S (1983b) Antagonistic effect of urea on oxygenation-linked binding of ATP in an elasmobranch hemoglobin. Life Sci 32:2157–2161
Weber RE, Wells RMG, Rossetti JE (1985) Adaptation to neoteny in the salamander *Necturus maculosus*. Blood respiratory properties and interactive effects of pH, temperature and ATP on hemoglobin oxygenation. Comp Biochem Physiol 80A:495–501
Weber RE, Heath ME, White FN (1986) Oxygen binding functions of blood and hemoglobin from the chinese pangolin, *Manis pentadactyla*: possible implications of burrowing and low body temperature. Respir Physiol 64:103–112
Weber RE, Jensen FB, Cox RP (1987a) Analysis of teleost hemoglobin by Adair and Monod-Wyman-Changeux models. J Comp Physiol 157B:145–152
Weber RE, Kleinschmidt T, Braunitzer G (1987b) Embryonic pig hemoglobins Gower I ($\xi_2\varepsilon_2$), Gower II ($\alpha_2\varepsilon_2$), Heide I ($\xi_2\vartheta_2$) and Heide II ($\alpha_2\vartheta_2$): oxygen-binding functions related to structure and embryonic oxygen supply. Respir Physiol 69:347–357
Weber RE, Hiebl I, Braunitzer G (1988a) High altitude and hemoglobin function in the vultures *Gyps rueppellii* and *Aegypius monachus*. Biol Chem Hoppe-Seyler 369:233–240
Weber RE, Lalthantluanga R, Braunitzer G (1988b) Functional characterization of fetal and adult yak hemoglobins: an oxygen binding cascade and its molecular basis. Arch Biochem Biophys 263:199–203
Weinberg SR, LoBue J, Siegel CD, Gordon AS (1976) Hematopoiesis of the kissing gourami (*Helostoma temmincki*). Effects of starvation, bleeding, and plasma-stimulating factors on its erythropoiesis. Can J Zool 54:1115–1127
Weinstein Y, Ackerman RA, White FN (1986) Influence of temperature on the CO_2 dissociation curve of the turtle *Pseudemys scripta*. Respir Physiol 63:53–63
Wells RMG (1979) Haemoglobin-oxygen affinity in developing embryonic erythroid cells of mice. J Comp Physiol 129B:333–338
Wells RMG, Brittain T (1983) Non-cooperative oxygen binding in the erythrocytes of pre-implanted sheep embryos. Comp Biochem Physiol 76A:387–388
Wells RMG, Weber RE (1985) Fixed acid and carbon dioxide Bohr effects as functions of hemoglobin-oxygen saturation and erythrocyte pH in the blood of the frog, *Rana temporaria*. Pflügers Arch 403:7–12
Westman K (1970) Hemoglobin polymorphism and its ontogeny in sea-running and landlocked Atlantic salmon (*Salmo salar* L.). Ann Acad Sci Fenn A, IV Biol 170:1–28
Wheeler TJ, Hinkle PC (1985) The glucose transporter of mammalian cells. Annu Rev Physiol 47:503–517
Whetton AD, Dexter TM (1986) Haemopoietic growth factors. Trends Biochem Sci 11:207–211
White FN (1969) Redistribution of cardiac output in the diving alligator. Copeia 3:567–570
Whitfield CF, Morgan HE (1973) Effect of anoxia on sugar transport in avian erythrocytes. Biochim Biophys Acta 307:181–196
Whitfield CF, Rannels SR, Morgan HE (1974) Acceleration of sugar transport in avian erythrocytes by catecholamines. J Biol Chem 249:4181–4188
Whitfield CF, Mylin LM, Goodman SR (1983) Species-dependent variations on erythrocytes membrane skeletal proteins. Blood 61:500–506
Whitfield CF, Coleman DB, Kay MMB, Shiffer KA, Miller J, Goodman SR (1985) Human erythrocyte membrane proteins of zone 4.5 exist as families of related proteins. Am J Physiol 248:C70–C79
Wickerson M, Stephenson P (1984) Role of the conserved AAUAA sequence; four AAUAA point mutants prevent messenger RNA 3′ end formation. Science 226:1045–1051
Widdas WF (1980) The asymmetry of the hexose transfer system in the human red cell membrane. Curr Top Membr Transp 14:165–223
Widdas WF (1988) Old and new concepts of the membrane transport for glucose in cells. Biochim Biophys Acta 947:385–404

Widmer HJ, Hosbach HA, Weber R (1983) Globin gene expression in *Xenopus laevis*: anemia induces precocious globin transition and appearance of adult erythroblasts during metamorphosis. Dev Biol 99:50–60
Wieth JO, Brahm J, Funder J (1980) Transport and interactions of anions and protons in the red blood cell membrane. Ann NY Acad Sci 341:394–418
Wilhelm DF, Reischl E (1981) Heterogeneity and functional properties of hemoglobins from south brazilian freshwater fish. Comp Biochem Physiol 69B:463–450
Wilkins NP, Iles TD (1966) Haemoglobin polymorphism and its ontogeny in herring (*Clupea harengus*) and sprat (*Sprattus sprattus*). Comp Biochem Physiol 17:1141–1158
Williams J (1982) The evolution of transferrin. Trends Biochem Sci 7:394–397
Willis JS (1979) Hibernation: cellular aspects. Annu Rev Physiol 41:275–286
Willis JS, Ellory JC (1983) Ouabain sensitivity:diversity and disparities. Curr Top Membr Transp 19:277–280
Willis JS, Ellory JC, Wolowyk MW (1980) Temperature sensitivity of the sodium pump in red cells from various hibernators and non-hibernator species. J Comp Physiol 138:43–47
Wilson RR Jr, Knowles FC (1987) Temperature adaptation of fish hemoglobins reflected in rates of autoxidation. Arch Biochem Biophys 255:210–213
Winter CG, Christensen HN (1965) Contrasts in neutral amino acid transport by rabbit erythrocytes and reticulocytes. J Biol Chem 240:3594–3600
Winterhalter KH, Di Iorio EE (1984) Influence of heme pocket geometry on ligand binding to heme proteins. In: Nicolau C (ed) Oxygen transport in red blood cells. Pergamon, Oxford, pp 1–13
Woffendin C, Plagemann PGW (1987) Nucleoside transporter of pig erythrocytes. Kinetic properties, isolation and reaction with nitrobenzylthioinosine and dipyridamole. Biochim Biophys Acta 903:18–30
Wood RE, Morgan HE (1969) Regulation of sugar transport in avian erythrocytes. J Biol Chem 244:1451–1460
Wood SC (1980) Adaptation of red blood cell function to hypoxia and temperature in ectothermic vertebrates. Am Zool 20:163–172
Wood SC, Johansen K (1972) Adaptation to hypoxia by increased HbO_2 affinity and decreased red cell ATP concentration. Nature New Biol 237:278–279
Wood SC, Johansen K (1973a) Blood oxygen transport and acid-base balance in eels during hypoxia. Am J Physiol 225:849–851
Wood SC, Johansen K (1973b) Organic phosphate metabolism in nucleated red cells: influence of hypoxia on eel HbO_2 affinity. Neth J Sea Res 7:328–338
Wood SC, Lenfant C (1976) Respiration: mechanics, control, and gas exchange. In: Gans C, Dawson WR (eds) Biology of the Reptilia, vol 5. Academic Press, London, pp 225–274
Wood SC, Lenfant C (1979) Oxygen transport and oxygen delivery. In: Wood SC, Lenfant C (eds) Evolution of respiratory processes, a comparative approach. Lung biology in health and disease, vol 13. Dekker, New York, pp 193–223
Wood SC, Lenfant C (1987) Phylogeny of the gas-exchange system: red cell function. In: Farhi LE, Tenney SM (eds) Handbook of physiology, Section 3. The respiratory system, vol 4. Gas exchange. American Physiological Society, Bethesda, Maryland, pp 131–146
Wood SC, Johansen K, Weber RE (1975) Effects of ambient P_{O2} on hemoglobin-oxygen affinity and red cell ATP concentrations in a benthic fish, *Pleuronectes platessa*. Respir Physiol 25:259–267
Wood SC, Lykkeboe G, Johansen K, Weber RE, Maloiy GMO (1978) Temperature acclimation in the pancake tortoise, *Malacochersus tornieri*: metabolic rate, blood pH, oxygen affinity and red cell organic phosphates. Comp Biochem Physiol 59A:155–160
Wright JW, Gubernick DJ, Reynolds TJ (1977) Intravascular, carcass and gut fluid changes induced by food deprivation in xeric and mesic adapted rodents. Comp Biochem Physiol 58:137–142
Wyman J (1964) Linked functions and reciprocal effects in hemoglobin: a second look. Adv Protein Chem 19:223–286
Wyman J (1972) Reflections regarding hemoglobin. In: Rorth M, Astrup P (eds) Oxygen affinity of hemoglobin and red cell acid-base status. Alfred Benzon Symposium IV. Munksgaard, Copenhagen, pp 37–49
Yamaguchi K, Kochiyama Y, Matsuura F (1962) Studies on multiple hemoglobins of eel II. Oxygen dissociation curve and relative amounts of components F and S. Bull Jpn Soc Sci Fish 28:192–198

Yamamoto K-I (1987) Contraction of spleen in exercised cyprinid. Comp Biochem Physiol 87A:1083–1087
Yamamoto K-I, Itazawa Y, Kobayashi H (1980) Supply of erythrocytes into the circulating blood from the spleen of exercised fish. Comp Biochem Physiol 65A:5–13
Yamamoto K-I, Itazawa Y, Kobayashi H (1983) Erythrocyte supply from the spleen and hemoconcentration in hypoxic yellowtail. Mar Biol 73:221–226
Yamamoto M, Iuchi I (1975) Electron microscopic study of erythrocytes in developing rainbow trout, *Salmo gairdnerii irideus*, with particular reference to changes in the cell line. J Exp Zool 191:407–426
Yoshida M, Tada Y, Kasahara Y, Ando K, Satoyoshi E (1986) Ca content of human erythrocytes-what is the true value? Cell Calcium 7:169–174
Young JD (1983) Erythrocyte amino acid and nucleoside transport. In: Agar NS. Board PG (eds) Red blood cells of domestic mammals. Elsevier, Amsterdam, pp 271–289
Young JD, Ellory JC (1977) Red cell amino acid transport. In: Ellory JC, Lew VL (eds) Membrane transport in red cells. Academic Press, London, pp 301–325
Young JD, Jarvis SM (1983) Nucleoside transport in animal cells. Review. Biosci Rep 3:309–322
Young JD, Ellory JC, Tucker EM (1975) Amino acid transport defect in glutathione-deficient sheep erythrocytes. Nature 254:156–157
Young JD, Jones SEM, Ellory JC (1980) Amino acid transport in human and in sheep erythrocytes. Proc R Soc Lond B 209:355–375
Youson JH (1981) The kidneys. In: Hardisty M, Potter IC (ed) The biology of lampreys, vol 3. Academic Press, London, pp 192–261
Youson JH, Lee J, Potter IC (1979) The distribution of fat in larval, metamorphosing and young adult anadromous sea lampreys, *Petromyzon marinus*. Can J Zool 57:237–246
Yu J, Steck TL (1975) Isolation and characterization of band 3, the predominant polypeptide of the human erythrocyte membrane. J Biol Chem 250:9170–9175
Zanjani ED, Kaplan ME (1979) Cell-cell interaction in erythropoiesis. In: Brown EB (ed) Progress in Hematology, vol 11. Grune & Stratton, New York, pp 173–191
Zanjani ED, Yu M-L, Perlmutter A, Gordon AS (1969) Humoral factors influencing erythropoiesis in the fish blue gourami, *Trichogaster trichopterus*. Blood 33:573–581
Zanjani ED, McGlave PB, Bhakthavathsalan A, Stamatoyannopoulos G (1979) Sheep fetal haematopoietic cells produce adult haemoglobin when transplanted in the adult animal. Nature 280:495–496
Zanjani ED, Lim G, McGlave PB, Clapp JF, Mann LI, Norwood TH, Stamatoyannopoulos G (1981) Hemoglobin phenotypes in genetically AA sheep fetuses transplanted with cells of genetically BB adult animals. In: Stamatoyannopoulos G, Nienhuis AW (eds) Hemoglobins in development and differentiation. Alan R Liss, New York, pp 263–274
Zapata A (1980) Splenic erythropoiesis and thrombopoiesis in elasmobranchs: an ultrastructural study. Acta Zool 61:59–64
Zapata A, Carrato A (1981) Ultrastructure of elasmobranch and teleost erythrocytes. Acta Zool 62:129–135
Zimmermann A, Schatzmann HJ (1985) Calcium transport by red blood cell membranes from young and adult cattle. Experientia (Basel) 41:743–745
Zinkl J, Kaneko JJ (1973) Erythrocyte 2,3-diphosphoglycerate in normal and porphyric fetal, neonatal and adult cattle. Comp Biochem Physiol 45A:699–704
Zurini M, Krebs J, Penniston JT, Carafoli E (1984) Controlled proteolysis of the purified Ca^{2+}-ATPase of the erythrocyte membrane. J Biol Chem 259:618–627

Subject Index

MIX
Papier aus verantwortungsvollen Quellen
Paper from responsible sources
FSC® C105338

If you have any concerns about our products,
you can contact us on
ProductSafety@springernature.com

In case Publisher is established outside the EU,
the EU authorized representative is:
Springer Nature Customer Service Center GmbH
Europaplatz 3, 69115 Heidelberg, Germany

Printed by Libri Plureos GmbH
in Hamburg, Germany